Introduction to General and Applied Entomology

Life History of the Indian Moon Moth *Actias selene*

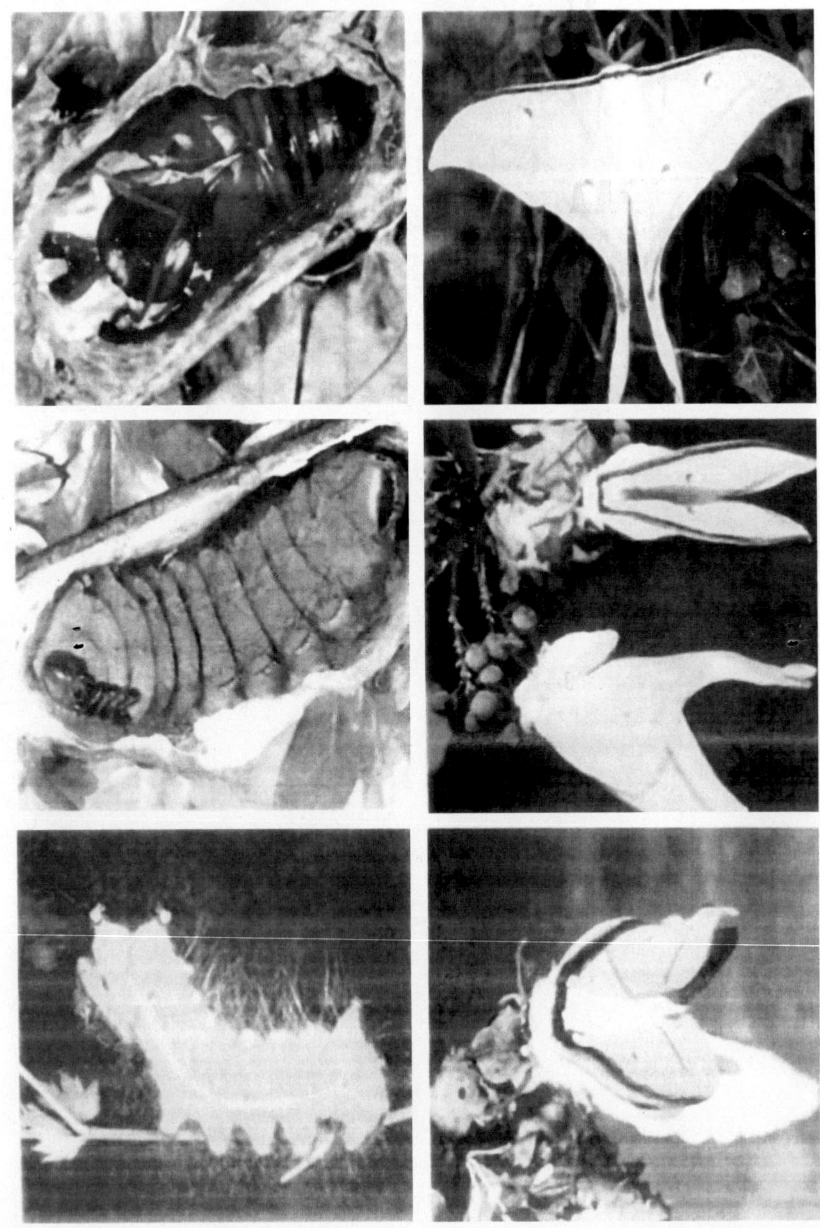

Introduction to General and Applied Entomology

3rd Revised Edition

V.B. Awasthi
M.Sc., Ph.D.
Former Associate Professor
Department of Zoology
J.N.V. University, Jodhpur (India)

Published by:

SCIENTIFIC PUBLISHERS (INDIA)
5A, New Pali Road, P.O. Box 91
Jodhpur 342 001, India
E-mail: info@scientificpub.com
Website: www.scientificpub.com

Print: 2025

Disclaimer: Whereas every effort has been made to avoid errors and omissions, this publication is being sold on the understanding that neither the editors (or authors of chapters in edited volumes) nor the publishers nor the printers would be liable in any manner to any person either for an error or for an omission in this publication, or for any action to be taken on the basis of this work. Any inadvertent discrepancy noted may be brought to the attention of the publishers, for rectifying it in future editions, if published.

Third Edition, 2012
Reprinted 2009
Second Edition, 2002
First Edition, 1997

ISBN: 978-81-72335-95-3
eISBN: 978-93-87307-37-7

Printed in India

PREFACE (3rd Edition)

With much pleasure I am presenting the third edition of my book "Introduction to General and Applied Entomology." I wish to thank the teachers and students both of this subcontinent and abroad for their whole hearted acceptance of the previous editions of the book. In this edition, some alterations have been made at appropriate places in the text. Likewise, 10 more pests of crops have been included along with their life histories and control methods. A good number of new illustrations have been introduced and the bibliography has been updated. In the part, dealing with behaviour, tropisms have been replaced by taxes etc. following the remarks of Wigglesworth.

The chapter, related to chemical control has also been revised keeping in view that a large number of chlorinated hydrocarbons, organophosphorus compounds, systemic insecticides, fumigants etc. have been either banned or restricted in use. Thus, for the control of pests, only those insecticides have been given that described in authentic books.

Further, only from the historical point of view, account of some of the chlorinated hydrocarbons and organophosphorus insecticides has been given under the foot note "banned", which were used in the past in large scale all over the world.

I hope the readers of Entomology will find the book extremely useful and welcome this latest edition with the same zeal as the earlier editions.

I wish to thank my friend Dr. Allah Noor Khan, Former Associate Professor of Agricultural Entomology, A.R.S., Mandore, Jodhpur for going through the proof and offering useful suggestions.

Jodhpur
5 April, 2009

V.B. Awasthi

PREFACE (1st Edition)

Although, there are a large number of standard and good books on the subject but they are either too exhaustive or too brief or mainly research oriented, and they deal, only one aspect of the subject, hence they do not fulfil the basic need of the post-graduate students of Entomology. Thus, a balanced account of general and applied Entomology both as required by the graduate students of Entomology of Indian Universities is hardly available.

This book provides a balanced and up to date account on morphology, anatomy, physiology of various organs and systems as well as developmental biology and different methods of control of insect pest and preventive measures for their hazards. The various appliances used in the pest management have also been given. The major pests of stored grains, cotton and fibre crops, sugar cane, oil seeds, pulses, sorghum, crops yielding cereal products, fruits and vegetables, polyphagous insects, household insect and insects of forestry and veterinary importance and their control have also been given together with information on the beneficial insects like honey bee, lac insect and silkworms. For the sake of convenience and their impact on the biology and control of insect pests, the ecology and behaviour have been dealt at the end; because, multiplication of a given species is maximum at a certain range of temperature and humidity and that controls too, the survival, feeding and dispersal of species. Ecological aspects of diapause and a brief account of the language of bees and their behaviour is also given.

A large number of illustrations have been given to make this fascinating subject easy and interesting. The principal aim has been to provide the basic information on insect biology, ecology and their control as well as benefits drawn out of them.

Since this book is an outcome of my comprehensive lectures hence information provided by several pioneer workers has been incorporated and it is not possible to thank them all individually, because their list is so long. But I must express my most humble and profound gratitudes to all of them. It is due to their hard work which enabled me to select and compile suitable information in such a manner that would be useful to the students of Entomology. As such, I do not claim of any sort of originality and in fact no one can claim originality in such a type of work.

I hope the students of Entomology, Forestry and Veterinary science will find this book useful for whom it is meant, and awake curiosity to know more and more on this subject and on these creatures which rival birds and flowers in their beauty. In the appendix, principal orders of insects have been given and a list of selected books covering a wide range of subjects is also given for further reading.

I am extremely grateful to the authorities of the Scientific Publishers, Jodhpur, particularly Shri Prahlad Rai Sharma and Pawan Kumar Sharma for their keen interest in my work and whole hearted support in the publication of the book.

I would like to thank my friend Shri A.P. Jain, Botanical Survey of India, Jodhpur for going through the proof and making necessary improvements. I also wish to thank my son Dr. Santosh for going through the manuscript, my wife Sneh Lata and daughter Shubha for the help in preparing the photo plates and labelling them carefully. Acknowledgement is also due to American Spring and Pressing Works Ltd. Bombay for sending the figures of sprayers and dusters of various types and allowing me to reproduce them as such.

I would appreciate if readers of the book will inform me of any errors or omissions that may come to their notice.

Jodhpur
November, 1996 **V.B. Awasthi**

CONTENTS

PART I - GENERAL ENTOMOLOGY

PART II (A) – APPLIED ENTOMOLOGY

Part-II (B): BIOLOGY AND CONTROL OF INSECT PESTS

Part - I

General Entomology

1

THE INTEGUMENT

The integument of the insects serves as exoskeleton and is ectodermal in origin. It performs several functions e.g. protection against moisture, dryness, diseases and natural enemies of many kinds. The integument also protects the soft tissues within, and serves as the place of attachment for the numerous muscles. The integument consists of (1) cuticle, (2) epidermis or hypodermis and (3) basement membrane (Fig. 1-A).

(1) Cuticle : It is secreted by the underlying epidermis and is a non-cellular layer. It covers the whole of the body and its appendages. The cuticle also forms the lining of the fore-and hindgut, tracheal system and parts of the reproductive tract. The rectal glands and anal glands are also lined by the cuticle.

COMPOSITION OF THE CUTICLE

The cuticle is composed of two main layers viz. (1) an outer epicuticle and (2) inner procuticle.

(A) Epicuticle : It is the outermost layer 0.03 μ to 4 μ in thickness and its detailed structure varies from insect to insect. It is very thin and nonchitinous in nature. It is composed of 4-layers. The first is a cement layer which is lipoprotein in character, second is wax or lipid layer which helps in preventing the loss of water through it. Thus it provides the water conserving properties to epicuticle. Hence waxes of the epicuticle play a very important part in protecting the insects from loss of water. The appearance of greasy layer on the body of cockroaches is due to the wax or lipid secreted by special epidermal cells. The third is a polyphenol layer responsible for the hardening of the cuticle and the fourth is the cuticulin layer, lipoprotein in nature.

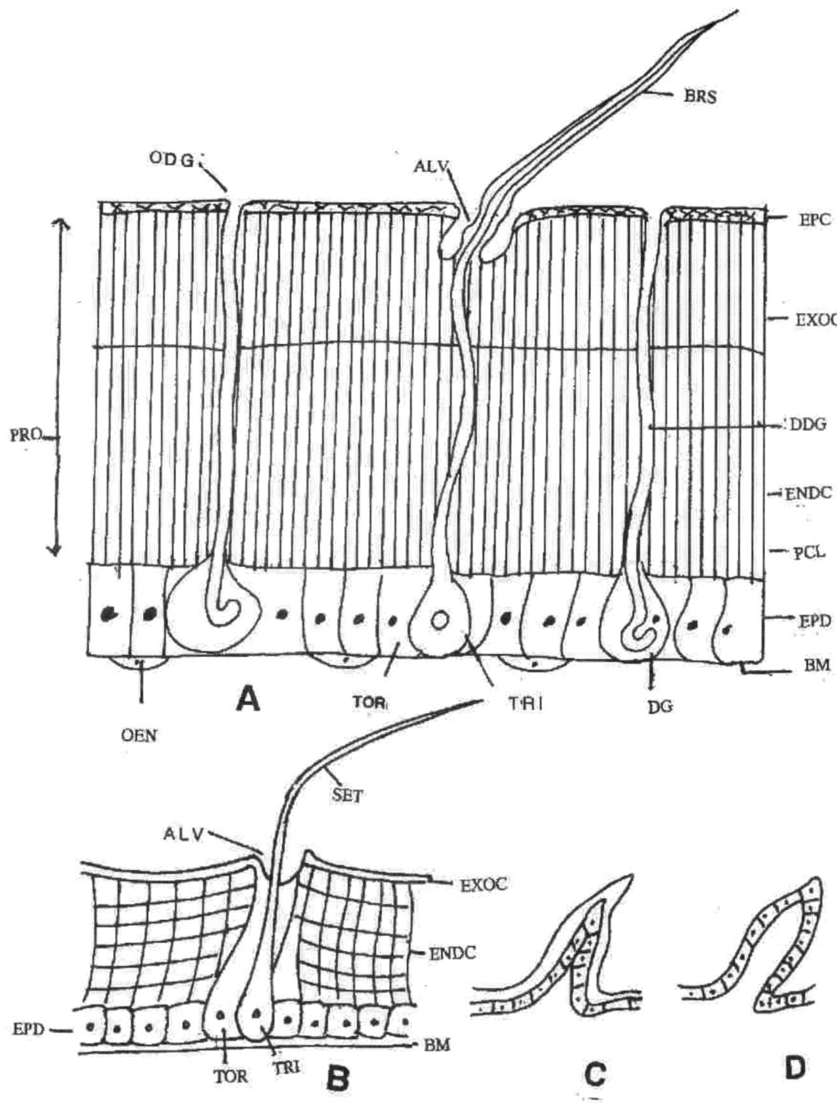

Fig. 1. A. Section of typical insect cuticle, B. A typical seta, C.D. Multi cellular process

Abbreviations: ALV - alveolus, BRS - bristle, BM - basement membrane, DG - dermal gland, DDG - duct of dermal gland, EPD - epidermis, EPC - epicuticle, EXOC - exocuticle, ENDC - endocuticle, ODG - opening of dermal gland, OEN - oenocyte, PCL - pore canal, PRO - procuticle, TOR - tormogen cell, TRI - trichogen cell, SET - seta.

(B) Procuticle : It is secreted by the epidermal or hypodermal cells. It is a thick layer which is composed of chitin, protein and a number of other substances. The procuticle shows a multi-lamillar structure, arranged more or less parallel to the surface. It is also traversed by pore canals running through all or only the outer layer of the procuticle. The pore canals lie perpendicular to the surface, and each one of them is initially occupied by thread like cytoplasmic extensions of the epidermis although at the later stage they may become filled with cuticular materials. These are very narrow, even less than 1 μ in diameter. They are not straight but are variously coiled. The pore-canals help in transporting the material to the epicuticle or outer procuticle layer e.g. the wax is secreted by way of the pore canals and crystallizes on the surface. Thus the insect is rendered water proof before moulting.

In the case of soft transparent membrane, the procuticle remains unchanged after the formation. But in other cases the outer part becomes changed into a hard, dark, sclerotized protein, known as exocuticle while the inner part which remains unchanged known as endocuticle. In this way the procuticle gets differentiated into two portions namely; (1) exocuticle and (2) endocuticle.

(1) Exocuticle : It is a thick layer which consists of mainly chitin and proteins. The proteins get tanned by phenolic substances to produce hard, brown material known as sclerotin. It is this sclerotin which gives rigidity to the hard parts like head capsule, segments of the limbs and other parts of the cuticle. In more flexible region of the integument, the exocuticle is absent or reduced or it may be altogether absent from the insects with a soft and thin cuticle.

(2) Endocuticle : It is the thickest layer of integument and also contains chitin and proteins. The proteins in the endocuticle are not tanned and thus this part of the cuticle is very soft and responsible for the extensibility of the integument and for combining toughness with flexibility.

The inner major constituents of the exocuticle and endocuticle are the chitin and the protein. The two substances i.e. chitin and protein are chemically combined to form a glycoprotein. The exocuticle is distinguishable from the endocuticle by its darker pigmentation and denser structure. It is this layer of the body wall which contains the hardening substances that forms the sclerites.

Chitin

According to Campbell (1929) the chitin is the nitrogenous polysaccharide making 25-60% of the body weight of the cuticle. On hydrolysis it yields acetic acid and glucosamine the nitrogenous polysaccharide. It is insoluble in water, alcohol, ether, dilute acids and dilute or concentrated alkalies but it dissolves with decomposition in concentrated mineral acids and sodium hypochlorite. When treated with KOH or NaOH at higher temperature, it is hydrolysed producing a substance known as chitosan and acetic acid.

Resilin

The protein of the cuticle occurs in many forms e.g. the ordinary protein of the cuticle, which is tanned to form sclerotin is called arthropodin. In addition to arthropodin, there is always present rubber like another protein called resilin. This is an elastic substance in which protein chains are bound together in a uniform 3-dimensional network so as to form a rubber like structure. Resilin may be deposited between the chitinous lamellae and provides the elasticity of the cuticle, or it may exist in pure state to form elastic hinges at the base of wings, and elastic tendons for muscles, connected to the wings.

Histological appearance of the cuticle varies in different insects and sometimes in the different part of the same insect e.g. some of the Coleoptera have a highly specialized structure but this holds good for all the insects, of course, with minor changes here and there.

Epidermis

The cell layer of the body wall is called epidermis or hypodermis. The cells are arranged in the single continuous layer and in most places they form a simple epithelium. In the growing stages of the insects, the epidermal cells are usually cubical or columnar in shape with the nuclei near their bases (Snodgrass, 1935). In adult insects, they become more or less indistinct and the cell areas are marked only by the presence of nuclei. There are specialized gland cells scattered amongst the normal epidermal cells. In addition to the gland cells, there are others which are concerned with the formation of cuticular sensilla. In the epidermis are also seen the fibres of the muscle attachment, the myofibrillae associated with the tonofibrillae that run through the epidermis and into the procuticle. It has also been suggested that the oenocytes which originate from the epidermis remain associated with this layer.

The epidermis secretes the greater part of the cuticle, it also produces the moulting fluid which dissolves the old cuticle before the immature insect moults. It also absorbs the digestion products of the old cuticle and repairs the wounds and differentiates so as to determine the surface pattern of the insect.

Basement membrane : The basement membrane, about 0.5μ thick forms the innermost lining of the body wall. It is very closely attached to the epidermis.

External processes of the integument

The external surface of the cuticle of insects is hardly smooth or bare, it shows numerous outgrowths in the form of spicules, spines, hairs and scales. The external processes of the body wall may, however, be classified into two categories viz. (1) cellular and (2) noncellular.

1. Cellular outgrowths

There are two types of cellular outgrowths viz. (1) unicellular (Fig. 1A) and (2) multicellular (Fig. 1, C,D.)

(1) *Unicellular outgrowths* : These outgrowths of the body wall are termed as setae. The setae form the body covering of the insects. A seta is a slender hair like hollow structure developed as extension of the exocuticle. It is produced by a single, usually large, epidermal cell or trichogen cell (Fig. 1B). The seta is attached by means of a small membranous ring of the body wall known as the setal membrane. Each seta arises from a cup like pit or alveolus formed by the depressed setal membrane. This membrane is produced by a second epidermal cell, the tormogen cell. Thus we see that the first epidermal cell forms the trichogen cell and the second adjacent cell that forms the articular socket for the first cell (i.e. trichogen cell), is termed as tormogen cell. At the base of the seta, there is a cylindrical internal cavity of the cuticle which is called the trichopore that contains the distal parts of the cells associated with the seta. The setae or macrotrichia are of about 5 types viz.—

(i) **Clothing hairs :** These covers the general surface of the body and appendages and may be plumose or branched e.g. Apoidea.

(ii) **Bristles** : When they are stiff or stout and called bristles as in the Tachinidae.

(iii) **Scales** : These structures are highly modified clothing hairs and are small, flat and scale like. They constitute the body covering of all Lepidoptera and several other insects.

(iv) **Glandular setae**: The setae which function as the outlet for the secretion of the epidermal glands are included under this group. Occasionally, these setae are stout or ridged known as glandular bristles or glandular setae, as in the case of the larvae of certain Lepidoptera. The secretions are produced in special gland cells that are associated with trichogen cell.

(v) **Sensory setae** : Very frequently, the setae of certain portion of the body, particularly those of the appendages are provided with sensory nerve cell lying in or just beneath the epidermis. The nerve cell is connected with the seta by a neural process and thus, they perceive various stimuli and become sensory in function.

(2) *Multicellular outgrowths* : These are hollow outgrowths of the entire body wall. They are covered by a layer of embryonic epidermal cells. They are fairly large and spiny in form (Fig. 1C, D).

(a) **Spurs** : These are set in a membranous ring around cuticle hairs and are movable. These differ from setae in the sense that these are thick walled, multicellular structures found on the tibiae of the legs.

(b) **Spines** : These are fixed to the surrounding cuticle and are multicellular. These are the outgrowths of the cuticle which are spine like setae and are produced by undifferentiated epidermal cells. These are occasionally seen on the hind tibiae of the Orthoptera.

2. Non-cellular outgrowths

These are also of various types

Microtrichia : These are fixed noncellular (hair-like) structures which are found on the wings of Mecoptera and Diptera. These are also found in the form of nodules, spicules, small spines, corrugations etc.

2

THE HEAD AND ITS APPENDAGES

The head is a compact box like structure which supports the mouth parts, antennae and eyes and within it lies the brain. The posterior surface of the head has a large opening, foramen magnum, which leads into the neck and thorax.

Orientation of the Head

According to the position of mouth parts the head may assume one of the three types of structures. If the mouth parts are directed downward and the cranium corresponds in position to the body segments the head is said to be "hypognathous" e.g. Orthoptera, if the cranium is turned upward on the neck so that the mouth parts are directed forward the head becomes "prognathous" e.g. Coleoptera. The third, "opisthognathous" type results from a deflection of the facial region, giving the mouth parts a posterior ventral position e.g. Hemiptera.

Segmentation of the head

The head has lost all the traces of segmentation and forms a strongly sclerotized capsule. However, it is a well known fact that at least six anterior segments are fused in the head. The primitive appendages of all except one of these segments are greatly modified in the form of eyes, antennae and mouth parts to perform the function of sight, feeling and ingestion of food.

Sclerites and sutures (Fig. 2A, B, C)

In the head, certain areas are strongly sclerotized known as sclerites. The sclerites are separated from each other by narrow lines termed as sutures. The sclerites and sutures represent the skeletal structures that can be recognized well in the insect head. In the special cases a number of structures may be modified or variously changed according to special requirements. In the head of a generalized insect e.g. grasshopper or locust the following structures are clearly recognized.

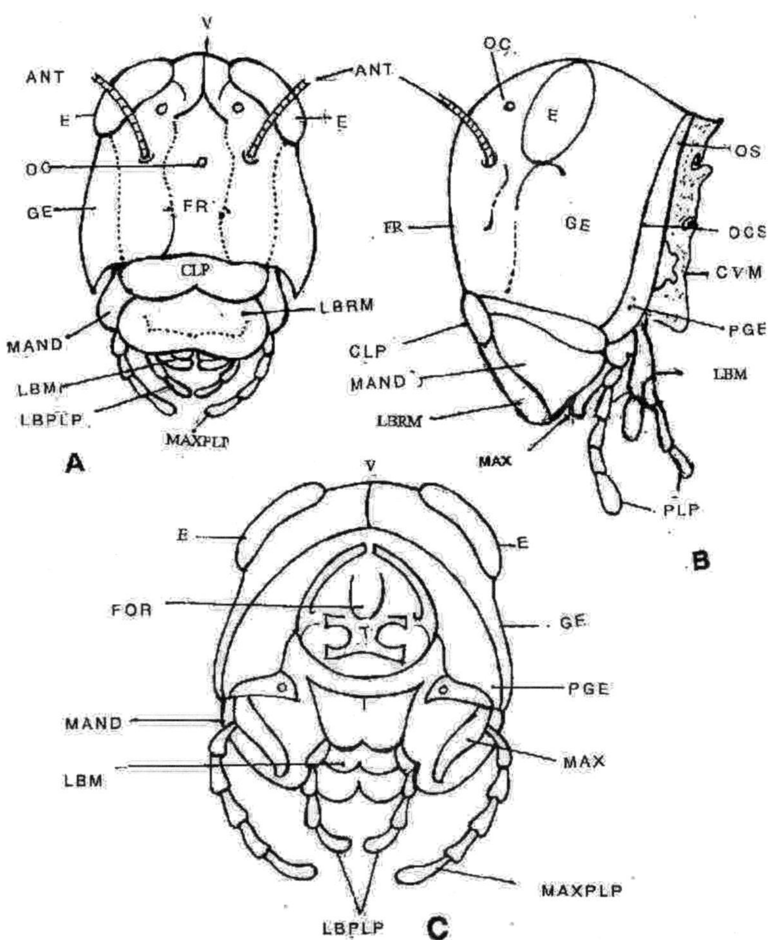

Fig. 2. A. Frontal view of a grasshopper head, **B**. Lateral view of a grasshopper head, **C**. Posterior view of a grasshopper head. *Abbreviations:* ANT - antenna, CLP - clypeus, CVM - cervicle membrane, E - eye, FR - frons, FOR - foramen magnum, GE - gena, LBM - labium, LBRM - labrum, LBPLP - labial palp, MAND - mandible, MAX - maxilla, MAXPLP - maxillary palp, OC - ocellus, OS - occipital sclerite, OCS - occipital suture, PGE - post gena, T - tentorium, V - vertex.

(1) The Frons

It is a large unpaired upper sclerite that occupies the frontal area of the head which extends from the frontal sutures to the clypeus and it laterally extends upto the eyes and to the genal suture running from the eyes to the anterior condyle of the mandible. The frons (FR) bears the median ocellus and the antennae although the antennae are said to belong to the vertex.

(2) The Clypeus

It is separated from the frons by the epistomial or fronto- clypeal or clypeal suture and the upper lip or labrum is attached along its anterior border.

(3) The Labrum

It is an unpaired sclerite movably articulated with the clypeus with the clypeo-labral suture.

(4) The Vertex

It is the upper part of the head, which lies between the compound eyes and behind the frontal suture. This area includes the dorsal portion of the two sclerites separated by the coronal suture or ecdysial cleavage line. These lateral sclerites are called the parietals, which extend back to the occipital sclerite and down each side of the head to include the compound eyes and the genae.

(5) The Occiput

It is a narrow, arching, sclerite lying behind the vertex and extending downward on each side to form postgenae. Behind the occiput lies a narrow rim which attaches the head to the neck membrane. A few small cervicle sclerites are present in the neck membrane.

(6) The Genae

These represent the parietals or lateral areas that extend below the eye and between the genal suture and the post genae.

(7) The post genae

The parts of the occiput below the level of the compound eyes and behind the occipital suture are known as post genae.

(8) The compound eyes

These are located in the lateral areas i.e. the parietals, and each eye is surrounded by a narrow ocular sclerite.

(9) The Ocelli

These are simple eyes, three in number. The median ocellus is located in the frons and the lateral ocelli are placed near the compound eyes.

(10) The Antennae

These arise in front of the eyes. Each is set in a membranous area, the margin of which is reinforced by a ridge, the antennal ridge or sclerite.

(11) The Tentorium

It is an endoskeletal structure which braces the head internally. The tentorium is composed of two pairs of invaginations which unite within the head to form a strong frame, overlying on the ventral nerve cord and supporting the foregut which pass above it. It has, anterior, posterior and dorsal arms. The anterior arms arise as invagination at each end of the clypeal suture, the posterior arms arise from the lower part of the post occipital suture and the dorsal arms extend from the anterior arms to the bases of the antennal sclerite.

(A) THE ANTENNAE

The insect head is provided with a pair of antennae. Each antenna is a slender structure composed of a number of parts. The antenna is made up of three regions viz., the first region called scape, the second pedicel and the remaining part is known as flagellum. They are of different form in different orders of insects. Some are undoubtedly modified for increasing surface area for receptors but many plumose antennae have no more receptors than filiform types. The common types of the antennae (Figs. 3A-I, 4A- E) are the following :—

(1) Filiform : These are long, thread like antennae e.g. grasshoppers. (Fig. 3E)

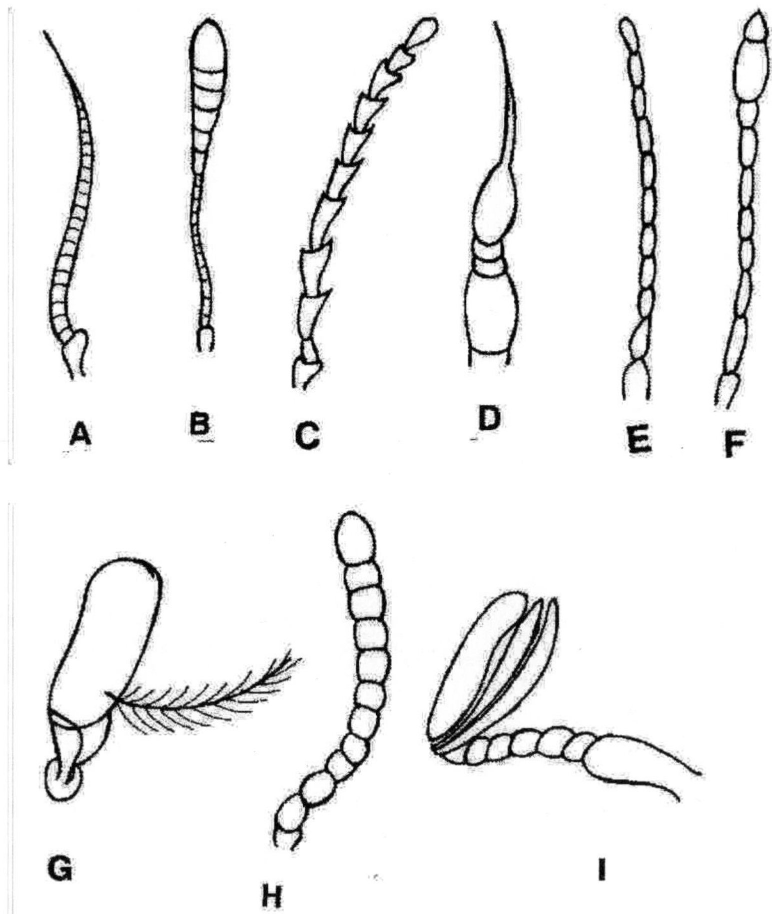

Fig. 3. Types of antennae : **A.** Setaceous, **B.** Clavate, **C.** Serrate, **D.** Stylate, **E.** Filiform, **F.** Capitate, **G.** Aristate, **H.** Moniliform, **I.** Lamellate.

(2) Moniliform : They are made up of globose segments and are beaded in appearance e.g. termites. (Fig. 3H)

(3) Setaceous : These are bristle like antennae where the size of the segments decreases from the base to the apex region e.g. Cockroaches. (Fig. 3A)

(4) Clavate : The last segment of the antenna is oval or like a cone or club (clubbed) e.g. Butterflies. (Fig. 3B)

(5) Capitate : They form a terminal large knob e.g. some beetles. (Fig. 3F)

(6) Lamellate : These have the terminal segment very much expanded on one side forming a broad plate e.g. some beetles. (Fig. 3I)

(7) Serrate : This type of antenna has saw-like teeth e.g. some beetles. (Fig. 3C)

(8) Pectinate : These have comb like side projections the monopectinate; when both sides have comb like projections they are known as bipectinate e.g. moths. (Fig. 4C)

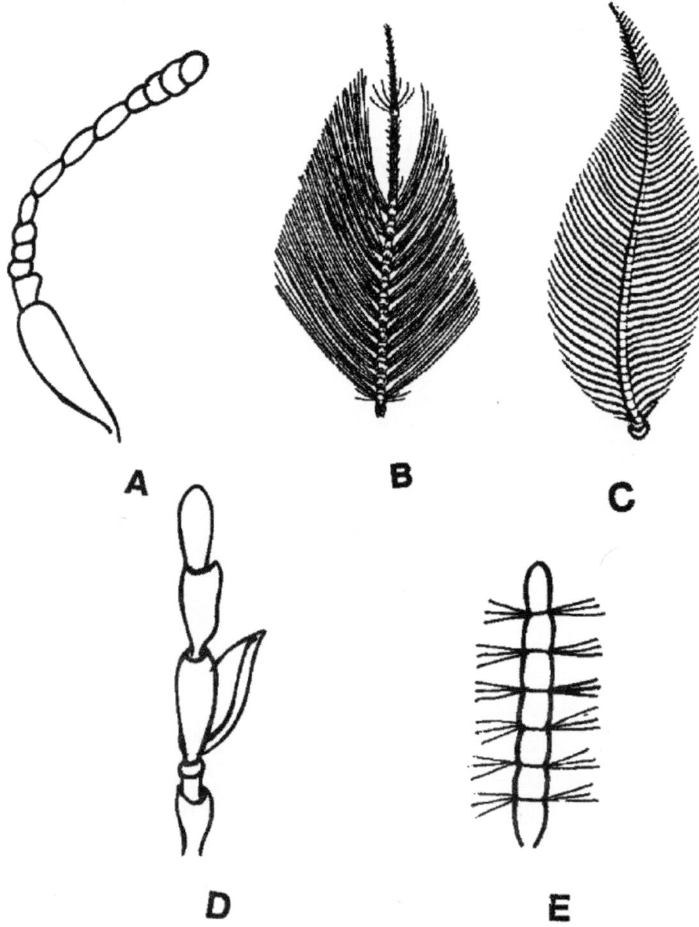

Fig. 4 Types of antennae : **A.** Geniculate, **B.** Plumose, **C.** Pectinate, **D.** Flabellate, **E.** Whorled.

(9) Plumose : These antennae have whorls of hairs coming off from each joint e.g. mosquitoes (Fig. 4B)

(10) Aristate : These antennae bear hairy bristles known as arista e.g. housefly, fleshfly etc. (Fig. 3G)

(11) Geniculate : It is bent sharply at the distal end of the scape e.g. wasps. (Fig. 4A)

(12) Stylate : These antennae have a hairy bristle called style at the end of the last segment e.g. Tabanus fly. (Fig. 3D)

(13) Flabellate : The antenna has a special apparatus known as Flabella (Fig. 4D)

(14) Whorled : The antennae having a whorl of bristles at every node (Fig. 4E)

(B) THE MOUTH PARTS

The appendages of the head which are situated around the mouth and which help the insects in the act of feeding are known as the mouth parts. The mouth parts (Figs. 5A, B; 6 A,B; 7 A, B, C.) in a generalized insect comprise of the following :

(1) Labrum (2) Epipharynx

(3) Mandibles (4) Maxillae

(5) Hypopharynx (6) Labium

(1) **Labrum** : It is a flap-like structure, attached to the clypeus and is capable of a limited up and down movement. It covers the mandibles in front and acts as upper lip and helps to pull the food into the mouth. It has a notch or cleft in the middle, which helps in holding the edge of leaf in position so that the mandibles can bite it effectively. Its ventral or inner surface is provided with gustatory organs.

(2) **Epipharynx** : It is inseparably attached to the labrum forming the inner, under or posterior face in the grasshoppers and crickets. Thus, it is continuous with the roof of the mouth and to the oesophagus. It is a sensitive area which contains the organs of taste.

(3) **Mandibles** : These are the most important structures in biting and chewing type of mouth parts. These are stiff, solid and compact structures articulating with the head possessing about three

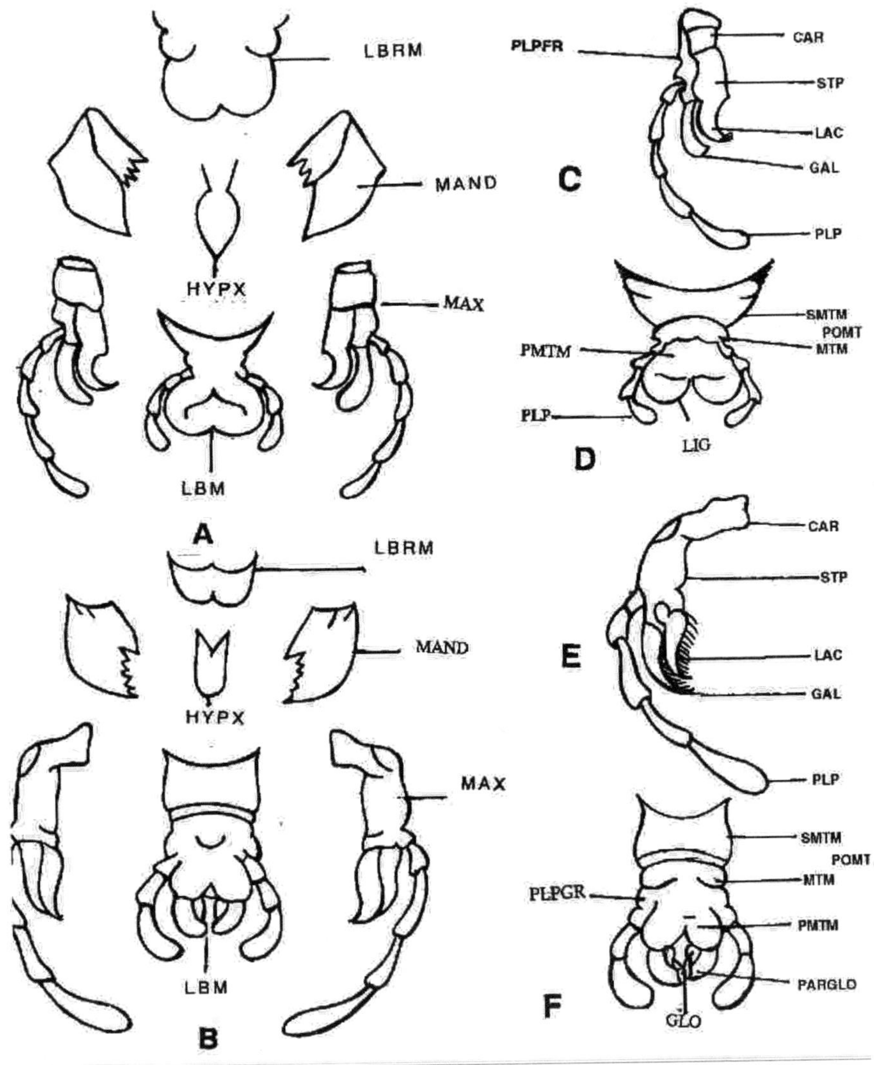

Fig. 5 A. Mouth parts of a grasshopper, **B.** Mouth parts of a cockroach, **C**. Maxilla of a grasshopper, **D**. Labium of a grasshopper, **E**. Maxilla of a cockroach, **F.** Labium of a cockroach.

Abbreviations: HYPX- hypopharynx, LBRM - labrum, LBM - labium, MAND - mandible, MAX - maxilla, CAR - cardo, PLPFR - palpifer, STP-stipes, LAC- lacinia, GAL-galea, PLP-palp, SMTM-submentum, MTM - mentum, PMTM - prementum, LIG - ligula, POMT - postmentum, PLPGR - palpiger, GLO - glossa, PARGLO - paraglossa faces and are shaped like a pyramid. The upper or outer surface is articulated and connected with a gena, the inner face is provided with large number of teeth which are quite wavy and strong and the teeth of the opposite mandibles are worked together. The hinder ends of the mandibles are articulated by a rounded process that fits into a socket of the genae, post genae or subgenae. They are acted upon by two powerful muscles, one closing against each other and the other pulling the other apart. The mandibles are regarded as first pair of jaws, and they move from side to side in a transverse plane instead of up and down. In addition to chewing the food, the mandibles also help in cutting of leaf or tissues or other objects on which the insect is feeding. In some insects, they carry the food whereas in other cases they act as organs of offence and defence.

(4) **Maxillae**: The second pair of jaws, which are much complicated in structure than the mandibles, also work side by side just like the mandibles. The exact shape of maxillae differs with the kind of food and mode of feeding. It is composed of cardo, stipes, and palpifer (Fig. 5C). The cardo is the first or proximal part and in many insects it is the only portion directly attached to the head. The stipes articulates with the distal border of the cardo and bears a lateral sclerite. The palpifer forms the maxillary palp which is the common structure of the maxillae. The maxillary palp is antennae like in shape, 1 to 5 or 7 segmented structure and functions as olfactory and gustatory organ. Maxilla is distally composed of two lobes viz. (1) an outer lobe galea and (2) an inner lobe lacinia. The galea is often of two segments and is variable in shape. The lacinia is also known as the blade; its inner border is spined or toothed and often modified for cutting, grasping and grinding the food.

As remarked earlier, maxillae are functionally a pair of sensory jaws where the lacinia helps the mandibles in holding the food when the mandibles are extended as well as assisting in mastication.

(5) **Hypopharynx** : It is a fleshy tongue, arising from the floor of mouth or preoral cavity and is usually attached to the inner wall of the labium. Its size and form is quite variable and it is more or less covered with hairs. The salivary glands open through it. It is a fine tube used for piercing in the act of feeding.

(6) **Labium**: It is formed by the fusion of a pair of appendages homologous to maxillae. The common name lower lip conveys good idea about its function because it encloses the mouth cavity from below or behind, and stands apposed to the upper lip. Out of all the

mouth parts, this is the most complicated one and is composed of two maxilla-like structures which have fused along the median line. Thus it consists of appendages resembling that of the maxillae. The labium is divided into two primary regions viz. (a) proximal-postmentum and (b) a distal-prementum (Fig. 5 D,F). The line of division between the two portions is known as labial suture. In some insects and many Orthoptera the postmentum is divided into a distal sclerite or mentum behind the labial suture and a submentum which is a proximal area of the original postmental plate. In larval forms and in some other cases the mentum is divided into two main parts, the prementum acts as the basis for the different appendages like labial palp, paraglossa and glossa.

On either side near the base of the prementum is the palpiger (Fig. 5F) which carry the labial palp. Labial palps are antenna-like pieces consisting of 1 to 4 segments and functioning as sensory organs. Similarly from the distal margin of prementum arises two pairs of lobes which collectively constitute the ligula. Thus, the ligula comprises of 4 unsegmented appendages viz. a median or inner pair of glossae and a lateral or outer pair of paraglossae. Usually the glossae are fused to form a median glossa or the prementum, or the prementum may bear a single median lobe which may be termed as ligula.

TYPES OF MOUTH PARTS

Insects feed on plants and animals in a variety of ways and their mouth parts have become modified for these purposes. Here, an account of only the representative of the more important types of mouth parts is given.

(1) The Piercing and sucking type (Fig. 6, B)

This type of mouth part is characterized by a tubular jointed beak. This tubular beak encloses several needle shaped stylets. The outer tube is formed by labium, which is simply a protecting structure for other parts. It has nothing to do with piercing the tissues or drawing up the liquid food. The mandibles and maxillae are principal components, sometimes supplemented or replaced by labrum- epipharynx and hypopharynx and are greatly elongated, slender structures which serve for piercing the skin of an animal or the epidermis of a plant. The labrum-epipharynx and hypopharynx also form a food channel or an inner tube in which the liquid food or sap is drawn. In some cases elongated 4-stylets formed by modification of two mandibles and two maxillae are present but no palps. The two maxillae are fused. The food channel and salivary duct is formed by maxillae as in Aphids, plant-bugs and bed bugs etc. In others six stylets: two mandibles, two maxillae, a labrum- epipharynx and a hypopharynx are present, and all of them play an important role in the piercing and sucking mechanism.

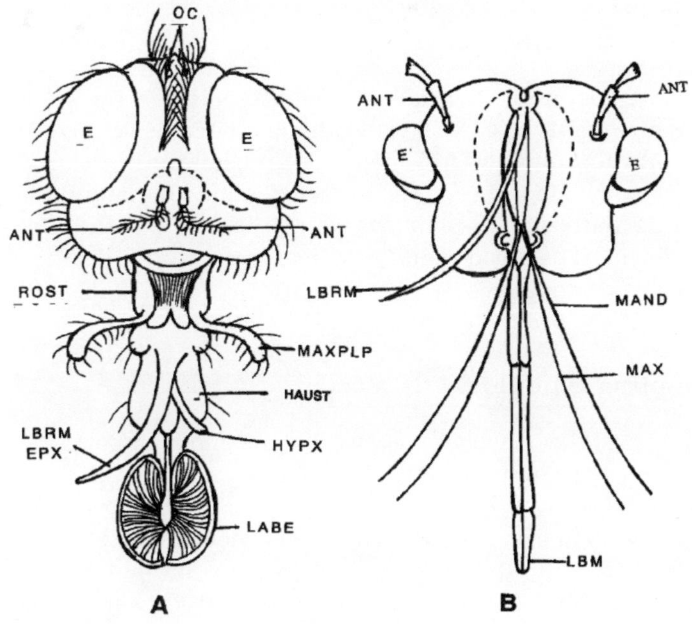

Fig. 6 A. Sponging type of mouth parts, **B.** Piercing and Sucking type of mouth parts.
Abbreviations - ANT - antenna, E - eye, HAUST - haustellum, HYPX - hypopharynx, LABE - labellum, LBRM - labrum, LBM - labium, LBRMEPX - labrum-epipharynx, MAND - mandible, MAX - maxilla, MAXPLP - maxillary palp, OC - ocellus, ROST - rostrum

(2) Chewing type (Fig. 7C)

This type is found in those insects which feed by mastication or by chewing the food. It is regarded to be a primitive type. In this type all the parts are marked easily. This type is usually found in the Orthopteroid orders. These consist of generalized mouth parts viz. mandibles, maxillae, labrum, labium, epipharynx and hypopharynx. The mandibles are hard, chitinous, toothed structures, fitted to work transversely and used for cutting, tearing and mastication of food. The maxillae are provided with powerful lacinia and galea and serve in cutting and grinding. Maxillary and labial palps are always associated with this type. This variety is the commonest type of mouth parts and are found in silver fishes, grasshoppers, crickets, termites, lice, beetles, wasps, many insects larvae and caterpillars. The mandibles may be adapted for grinding, chewing, catching, scraping etc.

(3) Rasping sucking type

This type is more or less intermediate between the chewing and piercing sucking types. The cone shaped beak is composed of the clypeus, labrum, elements of maxillae and labium. Within the beak are found the maxillae, hypopharynx and left mandible. Right mandible is reduced. The maxillae and the left mandible are stylet like in form. The food canal is between the labrum and the hypopharynx. Saliva passes between the hypopharynx and labium. e.g. Thrips of the order Thysanoptera.

(4) Siphoning type (Fig. 7B)

Very specialized variety, in which the most important working structures are the maxillae. Some portions of which are very much elongated and jointed to form a slender hollow tube which is coiled when not used like a watch spring underneath the head. Here the maxillary palps are rudimentary and the labrum is greatly reduced. Mandibles are absent. Labium is represented only by a very small basal plate and a long hairy or scaly 3-segmented labial palp. The proboscis is not in a position to pierce the skin or epidermis except in rare cases but feeding takes place by uncoiling the tube and extending it upto the exposed liquid and sucking it through the food channel. The food channel is formed by the locking of the right and left galeae of maxillae together and commonly the food is the nectar in the nectary of the flowers e.g. adult-moths or butterflies of the order Lepidoptera.

(5) Sponging type (Fig. 6A)

This type of mouth parts are found in housefly. There is a fleshy retractile proboscis on the lower side of the head and is known as labium. The basal segment of labium is known as rostrum whereas the distal one is called as haustellum. The terminal end of labium is termed as labellum and is specialized, and sponge like organ.

Labellum is traversed with a number of channels or furrows known as pseudotracheae which converge at one point in the centre of labellum. From this point the food is drawn up through the food channel into the oesophagus. The food channel is formed by labrum-epipharynx and the hypopharynx. The mandibles are absent. Maxillae are represented only by a pair of maxillary palps. There are only 2 stylets which are completely incapable of piercing the skin. These insects can only take liquid food or some solid substances like sugar which can be dissolved in saliva. Labellum is placed onto the liquid food when the pseudotracheae get filled with it by capillary action the food is sucked up from the central point into the oesophagus e.g. Houseflies, Fleshflies etc.

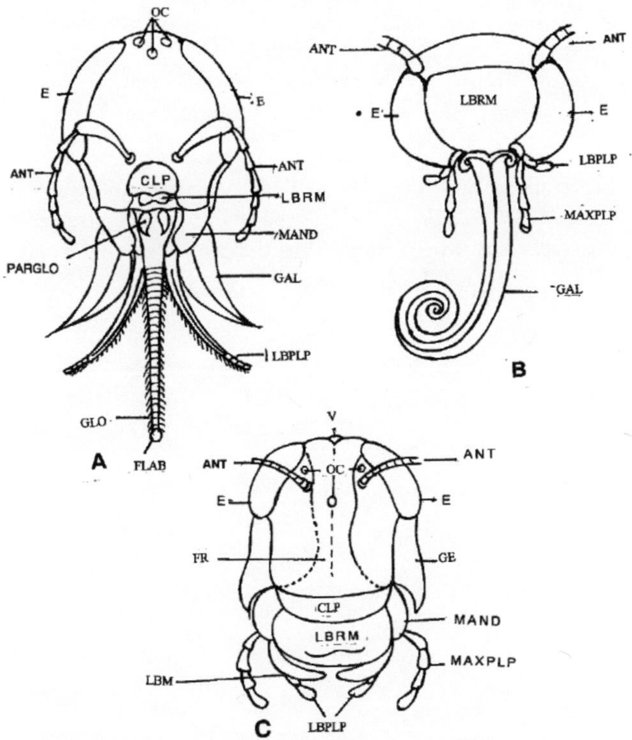

Fig. 7 A. Chewing-lapping type of mouth parts, **B.** Siphoning type of mouth parts, **C.** Chewing type of mouth parts.
Abbreviations : ANT - antenna, CLP - clypeus, E - eye, FLAB - flabellum, GE - gena, GAL - galea, GLO - glossa, LBM - labium, LBRM - labrum, LBPLP - labial palp, MAND - mandible, OC - ocellus, PARGLO - paraglossa.

(6) Chewing - Lapping type (Fig. 7A)

It is a combination in which the labrum and mandibles are like that of chewing type but the maxillae and labium are elongated and closely united to form a type of proboscis or lapping tongue. In this case both the pairs of palps are present, labial palps long and maxillary palps very small. A hairy flexible tongue is formed by glossae of the labium which are greatly elongated. The tongue can be quickly projected to reach the nectar and then contracted e.g. Honeybees, wasps etc.

(7) Degenerate type

In the larvae of certain Orders, the mouth parts become somewhat reduced although the mouth parts of the larvae are fundamentally the chewing type in general and that of the nymphs like adult. There are 2 categories in this case:

(1) *Hymenopterous, Trichopterous, Lepidopterous larvae* — In this case maxillae, hypopharynx and labium are associated or united forming a combined lower lip whereas the ligula and the hypopharynx form a median lobe. The labial palps are absent.

(2) *Dipterous larvae* — In the Orthorrhaphous Diptera mouth parts are of chewing type but in Cyclorrhphous Diptera like housefly and flower fly maggots, the true mouth parts are absent because the true head segments are invaginated into the throat and special mouth parts have been developed to serve during the larval life.

3

THE THORAX AND ITS APPENDAGES

A. THORACIC SEGMENTS

The thorax is the second region of the body of the insects. It is connected to the head by a membranous narrow neck region which permits a wide range of movement to the head. The thorax comprises of three segments viz. (1) prothorax (Fig. 8), (2) mesothorax (Fig. 9A) and (3) metathorax (Fig. 9B) (The wing bearing segments are called pterothorax). Each segment bears a pair of legs, and in all winged insects, the pterothorax support the wings. Within the thorax are the powerful muscles which operate the wings and legs. In the floor of the thorax medially lies the ganglionated ventral nerve cord.

(1) Prothorax (Fig. 8)

The prothorax in the locusts and grasshoppers is greatly expanded and overlaps the mesothorax. The outer part of the thorax consists of 3 types of sclerites viz. tergal, pleural and sternal, on the dorsal, lateral and ventral side, separated by sutures or connected with chitinous membranes. The tergum covers the segment like a bonnet and extends over the greater part of mesothorax. Only a small corner of each pleuron is visible which lies above the base of the leg and the lower margin of the proepisternum (Fig. 8, Eps 1). The pleuron of each thoracic segment is composed of two sclerites (1) an anterior-episternum and (2) a posteriorepimeron. The names of the sclerites are indicated by the profixes promeso or meta, pertaining to prothorax, mesothorax and metathorax e.g. proepisternum means episternum of the prothorax and so on.

In case of Acridids the tergum of thorax is called notum. The notum of the prothorax is called pronotum. The sternum is composed of two sclerites - a flat anterior and a pointed posterior. These are not easily recognized hence the entire area is termed as prosternum.

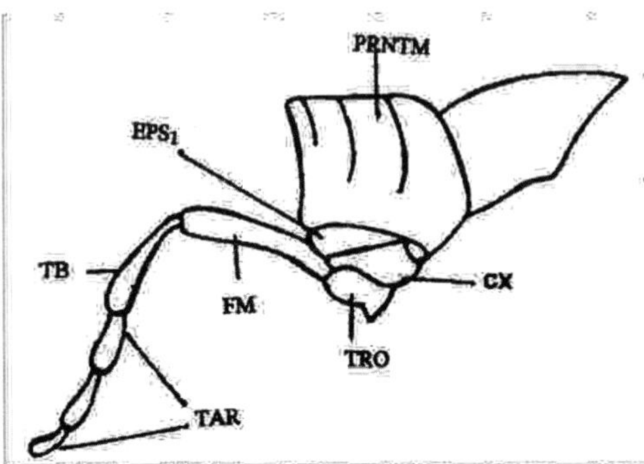

Fig. 8 Prothorax with leg attached
Abbreviations : CX - coxa, EPS1 - episternum of the prothorax, FM - femur, PRNTM - pronotum, TAR - tarsus, TB - tibia, TRO - trochanter.

The mesothorax and metathorax (Fig. 9 A,B)

In this case the two segments are firmly united together and form a strong box like structure. In all the wing bearing insects the mesothorax and metathorax supports the wings hence combined segments are usually called as pterothorax. The tergal plates of the pterothorax are distinct with strong transverse sclerotized areas for the attachment of the muscles as well as for the support of the various structures associated with the attachment and movement of the wings.

The pleurites of each side of the mesothorax and metathorax are firmly united to form a continuous lateral wall. The pleurite of each segment is composed of two sclerites which are separated by an oblique suture. These sclerites are capable of movement upside down as well as posteriorly. They are named according to the segment to which they belong. The sclerites of the mesothorax are termed as mesepisternum and mesepimeron and likewise of the metathorax as metepisternum and metepimeron (Fig. 9A, Eps 2, Epm 2, Eps 3, Epm 3). The spiracles are located on the sides of the mesothorax and metathorax. The mesothoracic spiracles (ii SP) are located in the membrane between the pro- and mesothorax whereas the metathoracic spiracles (iii SP) are located directly above the 2nd pair of legs. There is no prothoracic spiracle in present day insects. In this connection it may be mentioned that according to Snodgrass (1935) the thorax of post embryonic stages of all insects, except Diplura never has more than two pairs of spiracles and these two pairs are formed in the embryo on the anterior part of the mesothorax and metathorax respectively. He further remarked that in many insects, however, the thoracic spiracles migrate forward during development. The mesothoracic spiracles particularly are subject to this anterior migration and hence often occur on the sides of prothorax for which reason they are called "prothoracic spiracles".

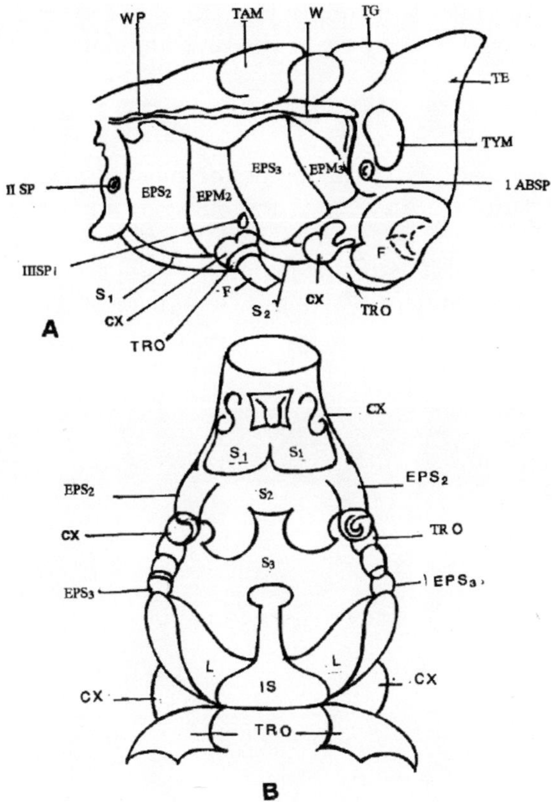

Fig. 9A. Lateral view of meso and metathorax, **B.** Ventral view of the entire thorax.

Abbreviations : CX - coxa, EPM2-epimeron of mesothorax, EPM3-epimeron of metathorax, EPS2-episternum of mesothorax, EPS3-episternum of metathorax, F-femur, 1S-sternum of the 1st abdominal segment, II SP - spiracle of 2nd thoracic segment, III SP - spiracle of 3rd thoracic segment, L - lateral lobe of the metasternum, S-1-prosternum, S-2 - mesosternum, S- 3 - metasternum, 1-ABSP - 1st abdominal spiracle, TYM-tympanum, TAM-tergum of mesothorax, TG - tergum, TRO - trochanter, TE-tergum of 1st abdominal segment, W - cut end of wing, WP - wing process.

The sterna of the meso- and metathorax consist of a flat plate which covers the ventral surface and is joined laterally with the pleura in front of the coxae. The anterior margin of the mesosternum is convex while the posterior edge is deeply notched. The metasternum is provided with a median quadrat lobe which fits into the notch of the mesosternum, while the posterior margin is deeply notched to receive the sternum of the first abdominal segment.

B. THE WINGS

The presence of wings is one of the most characteristic features of the insects and the insect's wings differ from that of other flying animals. It differs in the sense that the wings in the insects are the structures super added to the primitive motor equipment on their ancestors. The number of wings vary from two pairs to none. Primitive insects like silver fish, spring tails have no wings. Others such as lice and fleas have lost their wings due to the parasitic life. Flies and related insects have the second pair of wings modified into balancing organs known as halteres. Due to the various shape, size and texture, the classification of insects is based on the characters of the wings.

Origin of wings

There are several theories which explain the origin of wings in insects. The main theories are outlined below.

(1) **Tracheal gill theory** : This theory was put forth by Gegenbaur, Lubbock, Graber, Langversion, Woodworth and many others. According to this theory, the wings have been derived from thoracic tracheal gills or the plate like structures as shown in mayfly nymphs which have lost their original function and become adapted for the purpose of flight. It is claimed that the tracheal gills as they are basally articulated with the body and supplied with muscles, trachea, only require to become enlarged and adapted for flight. Tracheal gills, however, vary in position and may be developed from the dorsal aspect of the terga, from the sterna, or the pleura or at the apex of the abdomen and in head and even between the wings. At the same time, the wings differ in their mode of development and are not serially homologous with the tracheal gills. Further more, this theory also compels to assume that the ancestors of the winged insects were at least temporarily aquatic and thus acquired gills which subsequently developed into wings when these animals become air

breather. Lull (1940) mentions that the tracheal gill theory of the origin of the wings seems to have much evidence in its favour.

(2) **Paranotal theory** : This is the most widely accepted one and it is conveniently termed by Crompton (1916) as "paranotal theory". It is due to Miller (1873-75), Korscht, Heiden, Snodgrass, Packard etc. that it got full support. This theory has also been supported by recent workers like Sharrov (1966), Hamilton (1971-72), Wotton (1976), Rasnitsen (1976) etc. This theory postulates that the wings arose in the first instance as lateral extensions of the thoracic terga. It is a lateral fanlike tergal expansion which is known as paranotal lobe of the thorax. Snodgrass remarked that many of the carboniferous insects had in addition to wings a pair of small flat paranotal lobes projecting laterally from the tergum of the prothorax and these lobes according to him suggested that at an earlier period the wings themselves have developed from similar tergal lobes of the mesothorax and metathorax. These expansions i.e. paranotal lobes were very similar to those found on the prothorax of *Stenodictya lobata, Lemmatophora typica,* and other fossil insects. There is an inherent tendency towards the development of such structures in diverse groups of insects. During the course of their evolution it is believed that the tergal expansions become sufficiently large to function somewhat after the manner of paracheute in insects which possessed a tendency to glide. At a later stage, they acquired direct articulation with the tergal region, become supported with the trachea, and developed the power of independent movement.

Structure of the wing

Structurally, wing is a flattened double layered extension of the body wall and in its lumen there are nerves, tracheae and the body fluid. The wing development can be easily observed in any of the Hemimetabolous insect like *Lygaeus*, or *Dysdercus* wherein, the wing buds, appear in the second or third instar nymphs in the form of hollow, flattened outgrowths of the body wall along the latero-dorsal margins of the meso- and metathorax. Following each moult the wing buds become larger and larger and in the final moult the fully developed wings appear. In Holometabolous insects e.g. Lepidoptera, Diptera, Coleoptera and others, the wings develop internally (hence the name Endopterygota). The body wall invaginates and the wings are formed within the pouches underneath the cuticle. These invaginations are everted, usually during the last larval stage and the final moult discloses the pupa with the wing pads fully developed. If such a wing is cut across before it hardens fully, it will show the structures as shown in the Fig. 10.

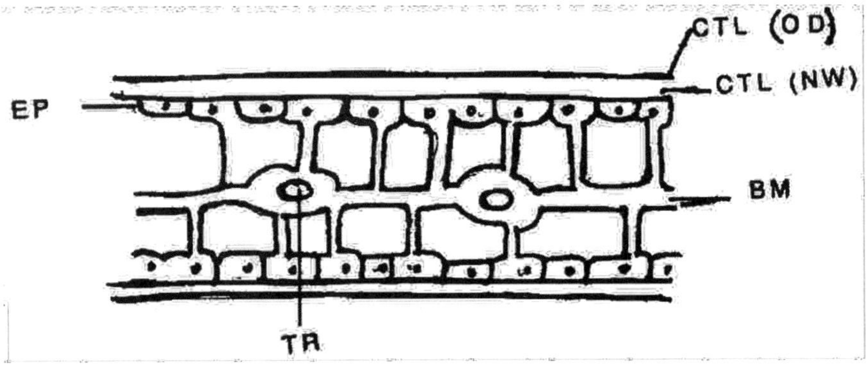

Fig. 10 T.S. of Wing
Abbreviations : CTL - (OD) old cuticle, CTL - (NW) new cuticle, BM - basement membrane, EP - epidermis, TR - trachea

As a result of the elongation of the hypodermal cells, the two basement membranes fuse and only at certain points the channels are left open. These longitudinal channels usually contain tracheae which determine the future pattern of the wing venation. In the fully developed wing the hypodermis is absent. Thus, a mature wing is a cuticular appendage in which the blood is seen circulating slowly between the membranes.

Articulation of wings (Fig. 11A)

Wings to be of any use in flying must have a well organized point(s) of articulation with the thorax. Since wings are mere outgrowths of the body wall, it is necessary that a point or points of flexure develop and that muscles be oriented a new, in order to give movement to these new organs. Such articulations are present and their complicated structures are represented in Fig. 11A. Muscles are attached to some of these axillary sclerites, but the main wing movements are brought about through the contraction and expansion of the pterothorax by enlarged longitudinal and dorsoventral muscles. In figure 11 B,C,D, are presented the probable movements of wings and thorax in flight. In B the wings are raised by the contraction of the tergo-sternal muscles, which draws the tergum and sternum closer together and thus elevates the wings on their wing process.

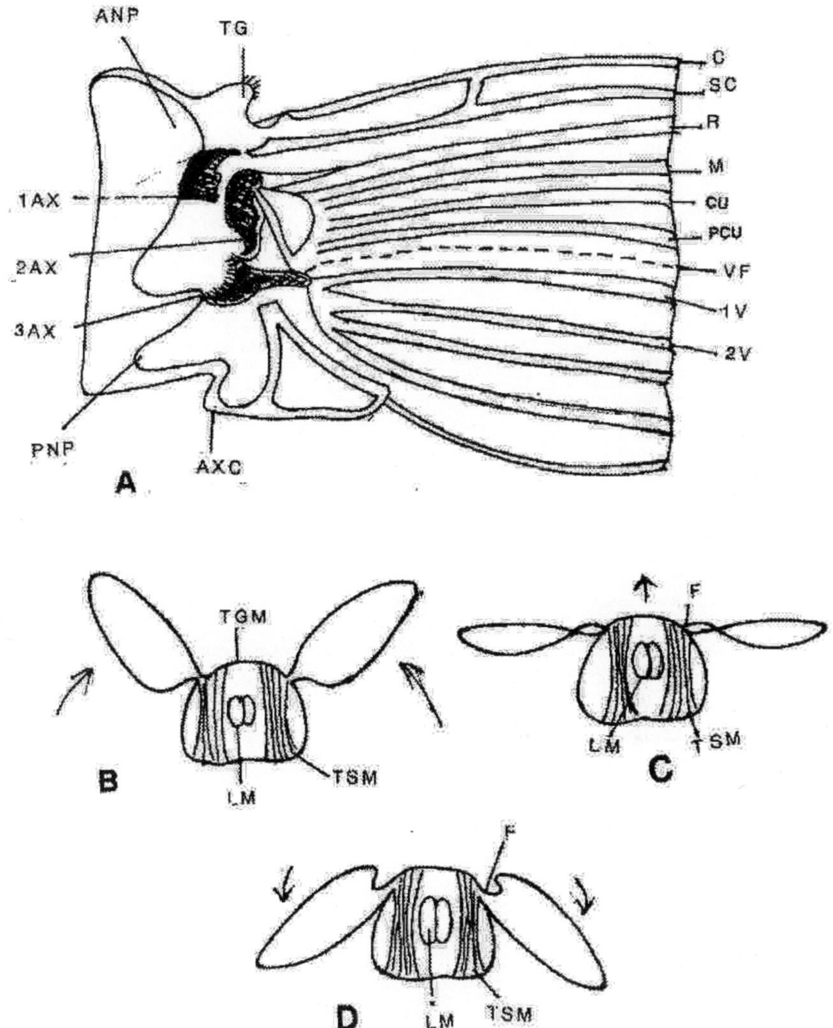

Fig. 11 A Articulation of wing **B, C, D. -** Wing movements
Abbreviations : ANP - anterior process, PNP - posterior process of thorax to which wings articulates, AXC - axillary cord, 1AX, 2AX 3AX - sclerites of wing base in the membrane which permits the movements of wings against the notal process, C - costa, SC - sub-costa, CU - cubitus, M - medial, R - radial, VF - vannal fold, 1V - vannal vein 1, 2V - vannal vein 2, PCU - postcubitus. TGM - tergum, TSM - tergo sternal muscles, LM - longitudinal muscles, F - flexure muscles.

In B and C, the contraction of the longitudinal muscles elevates the tergum and thus lowers the wings, while the flexer muscle causes torsion in C and its relaxation allows the wing margin to again elevate as in D.

Margins of the Wings (Fig. 12B) : The insect wings are mostly triangular in form and have three margins viz. (1) the outer or apical margin, (2) anterior margin or costa and (3) inner margin or anal margin.

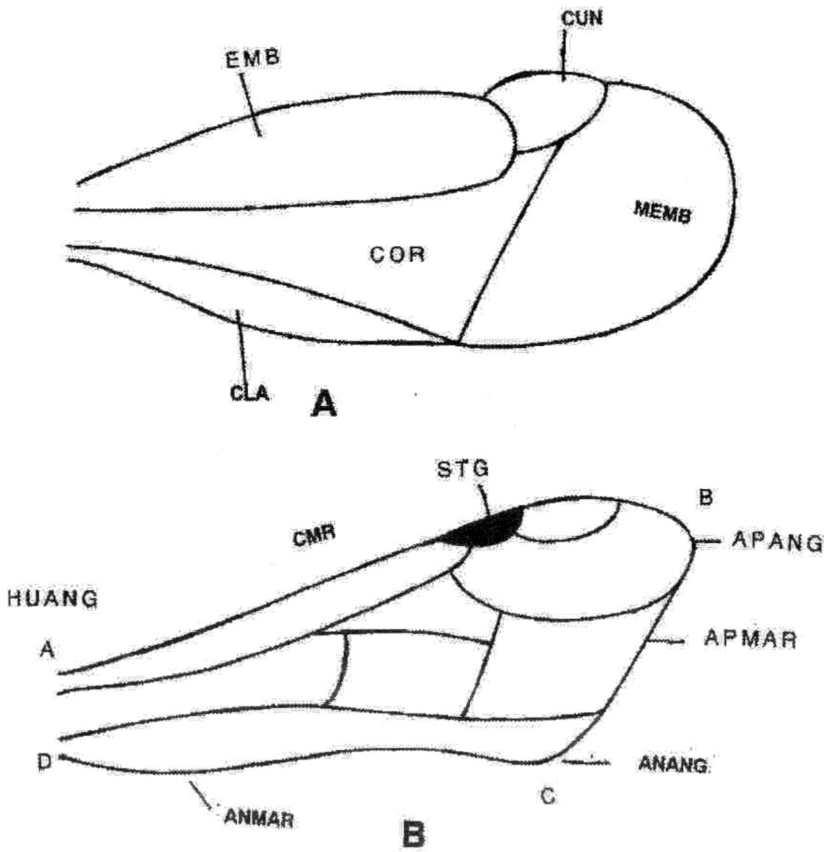

Fig. 12. A. Forewing of a Hemiptera, **B.** Wing of a Hymenoptera. *Abbreviations* : APANG - apical angle, APMAR - apical margin, ANANG - anal angle, ANMAR - anal margin, CLA - clavus, CMR - costal margin, CUN - cuneus, COR - corium, EMB - embolium, HUANG - humeral angle, MEMB - membrane, STG - stigma.

Angles of wings (Fig. 12 B) Similarly there are three well defined angles of wing viz. (1) humeral angle. This is the angle at the base of the costal margin of the wing, (2) the apex B or the angle between the costal margin and the outer margin, (3) the angle C between the outer margin and the inner margin which is known as anal angle.

Wing surface : The microscopic examination has shown that wings have fine hairs over its surface. When it is velvety, it is obviously clothed in those groups of insects e.g. Tricoptera. Psychodidae of Diptera they are closely covered with hairs. At the same time in Lepidoptera and Thysanura the wings are covered with scales.

Pterostigma or stigma (Fig. 12B) : In many insects near the costal margin of the wing a conspicuous opaque spot is found which is known as the stigma or pterostigma e.g. this pterostigma is found in the forewings of order Plecoptera and most Hymenoptera and in both the pairs of wings of Odonata.

Wing venation : The mature wing appears as thin membrane with longitudinal thickenings called veins. The arrangement of veins in the wing is termed as venation. Different orders have different types of venation. Much studies have been given to wing venation in order to determine the primitive type characteristics of the common ancestors of all Pterygota, because this may be used in establishing the evolutionary relationships. It was Cosmstock and Needham (1898, 1918) who evolved a method by which the venation of all the orders can be derived. Generalized system of venation was studied at the time of development and it was studied by the tracheae of the wings. The veins passing in the furrows are the concave veins and those of the ridges are termed as the convex veins. Convexity is shown by "+" sign and concavity by "–" sign. The wing venation must have evolved from a wing like the hypothetical form as shown in Fig. 13.

Hypothetical wing venation : (Fig. 13). The name applied to the longitudinal and cross veins and the abbreviations used to designate them are as follows :

1. The precosta (PC) is found in the fossil insects only and it is very small vein.

2. Costa (C+) : It is a convex vein running along the margin of the wing and is unbranched.

3. Subcosta (Sc-) : It is a concave vein with 2-branches SC1 and SC2.

4. Radial (R+) : It is a convex vein with two main branches R1 + passing straight along the margin and Rs- is concave vein passing also along the margin. The Rs branch is divided into 4 branches viz- R2 + R3, R4 + R5, thus there are altogether 5 branches of the radial. The Rs branch is known as radial sector.

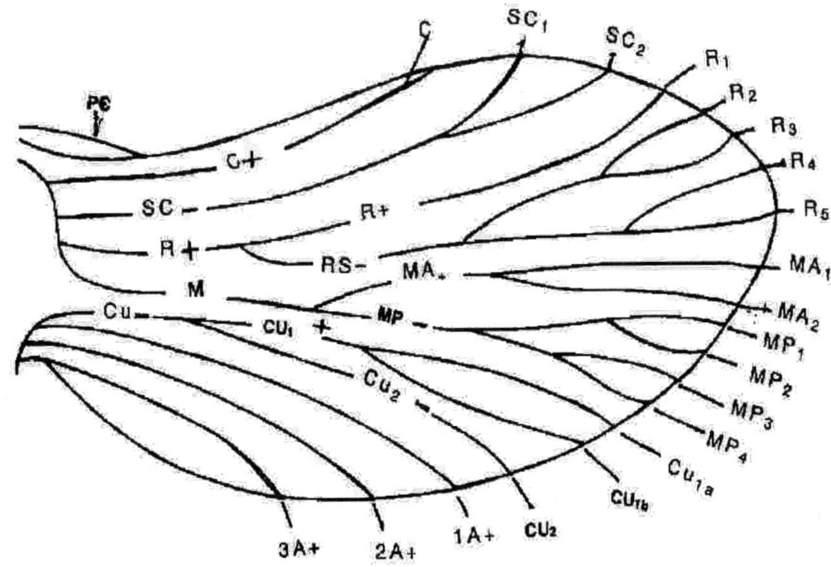

Fig. 13. Hypothetical plan of wing venation showing various veins. *Abbreviations* : 1A +, 2A +, 3A + branches of anal veins, MA - anterior media, MP - posterior media, MA1, MA2, MP1, MP2, MP3, MP4, - branches of the anterior and posterior media, RS - radial sector, CU1a - anterior cubitus, CUIb - posterior cubitus, PC - pre costa, C - costa, SC - sub costa, R + radials, M - media, CU - cubitus, SC1 - subcosta 1, SC2 - sub costa2, R1, R2, R3, R4, R5, - branches of radius, CU1, CU2, branches of cubitus.

5. Media (M-) : It is a concave vein and divides into two branches + median anterior (MA+) and median posterior (MP-). The former one divides into two branches MA1 and MA2 and MP which is a concave one, divides into 4 branches MP1, MP2, MP3 and MP4.

6. Cubitus (Cu-) : It is also a concave vein divided into two branches Cu1 + and Cu2-. The Cu1 divides into 2 branches anterior and posterior veins: Cu1a and Cu1b. Cu2- is a concave vein passing straight to the margin without any division.

7. Anals (IA+) : In the hypothetical wings there are 3-A, out of which the 1A+ is a convex and 2A ± may be a concave or convex vein. 3A + is always a convex vein.

There are certain small transverse radials which connect two longitudinal veins or their branches and they are called cross veins. The cross veins are as follows (Fig. 14) :

1. Humeral (h) : It connects the subcosta and the costa near the humeral angle of the wing.
2. Radial (r) : It connects radial sector with the radial.
3. Sectorial (s) : It connects R2+R3 with R4+R5.
4. Radiomedial (r-M) : It connects the radius and medial.
5. Median (M) : It extends between M2 and M3.
6. Medio-cubital (M-Cu): This cross vein connects media with cubitus.

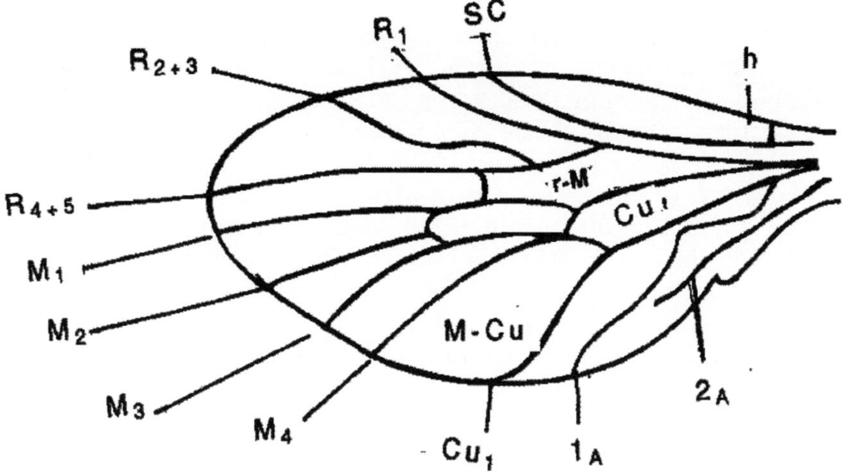

Fig. 14. Wing of a horse fly, *Anisopus punctatus* with all the cross veins. *Abbreviations* : 1A,2A - anal veins, CU1 - cubitus 1, Cu - cubitus, M-Cu - mediocubital, r - M - radiomedial, h - humeral, SC - subcosta, R1,-R2+3-radials, R4+5 - radial, M1, M2, M3, M4-branches of media.

Kinds of wings

In Orthoptera the fore wings are tough and leathery and are known as tegmina. It is a protective covering having no part in flying, it is the hind wings which are flight wings. In Hemiptera the wings are termed as hemelytra. In Heteroptera half of the wing is tough and the other half is membranous, in which case the distal part is membranous and the proximal part is toughened. It is again modified into different types. The apical half is known as corium and on the inner margins of it, there is a narrow ridge termed as clavus found in the family Lygaedae. In Capsidae out of corium a narrow triangular area is differentiated on the costal margin near the membranous area and is known as the cuneus. In another family Anthocoridae out of the corium another narrow area is differentiated and is known as embolium (Fig. 12A).

Cells in wings

There are a lot of cross veins in the wings. Due to their appearance, wings are divided into a number of spaces which are termed as cells (Fig. 14). The cells completely surrounded by veins are called closed cells, those that extend to the margin of the wing are called open cells. The names of the cells are designated from the veins forming inner margin of the cells and thus two groups of cells-basal and distal are formed. Distal cells are bounded by the branchings of the main veins while the basal cells are bounded by the main veins themselves. The cells formed by the basal portion of the radius is referred to by the term "Cell R" whereas the cell of the distal is termed as "Cell Rs". When two veins join together or fuse together, cell in between becomes obliterated. Thus, if two branches of radius R2 and R3 unite, the fusion is indicated by writing at the margin of the wings R2+3, which shows that branches R2 and R3 have fused together and the cell behind such a vein is termed as R3+4 and not cell R2. The cell R2 has been eliminated with the fusion of veins.

Modifications of veins

The veins of the hypothetical forms are not found in higher forms. This is due to certain modifications e.g. a complete median vein was present in Palaeozoic fossil insects but in the recent forms only in Ephemeroptera. In most cases M is divided into two named as MA and MP. MA atrophies completely in most cases and in the recent forms median is represented by MA unit, but it is designated by letter M only. In Odonata it is MA which is present and not MP. The anal veins are extremely variable and in wings with a reduced anal area, one or more are atrophied. On the other hand, in insects with a well developed anal area, the anal veins may be branched through the subdivision of 2A. The modification come into form by two ways only (1) by reduction of veins and by (2) addition of veins; generally reduction is brought about either by complete atrophy or complete degeneration of the veins or one or more branches and also by coalescence or joining of one vein by another vein e.g. due to atrophy a single well developed anal vein in the *Anisopus* is represented by R 2+3 or R 4+5 in the place of two originally separate veins viz. R 2+R3 or R4 + R5 (Fig. 14). Coalescence takes place in two ways : (1) one at the point of divergence and in this the divergent branches are pushed apart. The divergence thus becomes pushed forward and forward upto the margin and ultimately forms one vein,(2) the other way of coalescence is fusion at the apices.

Addition : It occurs due to the multiplication and due to the increase in the number of branches of the veins or due to the development of the secondary longitudinal veins between existing ones.

THE LEGS

Types of Legs

The insect thorax is composed of three segments viz., prothorax, mesothorax and metathorax. Each thoracic segment is provided with a pair of appendages - the legs. Each leg, possesses the following movable parts — (1) the coxa, (2) trochanter, (3) femur, (4) tibia and (5) tarsi. Pretarsus is present in some insects as 6th part (Fig 15). The coxa, femur and tibia are single pieces in each leg while the

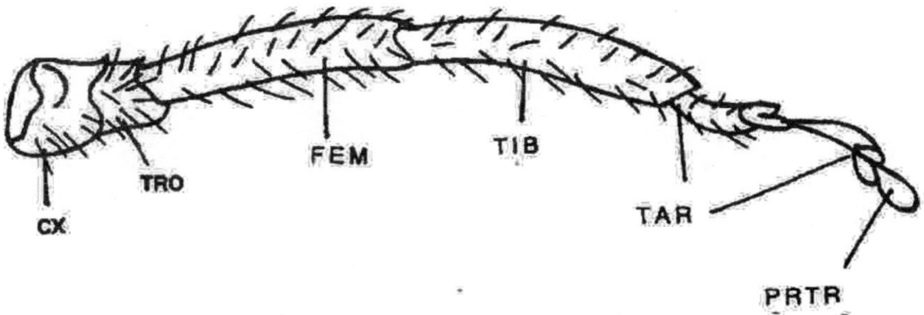

Fig. 15. Structure of an insect leg.
Abbreviations : CX - coxa, FEM - femur, PRTR - pretarsus, TIB - tibia, TAR - tarsus, TRO - trochanter.

trochanter is composed of two and the tarsal region 1-5 components except in Collembola where the tarsus is fused with the distal end of the tibia and called tibiotarsus. Two claws with a pad- the pulvillus, arolium or empodium in between, are attached in the distal end of the last tarsus and the entire structure is called pretarsus. For different purposes, the various segments of the legs are variously modified and they also bear a number of cuticular structures meant for a definite function.

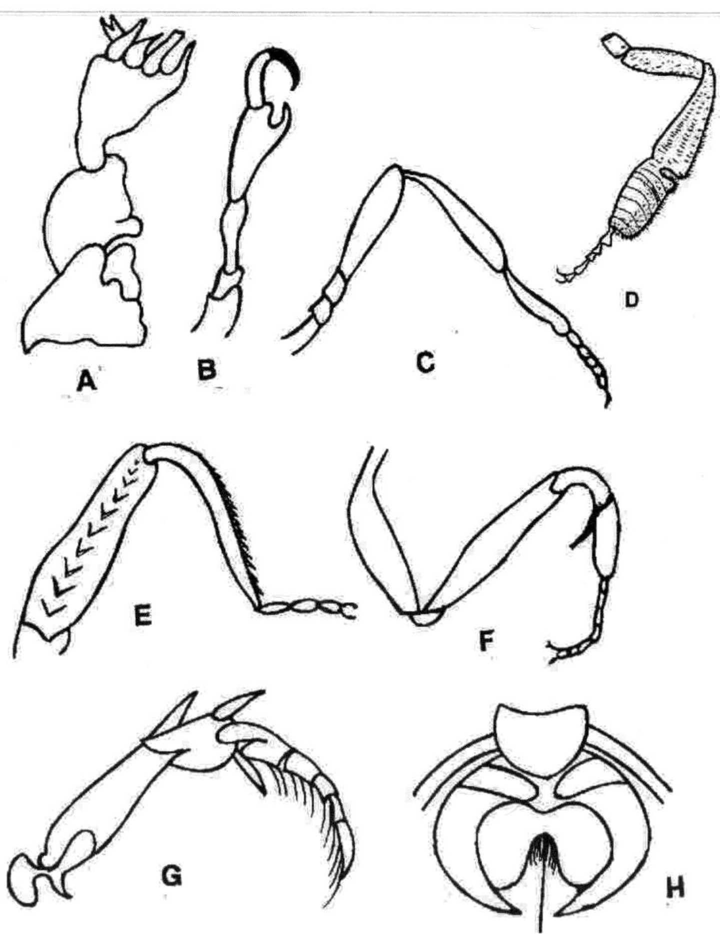

Fig. 16. Different types of legs. **A.** Fossorial, **B.** Scansorial **C.** Ambulatorial, **D.** Foragial, **E.** Saltatorial, **F.** Raptorial, **G.** Natatorial, **H.** Foot of housefly adapted to walk upside down (After Metcalf and Flint, 1962)

The legs in insects are highly modified and accordingly perform different functions like (1) walking - ambulatorial e.g. bugs, (2) running - cursorial e.g. tiger beetles and cockroaches (3) clinging - scansorial e.g. louse; (4) sound producing - stridulatorial e.g. crickets, (5) hopping - saltatorial - e.g. grasshoppers (6) digging - fossorial e.g. molecrickets, (7) swimming- natatorial e.g. water beetles (8) grasping - raptorial e.g. mantids etc. (Fig. 16 A-H).

4

THE ABDOMEN AND ITS APPENDAGES

The third division of the insect body, the abdomen, consists of ten or eleven segments (Fig. 17 A,B). In most insects, the actual number of segments present is less and it is not possible to recognize more than five or six segments. The reduction occurs at the posterior end of the body.

Each abdominal segment consists of a tergum or dorsal plate, a sternum or ventral plate, lateral areas of membrane connecting tergum and sternum and a spiracle on each side, usually situated in the lateral membrane. In Acrididae in the first abdominal segment is located a large tympanic membrane that covers the ear which is the characteristic feature of the family.

In most insects, the 8th and 9th abdominal segments are greatly modified to form reproductive appendages or the external genitalia. The 9th abdominal segment in the male insects and the 8th and 9th in the female insects form the genital apparatus.

Male genitalia

In recent years the genitalia of male insects have been studied in great detail because here we find structures that are of great aid in distinguishing species. Unfortunately the homologizing of these structures is not easy and no one has yet succeeded in presenting an adequate account of them. As a result a great varieties of names have been applied. The 9th abdominal segment is the genital segment in male insects in which also lies the male genital opening.

In a typical male Pterygote insect, the complete external genitalia are composed of (1) penis, (2) a pair of parameres or the inner processes and (iii) a pair of claspers that serves to grasp the female when mating and between which lies the aedeagus, within the aedeagus lies the penis. The penis is formed by the fusion of paired outgrowths and it bears the male gonopore. The original paired condition is found only in Ephemeroptera and in certain Dermaptera. The inner processes or parameres are present in several orders but it is not certain that they are homologous in all cases. The claspers or gonopods are derived from the appendages of the 9th segment (Figs. 17B, 18 A,B).

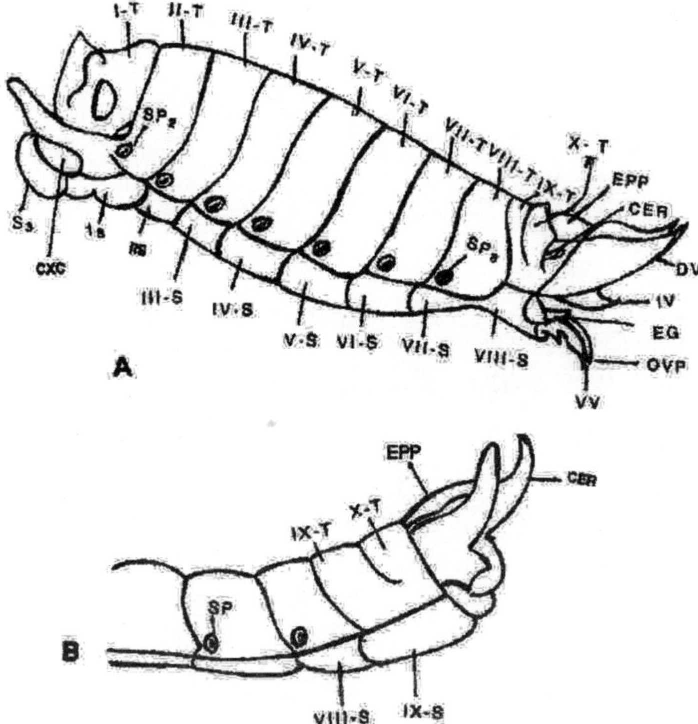

Fig. 17. A. Abdomen of a female grasshopper (lateral view), **B.** Terminal segment of a male grasshopper.
Abbreviations: CER - cercus, CXC - coxal cavity, DV - dorsal valve, EG - egg guide, EPP - epiproct, IV - inner valve, OVP - ovipositor, S3 - sternum of 3rd thoracic segment, I S-IX S 1 to 9th abdominal sterna, SP2 - SP8 2 to 8th spiracles, I-T to XI-T - 1st to 11th terga, SP- spiracle, VV - ventral valve.

In several apterygote and pterygote insects, the gonopods are unmodified limb bases bearing either long styli but perform no genital functions as in Thysanura, or short styli as in Dictyoptera. In many insects e.g. Ephemeroptera, Mecoptera and Trichoptera, the styli alongwith their limb bases form functional clasping organs. They are absent in Coleoptera.

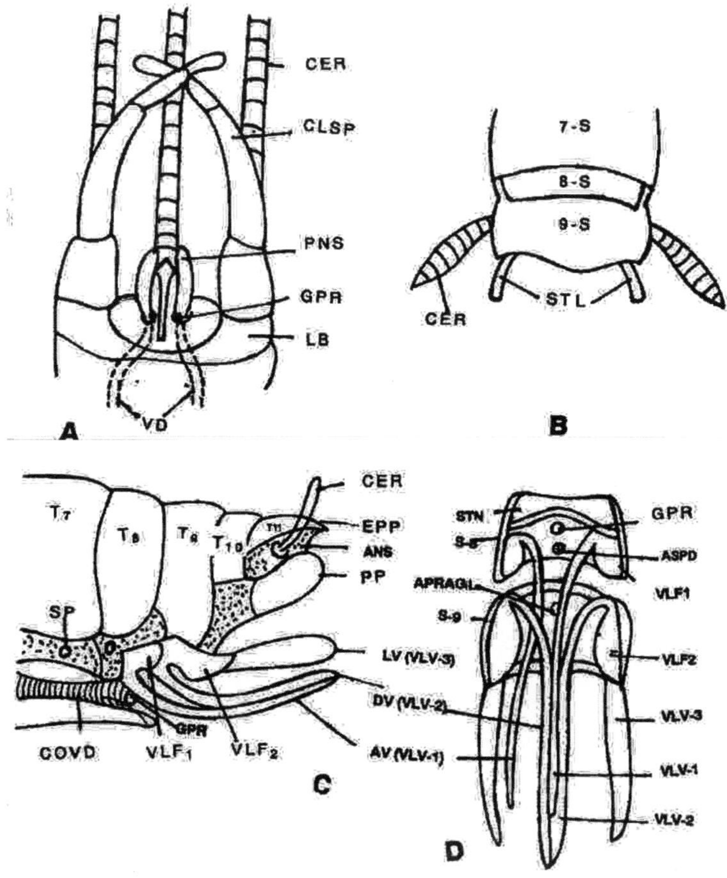

Figs. 18. A. Showing male genitalia of *Ephemera* **B.** Showing male genitalia of *Blatta*, **C.** Showing segmental relations of the parts of ovipositor of a pterygote insect, **D**. Ventral view of genital segment and ovipositor. *Abbreviations* : ANS- anus, CER – circus, COVD – common oviduct, EPP – epiproct, PP – para proct, LV (VLV3) – lateral valve, DV (VLV – 2) – dorsal valve, AV (VLV-1) anterior valve, VLF1 – valvifer-1, VLF2 – valvifer-2, ASPD- aperture of spermathecal duct., STN - sternum, SP - spiracle, STL - stylet, T - tergum, GPR - gonopore, CLSP - clasper, PNS - penis, LB - limb base, VD- vas deferens, 7-S - sternum, 8-S - sternum, 9-S - sternum, APRAGL- aperture of accessory gland.

Female genitalia

In all pterygote insects the ovipositor is composed of three pairs of valves (or valvulae) viz. (1) anterior or ventral, (2) posterior or inner and (3) lateral or dorsal. The first pair, anterior or ventral valves, arise from the valvifer-1 of the 8th segment. The 2nd, posterior or inner valves are the median outgrowths of the valvifer-2 of the 9th segment and the 3rd pair, lateral or dorsal valves arise from the valvifer 3 of the 9th segment. These three pairs of valves (valvulae) are variously modified in several insects (Figs. 17A, 18 C,D). In several Hymenoptera, the ovipositor is modified as a poison injecting sting and loses its original functions; the right accessory gland is modified as poison gland and the left accessory gland as alkaline gland. Both the glands open separately. The poison gland discharges its contents into a large poison sac which opens into a bulb like median stylet (Fig. 19).

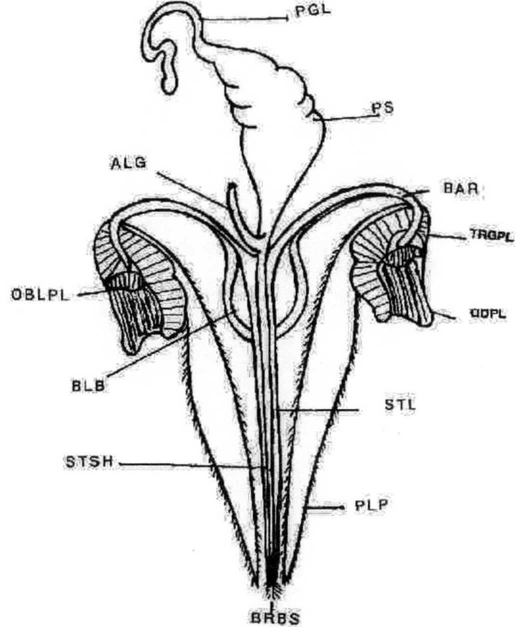

Fig. 19. Sting appratus of a bee.
Abbreviations : BAR - bar, ALG - alkaline gland, PGL - poison gland, PS - poison sac, TRGPL -traingular plate, OBLPL - obleque plate, QDPL - quadrate plate, BLB - bulb, STL - stylet, STSH - stylet sheeth, PLP - palp, BRBS - barbs.

In Coleoptera, Lepidoptera, Diptera and several other insects the valves (Valvulae) are either poorly developed or not at all developed. In these insects the apical abdominal segments form an extencile tube that functions as an ovipositor.

5

THE DIGESTIVE SYSTEM

ALIMENTARY CANAL

The alimentary canal in insects (Figs. 20 AB, 21 and 22) is a tubular structure of variable length arising as an infolded part of the body wall. The structure of the wall of the alimentary canal is therefore essentially the same as that of the integument. It is this part of insect anatomy which discharges the functions of digestion, absorption and egestion of the food materials. Associated with the alimentary canal are the organs of ingestion namely the mouth parts and other glands such as salivary glands, and hepatic caecae which pour their secretions into it. The entire alimentary canal extends from mouth to the anus in insects. This tubular structure is often longer than the body of the insect and therefore, variously convoluted upon itself. In fact, the length of the alimentary canal and the complexity of its structure depends on the food habits of the insects. The alimentary canal in insects is divisible into three primary regions on the basis of embryonic origin :

1. Stomodaeum
2. Mesenteron
3. Proctodaeum

1. STOMODAEUM

It is ectodermal in origin. The stomodaeum is also known as the foregut. In the primitive insects e.g. Collembola (Spring tail), Thysanura (Silver fish), etc. - the stomodaeum is an undifferentiated simple tube; however in many highly evolved forms, this part of the alimentary canal has become differentiated into four primary regions viz., (1) pharynx, (2) oesophagus, (3) crop, and (4) proventriculus.

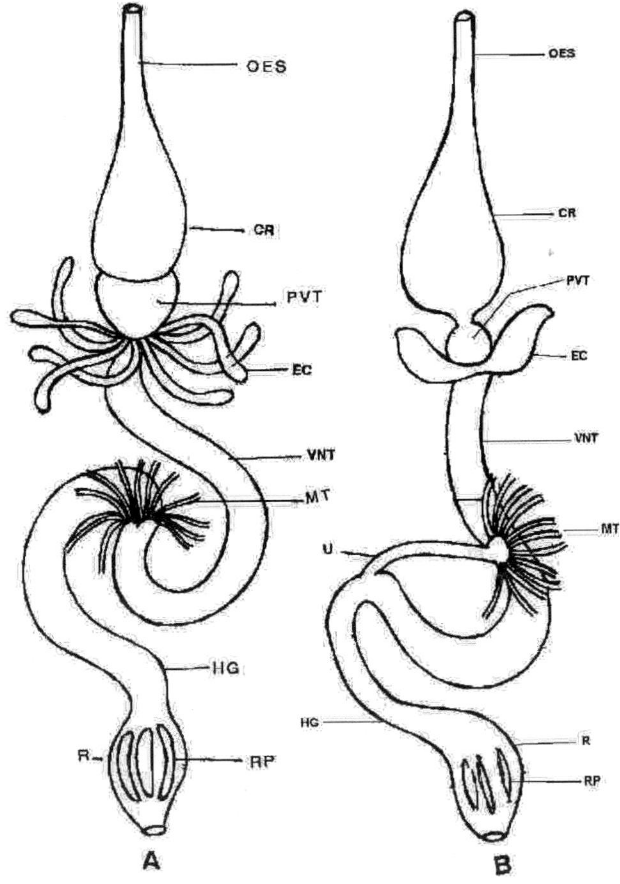

Fig. 20 A. Alimentary canal of *Periplaneta americana*, **B**. Alimentary canal of *Gryllodes sigillatus*.
Abbreviations : CR - crop, EC - enteric caeca, HG - hind gut, OES - oesophagus, PVT-proventriculus, R - rectum, RP - rectal pad, U-ureter, VNT - ventriculus, MT - Malpighian tubules.

1. **Pharynx** : This part of the stomodaeum is although structurally undifferentiated from the buccal cavity but it can be identified as that part of the foregut which is provided with dorsal dilator-muscles, originating from the frontal and dorsal area of the head wall. Usually the pharynx is a simple structure and occupies a

position between the brain and the sub-oesophageal ganglion, but in some insects it is divided into an anterior pharynx and a posterior pharynx.

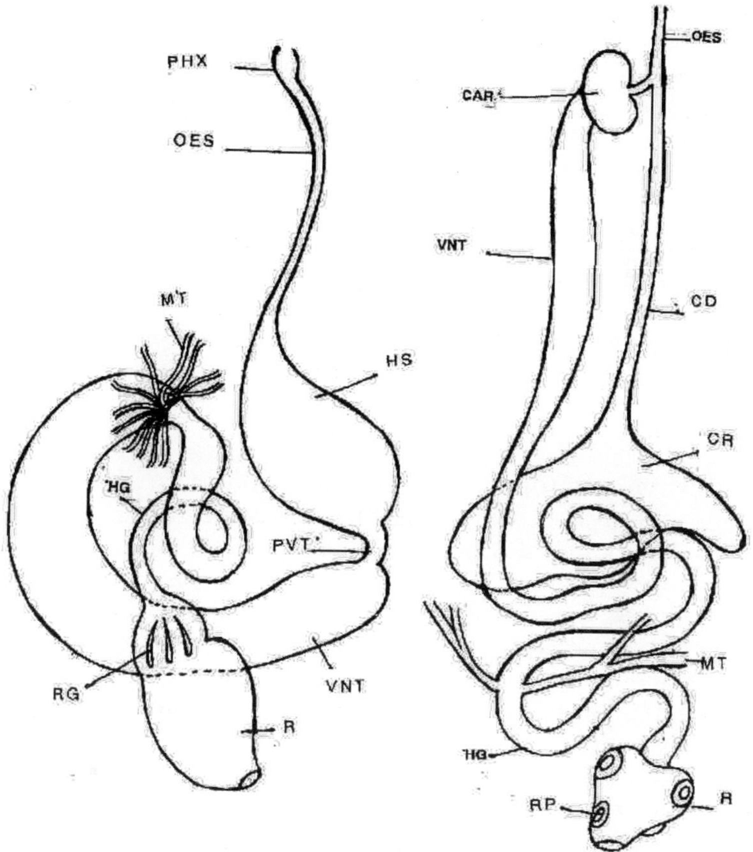

Fig. 21 Alimentary canal of Honey bee. **Fig. 22** Alimentary canal of House fly.

Abbreviations : PHX - pharynx, OES - oesophagus, HG-hindgut, HS - honey stomach, PVT - proventriculus, VNT - ventriculus, RG - rectal gland, R - rectum, MT - Malpighian tubule, CR - crop, RP - rectal pad, CD - crop duct, CAR - cardia.

The primary function of the pharynx is to receive the food ingested by the insect. However, in some insects having sucking mouth parts e.g. Lepidoptera, Hymenoptera, Diptera and Hemiptera the pharynx performs the function of a well developed sucking pump.

2. Oesophagus - It is without any morphological status and is merely a narrow part of stomodaeum following the pharynx. Its chief function is the conduction of food material from the pharynx to the crop. It is variable in length and the inner walls are longitudinally folded.

3. Crop - This part of the foregut in insects is nothing but an enlargement of the posterior part of the oesophagus. In some insects, it occurs as a lateral diverticulum of the oesophagus having either a simple sac or has a long slender tube with bladder like swelling at the end e.g. larvae of *Sarcophaga, Musca.* etc.

The chief function of the crop is storage, however, according to Sanford (1918) it is supposed that this region of stomodaeum also performs the function of digestion and absorption of food substances e.g. cockroaches. Although the view has been strongly criticised by Abbott (1926), Wigglesworth (1928), Treherne (1960) and others but according to Krishna and Saxena (1962) the crop of housefly does the function of digestion and absorption.

4. Proventriculus : The proventriculus or (Gizzard) is the terminal part of the stomodaeum. It is structurally the most highly specialized part of the alimentary canal and is best developed in Orthoptera, Coleoptera, Dictyoptera etc. It is usually absent or reduced in sucking insects e.g. Hymenoptera, Hemiptera and Diptera. In its simpler form as in many larvae and adults, this region is merely the narrow posterior end of stomodaeum, which is more or less invaginated into the anterior end of the mesenteron to form the cardiac valve.

Types of proventriculus

In insects with biting and chewing type of mouth parts the inner walls of the proventriculus develops strong cuticular plates or teeth e.g. Blattidae, Gryllidae, Tettigoniidae (Fig. 23A-D). In some biting and chewing insects e.g. Acrididae, the inner walls of the proventriculus are thrown into longitudinal folds. The surface of the folds are not strongly sclerotized, in grasshoppers they are provided with a few small marginal teeth. The number of longitudinal folds varies from 4 to 8 in different insects projecting into the lumen of the mid gut. In fact, this fold type of proventriculus is a simple condition while the presence of teeth or plates is a character of complicated conditions. According to Eidmann (1924) the proventricular armature itself is a primitive equipment of the insect's alimentary canal which has disappeared in most of the sucking type of insects.

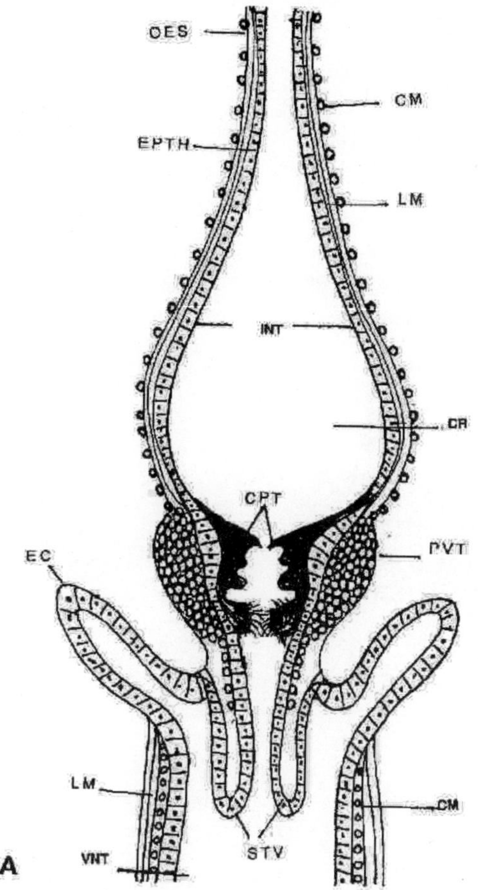

Fig. 23 A. Proventriculus with teeth and plates.
Abbreviations : CR - crop, CM - circular muscles, EC - enteric caecum, EPTH - epithelium, INT - intima, LM - longitudinal muscles, OES - oesophagus, CPT - cuticular plates with teeth, PVT - proventriculus, STV - stomodaeal valve, VNT - ventriculus.

FUNCTION OF PROVENTRICULUS

The function of proventriculus varies according to the structure of the organ, and the nature of food materials in different insects. In its simpler conditions it functions as sphincter between the crop and

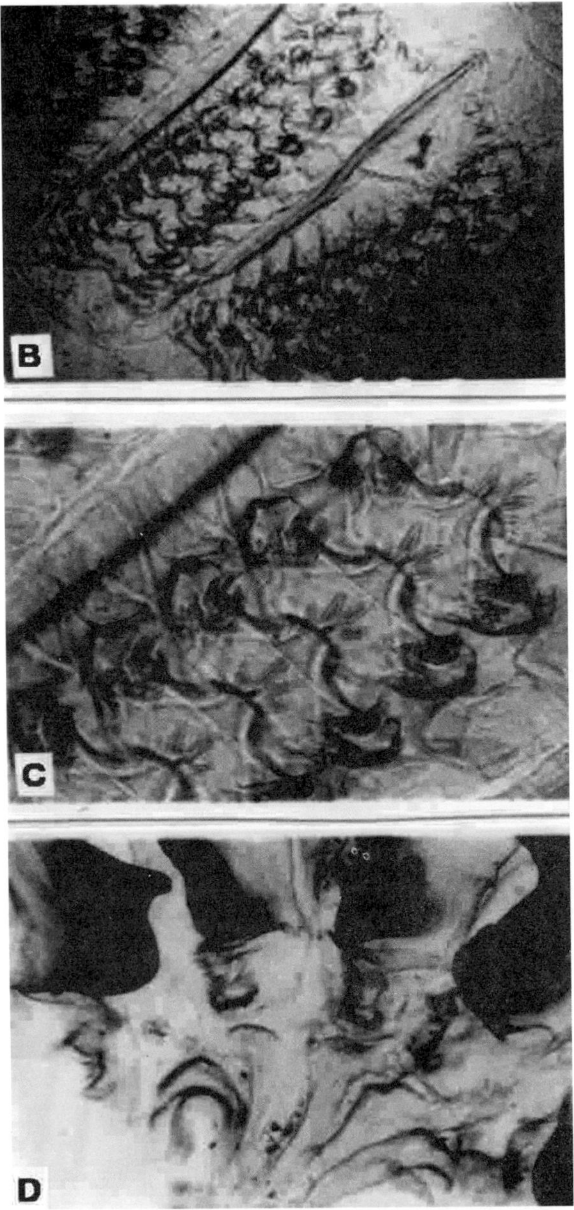

Fig. 23B Inner view of proventriculus of the house cricket, *Gryllodes sigillatus* **23C** A magnified part of B. showing arrangement of spines. **23D** Inner view of proventriculus of *Periplaneta americana*. (B.C.D. Awasthi, unpublished).

mid gut, to regulate the passage of the food from the foregut into the mid gut. However, with the development of complicated folds and teeth, the proventriculus acquires diverse functions; e.g. when folds are present - it may serve to retain the food in the crop where partial digestion may occur as in the Orthoptera and Coleoptera. When lamellae or spines are present, the proventriculus probably functions as a strainer to prevent larger pieces of hard indigestible matter in the food from entering the mid gut, such materials being later disposed off by regurgitation e.g. *Dytiscus*. Sometimes the teeth of the proventriculus help in breaking up the hard parts of the food and thus function as an organ of trituration or grinder and under such conditions this region is loosely termed as "Chewing Stomach". Work has been done experimentally to prove this function of proventriculus by several workers.

STOMODAEAL VALVE

The stomodaeal valve is essentially a circular fold of stomodaeal wall projecting into the ventriculus from the posterior end of the stomodaeum. The valve is composed of two cellular lamellae and is covered on each side by stomodaeal intima. The valve regulates the transport of food from proventriculus to the mesenteron and prevents movement of the food from mesenteron to the stomodaeum.

Histological details of the foregut reveal that it is composed of the following layers :

An innermost lining of intima which is the cuticular continuity of the body wall, a layer of epithelium and an inner layer of longitudinal muscle fibres and the outermost layer of circular muscle fibres (Fig. 24C).

MESENTERON

This is also known as the ventriculus or mid-gut. It forms the middle part of the alimentary canal in insects and functions for all the practical purposes as a "Stomach". It is endodermal in origin, and thus differs from both the stomodaeum and proctodaeum in the absence of an intima on its inner surface.

In general form, the ventriculus is tubular and is approximately uniform in diameter. In certain insects, the ventriculus is differentiated into two portions termed as cardia and the functional stomach, differing both histologically and functionally. The presence of cardia is clearly seen in many Dipterous insects like mosquitoes and flies.

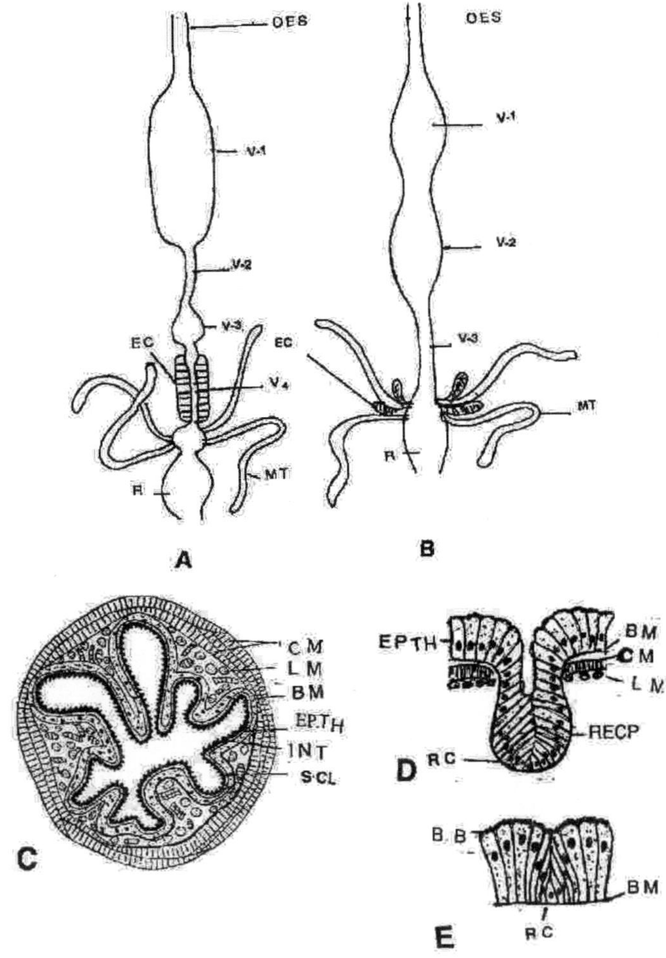

Fig. 24 A. Alimentary canal of a Heteropteran insect, **B.** Alimentary canal of a Homopteran insect. **C.** T.S. of foregut, **D-E.** Part of mid gut *Abbreviations:* BB- brush border, EC - enteric caecum, OES - oesophagus, MT - Malpighian tubules, R-rectum, V-1 - 1st ventriculus, V-2 - IInd ventriculus, V-3 - IIIrd ventriculus, V-4 - IVth ventriculus. CM–circular muscles, BM– basement membrane. LM–longitudinal muscles INT–intima, EPTH- epithelium, RC - regenerative cells, RECP - regenerative crypt, SCL - scales.

The regional differentiation of the ventriculus attains its climax of development in the Hemiptera (Fig. 24 A,B). In the Homoptera, it is usually divided into three distinct parts - an anterior sac like part, a middle crop-like enlargement and a posterior slender tubular part. In the Heteroptera the ventriculus is made up of 4 distinct regions. They are 1st, 2nd, 3rd, and 4th ventriculi. The first ventriculus usually is a saccular structure. This is followed by a fairly long tubular 2nd ventriculus. This is succeeded by a small bulbous 3rd ventriculus which finally opens into a short slender tubular 4th ventriculus.

In addition to the differentiation into distinct regions, the ventriculus unlike the stomodaeum has blind pouches varying in number and in length, and developed on different parts of this region. These are called the enteric caecae or caecal diverticula. The former are usually present in the anterior end of the ventriculus surrounding the stomodaeal valve. However, they may occupy certain other positions in the ventriculus as for e.g. in the anterior, middle and posterior ends e.g. Scarabid larva or as in *Blissus leucopterus*. The function of the caecal diverticula - specially in Heteroptera is, to enable the bacteria present in them to kill the foreign bacteria present in the alimentary canal of the insect and thus to provide a safe place for the multiplication of normal bacteria in the digestive tract.

HISTOLOGY

Histological details of the mesenteron show that the epithelial walls are characteristically thickened and the cells are more spongy than those of other parts of the alimentary canal (Fig. 24 D-E).

Since the ventriculus can be ``loosely'' regarded as the ``functional stomach'' of the insect, it is thus obvious that the maximum amount of digestion and absorption of the food as well as secretion of digestive enzymes occur in this region. Consequently, the epithelium of the ventriculus is made up of various types of cells. Some digestive in function, some absorptive in function, while others secretory in function. Sometimes the same cells may perform a double function i.e. digestive as well as absorptive, because of this multiplicity of activities in the ventricular region, there is always a constant degeneration and regeneration of epithelial cells. The structure of epithelium of the mid-gut varies greatly according to the state of the digestive process. Most of its cells are columnar in nature with granular or spongy cytoplasm. In addition to these large spongy cells that form most of the epithelium, there are other types of smaller cells occurring either singly or in groups between the bases of the larger cells. The larger cells exposed or projecting into the mid-gut lumen are the digestive cells i.e. the cells which secrete digestive enzymes and are responsible for the absorption and transmission of digestive materials. The smaller basal cells are regenerative cells (Fig. 24 D,E). These regenerative cells play an important role in the replacement of digestive cells when they are exhausted or injured due to secretory or other activities. The cells of the mid-gut epithelium rests on a membrane of the nature of connective tissue.

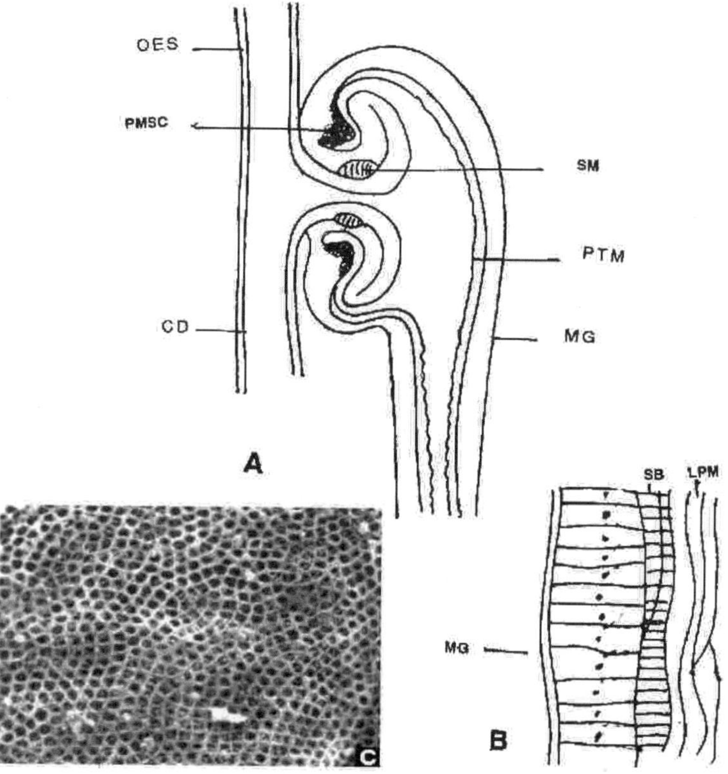

Fig. 25, A,B, Showing modes of origin of peritrophic membrane. **C.** Peritrophic membrane (From Elzinga, 1997).

Abbreviations : CD - crop duct, LPM - lamellated peritrophic membrane, MG - mid gut, OES - oesophagus, PTM - peritrophic membrane, PMSC- peritrophic membrane secreting cells, SB - striated border, SM - sphincter muscles.

The secretion of digestive enzymes takes place in two ways from the secretory cells : (1) Holocrine and (2) Merocrine. In the holocrine type of secretion of juices, which occur in Orthoptera and Coleoptera, the active epithelial cells disintegrate completely in the process of secretion and are replaced by the growth of new cells from special cell nests or nidi i.e. regenerative cells. In the merocrine type, the cells do not break down. They may discharge their secretion through the straited border. Occasionally, a special type of merocrine secretion known as vesicular secretion is described in insects where descrete droplets of fluids are eliminated through the striated border. In addition to culumnar and regenerative cells, goblet cells also occur occasionally in the mid gut epithelium.

The muscle fibres have a reverse order in the mesenteron i.e. an inner layer of circular muscles and an outer layer of longitudinal muscles. In addition to these, the ventriculus contains in most insects a thin peritrophic membrane surrounding the food contents (Fig. 25 A,B).

Peritrophic Membrane

The peritrophic membrane is present in majority of the insects both larval instars and adults. It has disappeared in most Hemiptera, perhaps because these take only liquid food but it is present in some corixids which feed on detritus and other solid matter. It has been also found to be present in Nepidae. It is said to be absent in blood sucking adults of Tabanidae. In the latter case, out of 17 species peritrophic membrane was found only in 14 species. It is detected in the adults of the flies, and mosquitoes (except *Anopheles*). It is also present in Muscidae. It is wanting also in the Carabidae and Dytiscidae, possibly because their digestion is largely extra intestinal and possibly because of holocrine type, in which the secretory cells break down completely during the process. It is permeable to the digestive enzymes and the products of digestion (see Fig. 25C).

The peritrophic membrane has been considered to arise in two ways (Wigglesworth, 1930, Waterhouse, 1953). In some Lepidoptera, Diptera, Dermaptera, it is secreted by the cells or a group of cells present at the junction of fore and mid-gut and extruded in tubular form by a muscular press in this region. While in other insects, it arises by the delamination from part or all of the general surface of the ventricular epithelium and a series of concentric membranes, formed by successive delamination are often present e.g. Phasmida,

Acridida, Ephemeroptera, Odonata, Coleoptera and Hymenoptera (See Fig. 25 A,B).

PROCTODAEUM

It is also known as the hindgut and is the posterior most part of the alimentary canal. Like stomodaeum, it is ectodermal in origin. In primitive insects, it is a simple tube connecting the mesenteron to the anus. However in insects the proctodaeum is differentiated into several distinct regions. The hindgut starts exactly a little anterior to the point where the bases of the Malpighian tubules are located. This point or region is normally constricted and is known as the "pylorus". The region into which the proctodaeum is differentiated varies in different insects. Normally, it is divided into anterior and posterior divisions, the latter being commonly referred as rectum. These two divisions of the hindgut are usually separated externally by a sharp constriction and internally by rectal valves. Although the anterior part of the hindgut is a simple tube of varying length, it is often subdivided into an anterior ileum and a posterior colon. These two parts although externally not well differentiated, differ widely in their histological details.

The posterior part of the rectum undergoes profound modifications in certain insects. In the larva of most Lepidoptera this region is provided with a cellular membrane lying beneath the muscle layer, between these two membranes the Malpighian tubules which start from the pyloric region of the hind gut traverse the course and they are thrown up into various convolutions and loops before they end or break up into a plexus of reuniting branches.

Further, in many insects the rectum has certain accessory structures known as rectal organs or anal glands. In most insects the rectal epithelium and intima of the rectal sac forms structures commonly known as rectal glands or anal glands. They appear sometimes as oval or elongated pad like thickening on the intestinal wall (rectal pads) e.g. Plecoptera, Odonata, Orthoptera etc. or in certain other cases as conical processes projecting out from the wall into the lumen of the intestine (rectal papillae e.g. Diptera, Fig. 26D). Anal glands are ectodermal glands opening near the posterior end of the rectum. They function principally as organs of defence by discharging substances having strong and repulsive odours or even explosive properties e.g. bombardier beetle.

PYLORIC VALVE

Two types of valvular structures are associated with the opening of the mesenteron into the proctodaeum. In some insects, these valves arise from the posterior part of the ventriculus and are known as ventricular valves in others they are developed in the pyloric region of the anterior end of the proctodaeum and constitute the proctodaeal valve.

Histology of the proctodaeum indicates that its structure resembles in essential details to that of stomodaeum: there is a well defined cuticular intima. The epithelium has flat columnar cells having no secretory function, an inner layer of circular muscle and an outer layer of longitudinal muscles

FILTER CHAMBER

In a majority of the Homoptera the two ends of the mesenteron and a part of the anterior end of the proctodaeum, all are brought into intimate proximity with one and other and are covered by a connective tissue sheath. This is known as the filter chamber. The chief purpose of this organ is to serve as a devise for allowing some of the excess of water and soluble carbohydrates of the food to pass directly from the anterior part of the stomach into the intestine by diffusion, while the protein and fatty materials are retained to be digested and absorbed in the ventriculus. The form, size and disposition of the organs of the filter chamber varies in different Homoptera (Fig. 26 A-C).

CEPHALIC GLANDS

In the majority of insects there are present cephalic glands or head glands or salivary glands (Figs. 27 A-D; 28, A,B,C; 29) of diverse nature that open either in the buccal cavity or in the pharynx. Some of them secrete a slimy substance i.e. saliva while in others they have also a number of digestive enzymes in their products. There are mainly four types of cephalic glands or salivary glands:

(1) Mandibular glands : These are small glands that open near to the base of the mandibles in Apterygota, Orthoptera, Isoptera, Dictyoptera, Coleoptera, Trichoptera and Hymenoptera (Bruntz, 1908, Suslov, 1912). In larval Lepidoptera these are quite large and secrete saliva, the normal salivary glands (labial) having become specialized for silk production in this group. In *Apis* the glands are larger in queens than in the workers and are very small in drones. In the queen honeybee they produce the "queen substance" and other pheromones concerned with colony control while in workers, they produce some saliva and serve to soften the cocoon at the time of emergence (Figs. 27D, 28C).

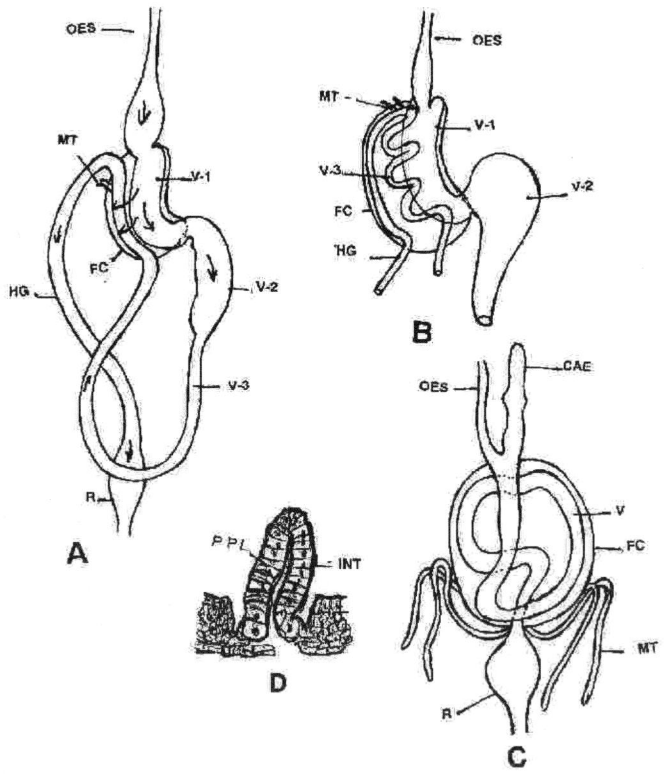

Fig. 26 A,B,C. Alimentary canal of Homoptera showing modifications of the filter chamber, **D**. Papilla like rectal pad of a fly (PPL)
Abbreviations : CAE - caecum, FC - filter chamber, HG - hind gut, INT- intima MT-Malpighian tubules, OES-oesophagus, R-rectum, V-ventriculus, V-1 - 1st ventriculus, V-2–2nd ventriculus, V-3 - 3rd ventriculus.

(2) Maxillary glands : These are present in some primitive group of insects e.g. Collembola, Protura, in the larvae of Neuroptera and certain Trichoptera. They open at the base of maxillae and lubricate the mouth parts. Much detailed information about these is not available.

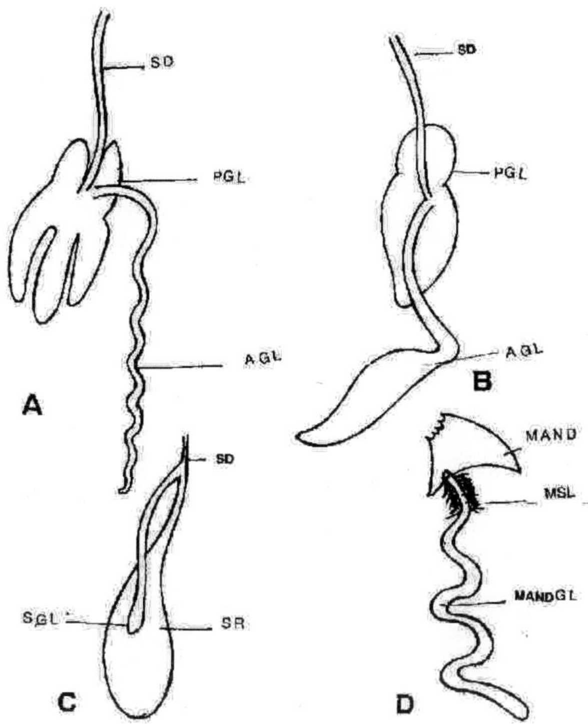

Fig. 27 A. Salivary gland of a Gymnoceratan Heteroptera, **B.** Salivary gland of a Cryptoceratan Heteroptera. **C.** Salivary gland of a termite, **D**. Mandibular gland.

Abbreviations : SD - salivary duct, PGL - principal gland, AGL - accessary gland, SGL - salivary gland, MAND - mandible, MANDGL - mandibular gland, MSL - muscles, SR - salivary receptacle or reservoir.

(3) Labial glands : The labial glands are commonly known as salivary glands (Figs. 27 A-C, 28A,B) and are paired structures, generally situated in the thorax on either side of stomodaeum. Their ducts combine to form a median salivary duct which opens on the labium usually near to the base of the hypopharynx. In many insects the ducts of labial glands are provided with taenidia in their cuticular lining and bear a close resemblance to tracheae; they are present in majority of the insects and assume a great variety of form and structure. Among Orthoptera and Dictyoptera, they are commonly very large and composed of a number of lobes, each lobe consisting of groups of glandular acini, a salivary reservoir is also present in many species in relation with each gland e.g. *Periplaneta americana.* In

Hemiptera, the salivary glands are differentiated into 1-4 lobes and a reservoir (accessory gland) is also present (Baptist, 1941). In adult Lepidoptera the labial glands form filamentous tubes. In the Diptera they are likewise tubular organs which in the Muscidae, may considerably exceed the total length of the body. Among Hymenoptera, labial glands are extremely well developed and assume great complexity. In the hive bee, they consist of two pairs of racemose organs, one pair being cephalic and other thoracic in position and their four ducts unite to form a common canal (Fig. 29).

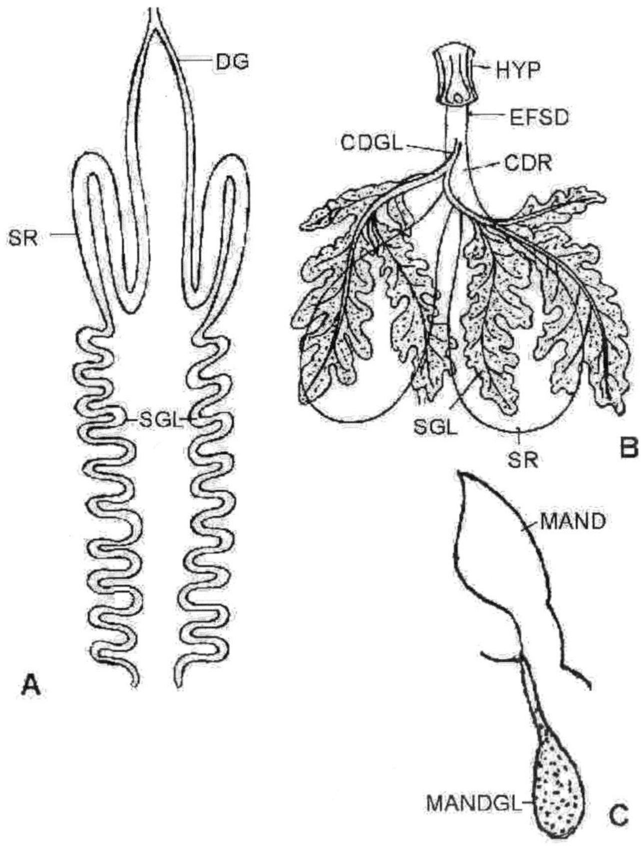

Fig. 28 A - Salivary glands of a caterpillar. **B**. Salivary glands of a cockroach. **C**. Mandibular gland of honey bee.
Abbreviations : CDR- common duct of reservoir. CDGL- common duct of gland, DG-duct of gland, EFSD-efferent salivary duct, HYP-hypopharynx, SR-salivary receptacle or reservoir, SGL- salivary glands. MAND-mandible, MANDGL - mandibular gland.

The labial glands secrete silk in Lepidopteran larvae but elsewhere their main function is to produce saliva. This watery secretion with an approximately neutral pH acts as lubricant in feeding either outside the body or in the foregut. In *Periplaneta,* the enzyme secreting cells are histologically distinct from other acinar cells (Day, 1951). In some blood sucking insects saliva contains substances which are poured into the host while feeding and inhibit the coagulation of its blood. The reaction to the bite of *Pediculus humanus* is also due to a secretion from one of the two pairs of salivary glands. The labial glands of Collembola may have the excretory function.

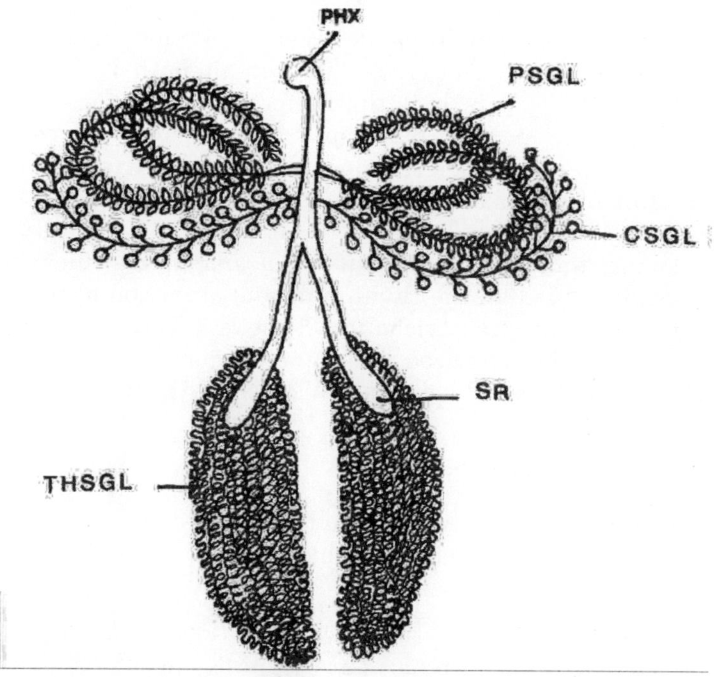

Fig. 29. Salivary glands of Honey bee.
Abbreviations : PHX–pharynx, PSGL-pharyngeal salivary gland, CSGL-cephalic salivary gland, SR- salivary receptacle, THSGL- thoracic salivary gland.

(4) Pharyngeal Glands

The pharyngeal glands occur in Hymenoptera (Fig. 29) and are well developed in worker honeybees. They are vestigeal in the queen bee and absent in male. There is one gland on each side of the head, each consisting of a long coiled tube to which numerous solid lobules are attached. The glands open at the base of the hypopharynx by separate ducts; in early days of life of the adult bee they produce brood food the "royal jelly" with which young larvae are fed and which plays some role in caste determination. It also produces diet of laying queens and drones. Later in life, when the bee begins foraging at about 3 weeks after emergence, the pharyngeal glands begin to secrete the enzymes amylase and invertase, becoming most active in this respect in bees about one month old. They are probably responsible for the invertase and amylase in the honey (Wigglesworth, 1965).

Royal jelly — *Apis* larvae destined to become queens receive a much higher proportion of `royal jelly' in their diet and receive it throughout their growth. It is this difference in diet which determines the difference in caste, but inspite of extensive analyses of royal jelly the nature of the determining factor is uncertain. Royal jelly contains 40% of dry matter comprising lipoproteins, neutral glycerides, free fatty acids (largely hydroxy fatty acids all with ten carbon atoms), sugar, amino acids, adenine, adenosine, guanosine, adenosine monophosphate, etc., all the B-vitamins and other components. It has a very high content of acetylcholine (1 mg./g.). The food given to the queen larvae is characterized by a much higher content of pantothenic acid, folid acid, and biopterine (Wigglesworth, 1965).

FUNCTIONS OF THE SALIVA

The functions of saliva secreted by the labial glands varies in different insects. The secretion of labial glands performs the following functions :

(1) It moistens the mouth parts and dissolves the food e.g. when a cockroach eats, its mouth parts are regularly moistened with saliva and in the feeding butterfly, a drop of saliva is extruded from the tip of the tongue.

(2) It also contains digestive enzymes which may act outside the insect body and which continues to act upon the food after it has been swallowed as in the case of amylase in the cockroach or invertase in butterflies and bees. The saliva of plant sap-sucking bugs commonly contains invertase and amylase as in Aphids but in Capsids and Jassids it contains lipase and protease.

(3) In the Hemiptera, Diptera and Siphonaptera which have mouth parts adapted for sucking the fluids, the saliva is discharged at the tip of the proboscis and its function may be to keep the tube moist and clean.

(4) The saliva contains anticoagulant in those forms which suck the blood of living animals.

(5) In many insects, the labial glands are not concerned with feeding and digestion but become modified for the production of silk or for other purpose. Thus, the saliva is used for moistening and lubricating the food, dissolving the dry sugar in Lepidoptera and Diptera and keeps moist piercing and sucking mouth parts in Hemiptera and prevents coagulation of blood, in blood sucking insects during feeding.

6

NUTRITION

Nutrition is the study of the food requirements of organisms, in this case, insects (Patton, 1963). House (1961, 1962) remarked that insect nutrition concerns more than mere dietetics. In its broadest sense, it involves many metabolic processes. It is a bond between physiologic and ecologic phenomena associated with natural selection and competition for food.

NUTRITIONAL REQUIREMENTS

The food constituents are classified into two major groups: essential and nonessential. The essential designates to those components that must be included in the diet, because they cannot be synthesized by the metabolic system of the insects. The nonessential indicates food materials that may have to be consumed to produce energy and that can be converted into a form in which they can be utilized through metabolic processes. The examples of essentials are vitamins, amino acids and some mineral salts. Although carbohydrates, proteins and fats are individually non-essential, but the animal must eat at least one of these non-essentials to maintain life.

The insects eat a wide variety of organic materials. Some insects are omnivorous, the majority show a considerable degree of specificity in their choice of food and many are so highly specific and specialized that they can develop only on a diet of one particular portion of a particular plant or animal. Insects can maintain life over relatively long periods of time on a diet of very limited nature or even without any food, but in order to grow and reproduce, insects like most other animals require in their food an adequate quantity of the following :

Water

The amount of water needed in the food depends upon the rate at which it is lost by the body, and this depends on one hand upon the properties of the cuticle, respiratory system and excretory system and on the other hand upon the drying power of the air. Although water is as essential to insects as to all other organisms, some can live in very dry environments as a result of being able to conserve the water resulting from the oxidation of food (Fraenkel and Blewett, 1944). Many insects, both larvae and adults are known to drink water under certain conditions. Several unidentified factors may also be needed in some insects.

Minerals

The various minerals which are found in insects take part in the regulation of the osmotic pressure of the intra- and extra cellular fluids, in the maintenance of an ionic balance suitable for the activity of the living cells and as cofactors in some enzyme systems and as an integral portions of others. Some salts are very essential and act as limiting factor in the growth of insects upon some diets, e.g. phosphorus acts as limiting factor in the growth of larvae of *Tribolium* and *Lucilia*. *Drosophila melanogaster* requires potassium, phosphorus, magnesium and sodium but not calcium except possibly in trace amounts. *Tribolium confusum* requires magnesium and potassium but not sodium or calcium. Manganese and magnesium, zinc and copper are probably essential in *Blatella germanica*.

Fats

Most insects can utilize fats and carbohydrates as sources of energy and for the synthesis of reserves of fat and glycogen. But fat is not required by all insects and is needed only in few insects where it acts as an accessory growth factor.

Some insects like pink boll worms and locusts need unsaturated fatty acids while majority of the insects do not require fat and thus it is not an essential factor.

Carbohydrates

Plant starch is the main source of carbohydrate but various types of sugars may be ingested in the normal diet from the plant sap, nectar and tissue. The utilization of carbohydrates varies with the species and depends on the ability to digest poly- and oligosaccharides to diffusible as well as absorbable forms and on the degree of absorbability of the molecule that is formed. The nutritive values of a great number of carbohydrates were rated for growth, longevity and fecundity. *Musca domestica* is one insect that needs no carbohydrate for growth. The presence of certain sugars in the larval diet of *Drosophila melanogaster* favoured their utilization by the adult. Utilization in adults may differ with sex. The carbohydrates in the order of utilization by insects generally include: dextrin, fructose, glucose, lactose, maltose, mannitol, raffinose, sorbitol, starch, sucrose and trehalose. Glucose and fructose are usually utilized well, sorbose and galactose not so well, and in general, pentoses are not used at all (Patton, 1963).

Proteins and amino acids

The proteins and amino acids are the principal components of the animal tissue. The nutritional value of proteins, as shown with protein or equivalent amino acid mixtures, varies for growth and also for reproduction. The amino acids needed for growth were determined in a number of insects. Generally insects like higher animals, need the well known 10-essential amino acids viz. arginine, histidine, isoleucine, leucine, lysine, methionine, phenylalanine, threonine, tryptophan and valine. These are the essential amino acids for growth and development of onion maggot, *Hylemya antiqua*, the pink boll worm, *Pectinophora gossypiella*, the beetle *Trogoderma granarium* and many other insects (House, 1962). These 10 amino acids were also needed for reproduction in *Aedes aegypti*. There are some exceptions concerning amino acid requirements for growth. For example, proline usually considered dispensable is essential in *Phormia regina* though it is synthesized by this insect. Usually dispensable amino acids must supplement essential ones for optimal nutrition of insects. Though not essential, iodoproteins and iodoamino acids promote growth and development of *Corcyra cephalonica*. Glycine can often be dispensed with but the remaining 10 amino acids listed above are the essential requirements in the diet of most insects.

Sterols

Insects require sterol for growth and development. Cholesterol and its short chain easters, 7-dehydrocholesterol, ergosterol, sitosterol and certain other phytosterols are satisfactory for many species. Usually however, cholesterol and some times a few other sterols are utilized best. For example, ergosterol was half, and sitosterol fully, as effective as cholesterol in the larvae of *Hylotrupes bajulus*. Recently, the essentiality of a sterol and the selective utilization of a number of steroids were determined in *Bombyx mori* and in two species of locusts. Cholesterol deficient diets for adult *Musca domestica* decreased the hatch but not the total production of eggs. The cholesterol is required by insects for larval growth and oogenesis and its function is similar to that of juvenile hormone. Cholesterol may act as a precursor to the various steroid hormones. However, there is much variation in detail in the sterol requirement of insects.

Nucleic acids

All insects appear to be capable of growing without nucleic acids in the diet, but an additional supply of RNA or DNA may often accelerate the rate of growth in various dipterous insects. The requirement for RNA in *Drosophila melanogaster* is primarily one for adenylic acid, essentially for adenine. *Agria affinis* utilized DNA as well as RNA. The beetle, *Sciobius granosus* needed RNA for growth. Except for this beetle, it appears that only flies and mosquitoes need nucleic acids (House, 1962).

Vitamins

Vitamins are organic molecules of diverse chemistry that are required in minute amounts for the growth and normal functioning of cells (Giese, 1962). They play an important part in metabolic processes in which they function as parts of enzymes. It is generally recognized that insects do not need fat soluble vitamins. Most insects require water soluble vitamins of B-series viz-biotin, choline-chloride, nicotinic acid, pantothenic acid, pyridoxine, riboflavin and thiamine for growth and development, though variations are found between species (House, 1962). Recent work determined the vitamin requirements of *Musca domestica, Calliphora erythrocephala,* and locusts. Several vitamins were shown to increase fecundity and fertility in adult tephritids, but vitamins had no effect on adult *Aedes aegypti*. Inositol was needed by two species of cockroaches, boll weevils, locusts and crickets. Though it is generally considered dispensable in insect diets. Similarly ascorbic acid (vitamin C) was found to be necessary for growth of *Schistocerca gregaria*. The latter insect needed inositol and β-carotene for normal pigmentation (House, 1962).

Micro-organisms as a source of accessory substances

The eggs of several insects such as *Drosophila, Calliphora, Lucilia, Aedes* and *Blatella* can be reared in sterilized media but this is possible only if the media are provided with all the necessary accessory factors. If these are deficient, then the infection of food with micro organisms e.g. in case of *Drosophila* yeast improves the rate of growth. Sterile *Lucilia* larvae will grow on beef muscle, they fail to grow on guinea pig muscle, but if this is infected with *Bacillus coli* or if yeast extract is added to it, normal growth takes place. *Lucilia* larvae cannot grow on sterile blood but growth is much improved by the presence of bacteria. Similarly a sterile blood medium is insufficient for larvae of *Culex* and *Aedes*, but it becomes adequate if infected with bacteria. In these cases there is little doubt that the micro organisms are synthesizing the accessory growth substances.

Types of micro organisms

The most commonly occurring micro organisms in insects are bacteria or bacteria like forms which are found in Blattaria, Isoptera, Homoptera, Heteroptera, Anoplura, Mallophaga, Coleoptera, Hymenoptera and Diptera. In addition, flagellates are found in wood eating cockroaches and termites, yeasts in Homoptera and Coleoptera and an actinomycetes in *Rhodnius*. In many cases, detailed information on the micro organism is not available.

7

DIGESTION AND ABSORPTION

Digestion in animals in general comprises the splitting up of the ingested food into simpler substances and the absorption of the essential nutrients into the body systems. In insects, likewise, as the ingested food passes through the alimentary canal, it is subjected first to the action of digestive enzymes. The result of this is the analysis of the food into products simple enough to be absorbed straight way into the general body system.

A number of factors govern the process of digestion in insects. Following are the important factors :

DIGESTIVE ENZYMES

The occurrence of digestive enzymes in an insect is governed by a variety of factors. One of the chief factors is the ability of the epithelial cells lining the alimentary canal to secrete these digestive juices. It is generally believed that the mid-gut is the chief site of digestion in insects, since this region shows the maximum enzymatic activity. However, there are evidences from many Orthoptera, Coleoptera, Diptera and a few Hymenoptera to prove that apart from midgut certain other regions of the alimentary canal, particularly the crop also acts as a site of digestion.

Another factor which governs enzyme secretion and occurrence, is the presence or absence of food inside the alimentary canal. It has been suggested that enzyme secretion is continuous in a majority of insects and to those larvae which are immersed in or always close to their food and is intermittent in many carnivorous forms and other insects which feed at definite intervals. In the former group i.e. where enzyme secretion is continuous, the enzyme secretion is not guided by any stimulus, arising from the presence of the food (such food stimuli being known as "Secretogogues") or from nerves or hormones. In the latter group of insects, however such secretogogues or other stimuli may be present.

The bulk of food material available to insects is constituted by the three main classes of foodstuffs viz: carbohydrate, protein and fat in proportions depending on the diet of the different species. Therefore, a corresponding diversity of enzymes capable of hydrolysing these substances will be required by most of the insects.

1. Carbohydrases

Carbohydrates exist as monosaccharides, oligosaccharides or polysaccharides, depending upon the number of simple sugar molecules in the structure. The carbohydrates are split first into their monosaccharide components - i.e. hexoses before they are assimilated (Patton, 1963).

According to Wigglesworth (1965, 1972) a wide range of carbohydrases are commonly present in insects e.g. the gut of the cockroach, *Blaberus* contains amylase but no cellulase, together with an α-glucosidase (sucrose and maltose), a β-glucosidase (cellobiose), an α-fructofaranoside (raffinose and sucrose), an α-galactosidase (melibiose) and a β-galactosidase (lactose). In *Byrsotria fumigata* in addition to above enzymes, cellulase and chitinase are also present.

The cellulase has also been found in the mid-gut and cellobiase throughout the gut and ventral labial glands of a plant louse, *Pseudocaecilius elutus* (Sinha and Srivastava, 1970). The cellulose is converted into cellobiose by cellulase and the cellobiose is converted into glucose by cellobiase. Termites, silverfishes, beetles, mulberry silkworms and a few cockroaches either secrete a cellulase in the gut or depend upon microorganisms for the digestion of cellulose (Patton, 1963; Wigglesworth, 1965).

Amylase (a group of three enzymes viz. α-amylase, α- oligosaccharase and maltase) hydrolyses starch as well as glycogen (Patton, 1963).

2. Proteinases

The digestion of proteins in insects appears to follow the same general rule as in other animals i.e. proteins are broken into peptones and polypeptides. And these are ultimately hydrolyzed to their constituent aminoacids. The insects proteinases have been found generally to be more active in neutral or alkaline media than in acidic.

3. Lipases

Enzymes capable of hydrolyzing neutral fat to glycerol and free fatty acids have been demonstrated in a wide variety of insects. Most of the lipases show maximum activity at alkaline pH values and under these circumstances the fatty acids released would form soaps, which would help to emulsify undigested fat and thus, facilitate further action of the hydrolytic enzymes.

Concludingly, it appears that all the three main classes of foodstuffs are generally digested by the enzymes and further that, a number of species have developed special adaptations to special feeding habits which enable them to digest animal products such as chitin, keratin, beeswax and plant products such as cellulose and lichenose. The enzyme activity also depends greatly on the feeding habit e.g. in carnivorous insects i.e. blow flies, proteinases and lipases are strongly developed while carbohydrases activity is feeble. Similarly in the phytophagous insects carbohydrases predominate over the proteinases. But it must be noted that if an insect does not feed on carbohydrates even then the activity of carbohydrases is present in the alimentary canal and likewise the activity of proteinases and lipases. Therefore it appears that high or low activity of a particular enzyme(s) is governed to some extent, by the types of food material, which an insect takes in its diet.

The activity of the digestive enzymes on a particular nutrient(s) (substrate(s) is influenced by a variety of factors chief of them being: (1) pH, (2) Temperature, (3) Presence of activators and inhibitors.

(1) pH - ENZYME RELATIONSHIP

Every digestive enzyme has a certain range of pH or hydrogen ion concentration within which its activity is optimum. The pH at which the enzyme activity is maximum is known as the optimum pH. It is at this pH that the maximum digestion of the corresponding substrate (food) is accomplished. If by any chance the pH falls below or rises above the optimum level, the digestion of corresponding substrate (food) is greatly influenced. Therefore, the role of pH on the enzymatic activity is a very important factor. For an efficient digestion of food, the pH in the different parts of the gut should remain within an optimal range, so that almost all the digestive enzymes are capable of acting well on their corresponding substrates. This pH- enzyme relationship varies for different enzymes, in different insects e.g. the enzyme amylase responsible for the digestion of the starch varies widely in its pH optima from about pH 5.5 to about pH 9.5 in different insects (Shinoda, 1930).

2. TEMPERATURE

It has been clearly shown by numerous studies that temperature is another important factor affecting enzymatic activity and thereby controlling the ability of the insect to digest food substances. It has been demonstrated in a number of insects that 37°C is the optimal temperature in which most of the enzymes are appreciably active. However, in certain insects e.g. *Blatta,* some of the enzymes particularly amylase has been found to be most active at 50°C (Day and Powning, 1949). Rise and fall of temperature above or below the optimum level correspondingly affects digestion.

3. ACTIVATORS AND INHIBITORS

In addition to pH and temperature, activity of the enzymes is also governed by certain activators or inhibitors. Activators or inhibitors are certain chemical substances which can accelerate or inhibit the enzymatic-activity, respectively. It has been seen that the Cl-ion-of NaCl inhibits the amylase activity in *Blatta* (Wigglesworth, 1927). Similarly Schlottke (1937) has reported the inhibitory properties by the Cyanids and Cystone (amine) of Protease in *Periplaneta americana.*

It may be mentioned here that although in majority of the insects digestive enzymes are secreted by the individuals themselves, yet there are forms where certain micro organisms, as stated earlier, present in the gut secrete the necessary enzymes and thus aid the insect in the digestion of food.

ABSORPTION

As soon as the food splits up into simpler products the next step is the study of the insect gut physiology and ability of these organs to absorb their products. Evidence for absorption studies depends mostly upon the accumulation of the absorbed quantities of materials in the cells. But there is no doubt that many materials are passed through the epithelial cells of the gut without accumulation. Thus it is clear that absorption is an active and often a highly selective process, certain groups of cells being specially concerned with the uptake of specific material from the lumen.

Carbohydrates

The carbohydrates are always broken down to monosaccharides before absorption by the insect's gut. Usually the maximum absorption of carbohydrates occurs in the midgut. Often the absorbed monosaccharides are rapidly converted to glycogen which accumulates in the mid-gut cells. It is well demonstrated in mosquitoes, *Prodenia* caterpillars etc. Sugar absorption may occur in the hindgut also as in the case of larva of *Aedes* sp. (Diptera).

Fats

Generally fat absorption does not occur in the foregut. However, in certain insects e.g. *Blatta orientalis* it has been reported that crop is the site of absorption of the fats. In the mid-gut the capacity to absorbs fat varies in different regions for instance in the Blowfly larvae the large anterior end and posterior section of mid gut, absorbs fat readily whereas the middle section normally does not absorb fat. In *Aedes* larvae anterior part of mid-gut absorb fat while in *Culex* larvae the middle region is the site of absorption. Absorption of fat by hindgut has been reported in some adult Hymenoptera also.

Proteins

Information on the protein or amino acid absorption is poor. In certain insects like *Aedes* larvae the absorption is in the posterior region of the mid-gut while in *Rhodnius* it occurs in the middle part of the mid-gut. In the case of *Tenebrio* larvae some of the proteins get transformed into amino acids and are then absorbed in the rectum.

Inorganic Materials

A number of insects have been experimented upon to investigate the absorption of certain inorganic materials such as Fe, Cu, Ca, Ba, Sr, Po4, Iodine, NaCl. acids and alkali, etc.

Iron is found to be absorbed in the gastric caecae in *Periplaneta americana* and in the mid-gut region in *Culex* larva and adults, also in the larvae of *Musca* and *Stomoxys*.

As regards Cu, *Lucilia* larva has been found to absorb it into two separate zones of mid-gut, while the adult of the same species absorb these inorganic materials through the mid-gut. Also other inorganic materials like Ba, Sr, Phosphates etc. have all been usually found to be absorbed by mid-gut. As far as iodine, acids and alkali and NaCl are concerned the site of absorption varies widely in different insects. However, it has been reported that these substances are often absorbed in the hind-gut.

It may be mentioned here that other substances such as dyes, pigments and even certain poisons are absorbed in the mid-gut.

8

THE CIRCULATORY SYSTEM

The circulatory system in insects is of open or lacunar type. The blood circulates in the haemocoel. In insects this system plays a very insignificant role in respiration. There is only one palsatile vessel.

1. COMPONENTS OF THE CIRCULATORY SYSTEM

The diaphragms and sinuses

The body cavity or haemocoel is divided into three sinuses by means of diaphragms (Fig. 30A). (1) The dorsal diaphragm extends across the abdominal cavity above the alimentary canal. The blood space thus formed is known as dorsal or pericardial sinus. It encloses the heart. (2) The ventral diaphragm stretches across the abdominal cavity just above the ventral nerve cord. The haemocoel thus formed is known as ventral or perineural sinus. (3) Between the dorsal and ventral sinuses is a large cavity known as visceral or perivisceral sinus which contains the main internal organs.

Pairs of alary muscles which arise from the terga spread out fan like over the surface of the dorsal diaphragm. Thus beneath the alary muscles lies the dorsal diaphragm. The alary muscles are attached to the body wall along the lateral parts of the dorsum. They are believed to function more or less as elastic sheets which bring about contraction and expansion of the heart.

Dorsal vessel

The dorsal vessel (Fig. 30 B,C) runs along the mid dorsal line just beneath the integument for the whole length of the body. It is a continuous tube whose posterior end is closed and is always open at the anterior or cephalic extremity. The dorsal vessel is divided into two regions : a posterior heart and an anterior aorta. The heart is perforated while the aorta is an unperforated tube. The wall of dorsal vessel in the heart and aorta is contractile.

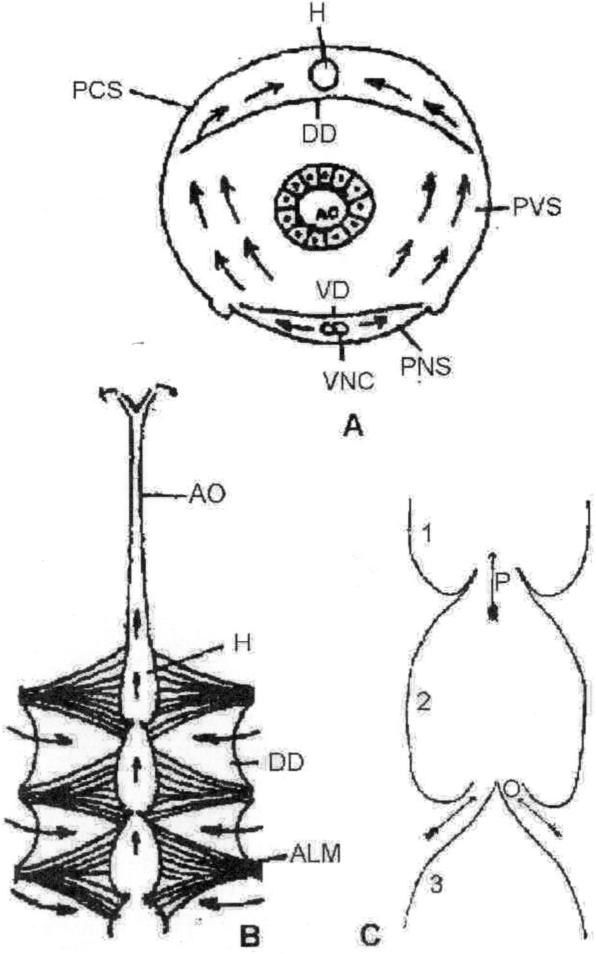

Fig. 30 A. Transverse section of abdomen, **B.** Dorsal vessel showing a few chambers of the heart, alary muscles and dorsal diaphragm, **C.** Valves of the heart. 1,2,3-chambers of the heart; 1,2-at the moment of systole, 2,3- at the moment of diastole (Fig. C From Imms, 1963, 1964)
Abbreviation: AC - alimentary canal, AO - aorta, ALM - alary muscles, DD - dorsal diaphragm, H - heart, O - ostium, P - intraventricular passage, PCS - pricardial sinus, PNS - perineural sinus, PVS - peri visceral sinus, VD- ventral diaphragm, VNC - ventral nerve cord.

Heart

The heart is usually confined to abdomen but may extend in some insects upto the prothorax. In most insects the heart is a continuous tube, rarely it is divided into chambers. Its segmentation is only shown by the presence of paired incurrent ostia and the alary muscles. It is composed of a single layer of cells with large nuclei and straited muscle cells.

The blood enters the heart through lateral incurrent ostia, a pair of which is situated at each constriction. The wall of the heart is reflected inwards and forwards at each ostium to form an auricular valve, which precludes the return flow of the blood into the dorsal sinus. In many insects each pair of auricular valve also prevents the backward flow of the blood in the heart itself.

The aorta

The aorta is the anterior prolongation of the dorsal vessel (Fig. 30B). It functions as the principal artery of the body. Its junction with the heart is frequently marked by the presence of aortic valves. The aorta extends forwards through the thorax to terminate in the head near the brain. In some insects its anterior extremity is an open funnel-like mouth but it may instead divide into two or more cephalic arteries each of which may subdivide into smaller vessel.

Accessory pulsatory organs

In addition to dorsal vessel there are often other sac like pulsating structures connected with the haemocoel which are concerned with maintaining circulation through the appendages. In the mesothorax and sometimes in the metathorax there is a pulsatile organ concerned with the circulation through the wings. In Odonata the blood space opens through a terminal ostium into an ampulla at the end of a dorsal diverticulum of the aorta. Contraction of this ampulla drives blood into dorsal vessel when it relaxes the ostium opens driving blood in from the sac beneath the tergum and hence indirectly from the wings.

In many Lepidoptera, Orthoptera, Coleoptera, Hemiptera and Diptera, pulsatile organs are well developed and they may be present beneath the tergum of mesothorax, metathorax or in legs or at the base of antennae and wings.

2. COMPOSITION OF BLOOD

The blood or haemolymph in insects is composed of liquid plasma or serum and blood cells or haemocytes. It makes up 15-75% of the volume of the insect and the plasma contains about 85% water. It is a colourless fluid but occasionally it appears with a greenish, yellowish or brownish tinge. The plasma is slightly acidic in insects in general. The pH of insect blood varies considerably depending upon the age of the insect, it may be slightly acidic or alkaline as in the case of *Bombyx mori*. The chemical composition of the insect blood reveals the presence of proteins, lipids, sugars (glucose in case of honeybee and trehalose in others), uric acid, a number of pigments, organic acids, bicarbonates, phosphates etc. Almost all amino acids are found in the insect haemolymph. Some 19-free aminoacids have been reported by Duchateau and Florkin (1958). The plasma of *Periplaneta* has more sodium than potassium but this is not true of all insects, the ratio of sodium and potassium varies from insect to insect. On the basis of sodium and potassium ratio the blood has been classified into two types (Duchateau *et al.,* 1953). In Odonata the concentration of sodium is greater than the concentration of potassium and magnesium, in Lepidoptera, Coleoptera etc. the plasma has a low concentration of sodium and potassium, and magnesium have a high concentration.

Haemocytes (Fig. 31 A-G)

Several attempts have been made to classify haemocytes but the account given by different workers is not uniform and is very ambiguous. The haemocytes have been classified from 3 to 9 types e.g. Mori (1979) has classified into 3 types, Jones (1962, 1964) into 4-types, Wigglesworth (1965) into 5-types, Price and Katcliff (1974) into 6-types and Arnold (1974) into 9-types. As remarked by many workers, the various types of haemocytes are the functional or developmental stages of a single cell type. Three main types i.e. (1) prohaemocytes or proleucocytes (2) plasmatocytes and (3) granulocytes are found in all insects whereas (4) spherules, (5) cystocytes, (6) oenocytoids and (7) adipohaemocytes are found only in certain orders of insects.

(1) Prohaemocytes (Fig. 31A)

These are also called proleucocytes. They are small, spherical cells with large nucleus. The cytoplasm is strongly basophilic and is rich in RNA. The mitochondria give a positive test for phospholipids. The prohaemocytes divide and give rise to other types of cells.

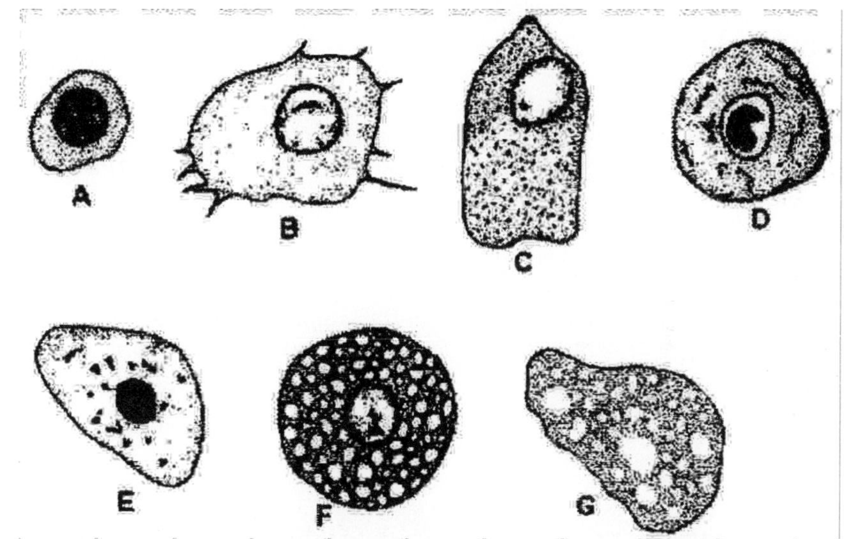

Fig. 31. Diagrams showing different types of haemocytes. **A**. Prohaemocyte, **B**. Plasmatocyte, **C**. Granulocyte, **D**. Oenocytoid, **E**. Cystocyte **F**. Spherule, **G**. Adipohaemocyte (From R.F. Chapman, 1972)

(2) Plasmatocytes (Fig. 31B)

They are the most abundant haemocytes; polymorphic-many of them are amoeboid others are rounded, oval, spindle-shaped or fusiform. They have basophilic vacuolated cytoplasm. They occasionally contain inclusions of neutral mucopolysaccharides.

(3) Granulocytes (Fig. 31C)

These are derived from the plasmatocytes. They are phagocytic and contain acidophilic granules in the cytoplasm. They are the largest of the three haemocytes. They are rounded or oval in outline most of the granules give a positive test for mucopolysaccharides or mucoproteins.

(4) Oenocytoids (Fig. 31D)

They are found in Coleoptera, Hemiptera, Diptera and Lepidoptera. They have been considered to be derived from oenocytes and not from the prohaemocytes.They are large, rounded or oval in form, the cytoplasm is basophilic and nucleus is eccentric in position.

(5) Cystocytes (Fig. 31E)

Also known as coagulocytes, connected with the coagulation or clotting of haemolymph. They have a yellowish-hyaline cytoplasm containing scattered granules. The plasma membrane is not easily discernible.

(6) Spherules (Fig. 31F)

They are spherical, oval or spindle-shaped and are found in Lepidoptera and Diptera. They have coarse spherules that occupy the whole cell. They give a positive test for mucopolysaccharides but not for proteins and lipids.

(7) Adipohaemocytes (Fig. 31G)

They have been found in all the Lepidoptera and Diptera so far studied and in some representatives of various other groups. They are charaterized by refringent fat droplets and other inclusions.

Functions

A number of functions are performed by the haemocytes. Some of the important functions are : phagocytosis, encapsulation, wound healing, coagulation, connective tissue formation, secretion and metabolism, immunity, storage and transport of food and detoxication etc.

Circulation of blood

The heart is the principal pulsatory organ. The circulation of blood is maintained by the rhythmic contractions of the heart which is composed of contractile tissues. The motion used in pumping the blood is a peristaltic wave that originates at the posterior end of the vessel and runs in anterior direction. During diastole (expansion) the blood from the pericardial sinus is drawn in the heart chamber through the incurrent ostia and during systole (contraction) the expanded tips of these ostia get closed and act as valves, so that the blood is prevented from returning through these into the pericardial sinus and is propelled forward. The blood propelled in such a manner in the aorta is discharged into the cephalic haemocoel. From the head and thoracic haemocoel the blood flows backwards into the perivisceral sinus or abdominal haemocoel. The backward flow of blood is probably assisted by the undulatory movement of the ventral diaphragm.

The accessory pulsatile organs present at the bases of the appendages cause the blood to flow into the antennae, legs and wings. Finally, the blood is returned to the pericardial sinus from the perivisceral sinus.

HEART BEAT

(1) Systole

The contraction phase of the heart beat results from the contractions of the muscles in the heart wall which start posteriorly and spread forward as a wave.

(2) Diastole

The relaxation phase results from relaxation of the muscles assisted by the elastic filaments supporting the heart and in some cases by the contraction of the alary muscles.

(3) Diastasis

This is the third phase in the heart cycle in which heart rests in the expanded condition. Increase in the frequency of the heart-beat results from reduction in the period of diastasis.

Control of heart-beat

The control of heart beat is yet not fully understood. A variety of factors regulates the heart beat and it may be myogenic or neurogenic. Krijgsman (1952) has suggested that the heart muscles contract myogenically but often has neurogenic pacemaker. The heart of *Periplaneta* is innervated by the neurosecretory axons. Similarly, the heart of Calliphora is innervated myogenic type (Normann, 1974). Pharmacological evidences are in the favour of a neurogenic origin of the heart beat. It has been suggested by a large number of workers that corpus cardiacum seems to play an important role in controlling the beating of insect heart (Awasthi, 1983). Cameron isolated

O-diphenol which according to him regulates heart beat, the O-diphenol was reported to be similar to that of adrenalin. Davey and Ralph demonstrated that the substance increasing the heart beat is a peptide that acts on pericardial cell. It has been reported that the active substance which increases the heart beat is similar to adrenalin. Further, it was remarked that the cardioregulatory substance extracted from the pericardial cells was indol alkylamine. The pericardial cells can be stimulated by the corpus cardiacum extracts to release an indol alkylamine, the active substance responsible for heart beat acceleration. Some workers have obtained serotonin in the brain and corpus cardiacum of *Periplaneta*. Others have extracted cardioregulatory substances from ganglia of the ventral nerve cord, metameric neurohaemal organs and corpus cardiacum. Numerous substances of diverse chemistry have been extracted from the nervous system and endocrine organs that have heart regulatory properties. On the basis of information available (see Awasthi, 1983) it can be said with some certainty that endocrines are one of the important sources of heart regulatory factors which control the heart beat directly or indirectly.

9

THE EXCRETORY SYSTEM

In insects, the function of excretion is chiefly concerned with the elimination of several types of substances such as mineral salts and water, and products of the metabolism of nitrogenous substances, sulphur and phosphorus; certain complex compounds that are produced incidentally as a result of certain accidents or as a by product in case of certain chemical changes and the acids and alkalies. Thus, it is a process by which the internal environment of the insects is regulated at a constant level by removing surplus and injurious materials.

In general, the organs responsible for the elimination of the substances, mentioned above are known as excretory organs. There are several types of excretory organs in insects chief of them are Malpighian tubules.

1. **The Malpighian tubules** (Fig. 32 A-D) : Malpighian tubules are found in all insects but are absent in Collembola and the family Aphididae of the order Homoptera. They are long, slender, blind tubules, which are often convoluted. They originate at the junction of mid and hind gut (the pyloric region) and have been regarded ectodermal in origin. But according to Savage (1956) they are endodermal in *Schistocerca gregaria.*

These tubules in insects in general are free at their distal ends but in some insects the distal ends of these tubules are attached to the rectum. The reassociation of Malpighian tubules is known as cryptonephridial condition that constitute cryptonephridial system in insects. All the Lepidopteran and Coleopteran insects have a cryptonephridial condition. The cryptonephridial condition helps in the conservation of water and other useful ions by absorbing them from the faecal matter.

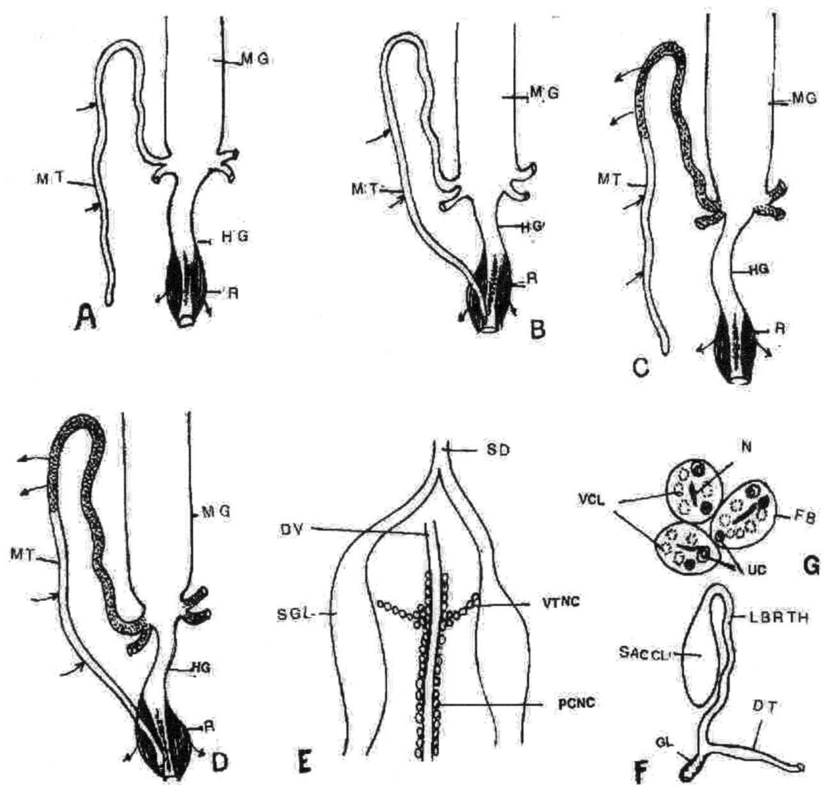

Fig. 32. **A**. Free floating Malpighian tubules as found in Orthoptera., **B**. Re-association of Malpighian tubules as found in Coleoptera, **C**. Malpighian tubules as found in Hemiptera, **D**. Re-association of Malpighian tubules as found in Lepidoptera, **E**. Pericardial and ventral nephrocytes **F**. Labial gland of Collembola, **G**. Urate cells.

Abbreviations : DT- duct of gland, DV- dorsal vessel, FB- fat bodies, GL- gland, HG- hind gut, LBRTH- labyrinth, MG- mid gut, MT- Malpighian tubule, N- nucleus, PCNC- pericardial nephrocyte, R- rectum, SD- salivary duct, SGL- salivary gland, SACCL- saccule, UC- urate cells, VCL- vacuoles, VTNC- ventral nephrocytes.

On the basis of arrangement and their nature, Wigglesworth (1931) has broadly classified Malpighian tubule system into 4-types viz. - A, B, C and D. The type A is the basic one and the remaining three B, C and D represent the variations met within different insect orders. In the basic type (A) the tubules are fairly large, highly coiled and their distal ends float freely in the haemolymph. Oftenly the tubules are associated with alimentary canal and richly supplied with tracheoles and muscle fibres.

In the majority of Orthopteroid insects (Fig. 32A) no visible difference exists in the appearance of the Malpighian tubules over their entire length. In the 2nd type, the Malpighian tubules differ from the first in that the distal ends of the tubules are attached to the rectal wall and are enveloped by a thin sheath (Fig. 32B). As remarked above, the tubules of this type, have been termed as cryptonephridic or cryptosolenic and are found in Coleoptera and Lepidoptera. In Coleoptera, (Fig. 32B) there exists no visible differentiation over the entire length but in Lepidoptera (Fig. 32D) the Malpighian tubules are clearly differentiated into two zones. The distal part is filled with a clear fluid and the proximal part has solid particles. In the order Hemiptera (Fig. 32C) each tubule consists of two segments, an upper or distal containing 2/3 of the tubules, made up of a granular epithelium containing clear fluid in the lumen and a lower or proximal segment containing 1/3 of the tubule, in which the epithelium is almost free of granules and the lumen is filled with uratic spheres.

The number of Malpighian tubules varies from group to group. The number of these tubules is 4-in Anoplura, Thysanoptera, Hemiptera, Diptera and Siphonaptera; 4-6 in Psocoptera and Coleoptera; 2-8 in Isoptera, and Lepidoptera; Mecoptera and Trichoptera have 6 Malpighian tubules. In Dermaptera 8-20, in Ephemeroptera 8-100, in Odonata 50-200, in Orthoptera 30-200 and in Hymenoptera 6-20. In mosquitoes the number of Malpighian tubule is five and in coccoidea and parasitic Hymenoptera only 2. Further, in some insects for example in Gryllidae, all the Malpighian tubules converge to open into a common ureter. The Malpighian tubules are white, yellow or cream coloured and they may have a smooth or beaded structure.

Histology

The Malpighian tubules are composed of 2-8 cells and are of various shapes and sizes. They are either provided with distinct cell boundaries or are syncitial. The cells are either smooth, or have a brush or honeycomb border towards the lumen. The cells rest on a thin basement membrane covered by a peritonial membrane occasionally containing fine muscle fibres which bring about the peristaltic movement of the tubules. The movement caused by the muscles of the tubules assures the mixing of the contents. The tubule carries the contents towards the hind gut for final disposal.

Stimulus for urine formation

Recent evidences suggest that the neurosecretory cells of the pars intercerebralis control water excretion. Removal of the neurosecretory cells or pars intercerebralis in a Coleopteran, *Anisotarsus*, caused retention of water. Pilcher (1970) has suggested that the rate of secretion by the isolated Malpighian tubules of *Carausius morosus* is accelerated by a diuretic hormone which can be extracted from the brain, corpora cardiaca and suboesophageal ganglion. According to Maddrell (1962) the isolated system of Malpighian tubules immersed in the haemolymph from an unfed *Rhodnius*, show no signs of secretory activity. When haemolymph from a newly fed *Rhodnius* is administered, it causes a rapid flow of urine. Thus it appears that the factor or stimulus responsible for urine formation is a hormone derived from the central nervous system. Maddrell (1962) further suggested that the cells of metathoracic ganglia induces much activity whereas the cells of the brain comparatively less.

Function of the Malpighian tubules

The Malpighian tubules chiefly excrete uric acid which is usually accompanied by water. The process of excretion takes place as follows:

The uric acid collected in the body cells diffuse into the blood which finally circulates round the Malpighian tubules. In the blood, uric acid combines with sodium and potassium carbonates to form sodium and potassium urates. These two salts are soluble in water and are easily absorbed by the distal part of Malpighian tubules. In this way, the urate salts are secreted into the lumen of the Malpighian tubules. Within the proximal part of tubule in alkaline pH, by the action of carbonic anhydrase, the urate salts are broken down to regenerate water, bicarbonates and uric acid. The water and bicarbonates are reabsorbed and free uric acid is precipitated. From here the uric acid solution loosely termed as "urine" is passed into the proctodaeum and is finally excreted through the anus.

It may be mentioned that this method of excretion involves a constant supply of water and in the majority of the insects water is a very important constituent and cannot be lost so easily. It is also probable that the basic solvents such as Na and K salts are equally valuable. Different methods have therefore been evolved which conserve these compounds by extracting the water and bases from the uric acid and returning them into the blood. This results in the excretion of only free uric acid made up of certain urates, ammonia and nitrogen. One method characterized by earwigs and grasshoppers is the development of the absorptive areas in the rectum. These areas extract the water from the excrement and return it to the blood. In certain other forms e.g. *Tenebrio* the tips of the Malpighian tubules at the reassociation end are enveloped by a membrane and attached to the rectum. In such cases, both the membrane and the absorptive areas in the rectum extract the water and return it into the blood. Very often the water absorbed is also returned to the Malpighian tubules again, consequently the same water is used over again and again. A second method in conservation of water is the division of Malpighian tubules themselves into two, excretory and absorptive zones as already stated.

In aquatic insects which are not compelled to conserve water, excretion of uric acid accompanies a good deal of water. However, in these forms continuous excretion of water is liable to deprive the insects of essential ions such as chlorides, K and Na. In order to eliminate this deficiency, some aquatic larvae have a body covering which is highly impermeable to salts. In addition to this, there are certain larvae where the rectum absorbs the salts from the urine before the latter is discharged. In certain other aquatic larvae e.g. *Aedes* the anal papillae can absorb the substances present in the fresh water. The gills of many larvae have a similar function.

Composition of urine

The urine, which the Malpighian tubules secrete is variable. It may be a clear liquid or a thick, pasty, suspension (e.g. meconium). The composition of urine depends on the nature of food. It consists of water, inorganic salts, chlorides and phosphates of sodium, potassium and magnesium. A yellowish or greenish pigment (Entomouro-chrome), riboflavin in the cells of some and kynurine in other insects, is also often present.

Nitrogen is excreted chiefly as uric acid, but presence of urea has been reported in good amount in the urine of cloth moth, and meat eating maggots (Diptera) eliminate much of their nitrogen as ammonia. Likewise, a number of aquatic insects such as larvae of

Sialis of Odonata, and Trichoptera also eliminate the end products of nitrogen metabolism as ammonia.

The hydrogen ion concentration of the insect urine mainly depends on the diet and the period of feeding. The meconium of *Cilerio* (Lepidoptera) is acidic 5.8-6.3. In *Rhodnius*, the pH in the first day of blood meal is 7.8-8.0 i.e. alkaline and contains chlorides of sodium and potassium that was in the ingested blood. This urine also contains urea, bicarbonate, sulphate and uric acid. Later, the urine becomes acidic i.e. pH 6.0 and is more concentrated. At this stage it appears yellow due to the yellow pigment or greenish pigment the "entomourochrome". The entomourochrome differs chemically in different species.

Excretion of nitrogen takes place in the form of uric acid and crystalline spheres and the percentage of free uric acid varies from 80% to 90%. The remaining uric acid appears to be the sodium urate or potassium urate. Besides the uric acid, sodium and potassium urates, and ammonia in aquatic insects, some insects excrete allantoine, xanthine and hypoxanthine.

The allantoine, which is excreted by some insects is further broken down to ammonia and carbon dioxide by the activity of uricase, allantoinase, allantoicase and urease:

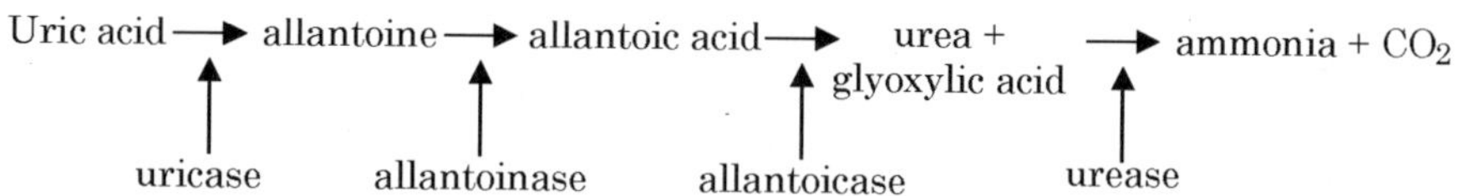

Excretion of ammonia has been regarded as primitive feature and uricotelic excretion has been achieved by the successive loss of the uricolytic enzymes. The Collembolan *Xenylla welchi* contains uricase, allantoinase and allantoicase and produces both allantoic acid and urea.

In several other insects, there are certain structures other than Malpighian tubules which perform the function of excretion viz. :

1. **Labial glands** (Fig. 32F): In Collembola and Thysanura where the Malpighian tubules are absent, the labial glands are excretory in function. They have an upper sac like part, a coiled labyrinth with a gland and excretory duct, which opens at the base of labium. The labial glands take up dyes from the blood and eliminate outside.

2. **Nephrocytes** : In some insects special cells, the nephrocytes perform the function of excretion. These cells are either pericardial cells distributed along the dorsal heart or garland like string of cells as shown in Fig. 32E.

3. **Urate cells** : Certain insects are provided with special cells in the body which become laden with crystalline spheres of uric acid. These cells are termed as urate cells (Fig. 32G). They serve the function of excretion of uric acid and often of pigments also. These urate cells are regarded as `Storage kidneys' and are scattered throughout the insect body. These cells collect the uric acid in solid form throughout the larval life and the excretory products are finally transferred to the Malpighian tubules in the pupal stage for elimination e.g. social Hymenoptera.

4. **Excretion by the gut :** According to Waterhouse and Day (1953) in Collembola the midgut performs the function of excretion. In case of aphids the excretory materials and the pigments are eliminated by the digestive tract. In *Periplaneta* the uric acid is present in the form of crystals and not in liquid form in the wall of hind gut, accordingly the hind gut does the function of excretion to some extent in this insect. Some insects excrete ammonia which appears to pass directly into the digestive tract without going to the Malpighian tubules e.g. in blowfly larvae the ammonia is produced in the midgut and is then reabsorbed and passed via blood to the hind gut. In aquatic insects, the ammonia may be secreted directly into the rectum, from the latter it is eliminated to the outside.

10

THE RESPIRATORY SYSTEM

In insects respiration takes place through a system of internal tubes known as tracheae, the finer branches of which convey air directly to the tissues. The tracheae open along the side of the body by a series of apertures, termed spiracles.

(A) DEVELOPMENT AND STRUCTURE OF A TRACHEA

Studies on the development of tracheae show (Fig. 33A) that these tubes arise as simple invagination of the ectoderm along the side of the body in insects. The external openings of these invaginations become the spiracles. These invaginated tubes later extend inward and branch into ramifying structures that ultimately become a well developed tracheal system in insects. The walls of the tracheae contain the same contents as that of the body wall but in the reverse order.

The characteristic feature of an insect trachea is its closely ringed appearance (Fig. 33B) resulting from the presence of folds or the thickenings of intima in the form of minute circular spiral ridges. (Snodgrass, 1935). These are known as taenidia. The taenidia prevent the collapse of the trachea if the pressure within is reduced.

Modifications of the trachea

The tracheal tubes thus formed are rarely uniform in diameter. In places the tracheae are expanded or widened to form thin walled swellings known as air sacs. The taenidia are either absent in the air sacs or poorly developed and are often irregularly arranged. Consequently the air sacs will collapse under pressure and they play very important part in ventilation of the tracheal system. The air sacs vary in size from minute vesicles to large bags and may be widely distributed in the body, head and appendages of the insect e.g. in *Melanoplus* there is a pair of large thoracic air sac and 5-pairs in the abdomen. In certain forms such as *Musca, Apis, Bombus* etc. the air sacs attain greatest development and they occur as dilations of the main trachea (Fig. 33 D,E).

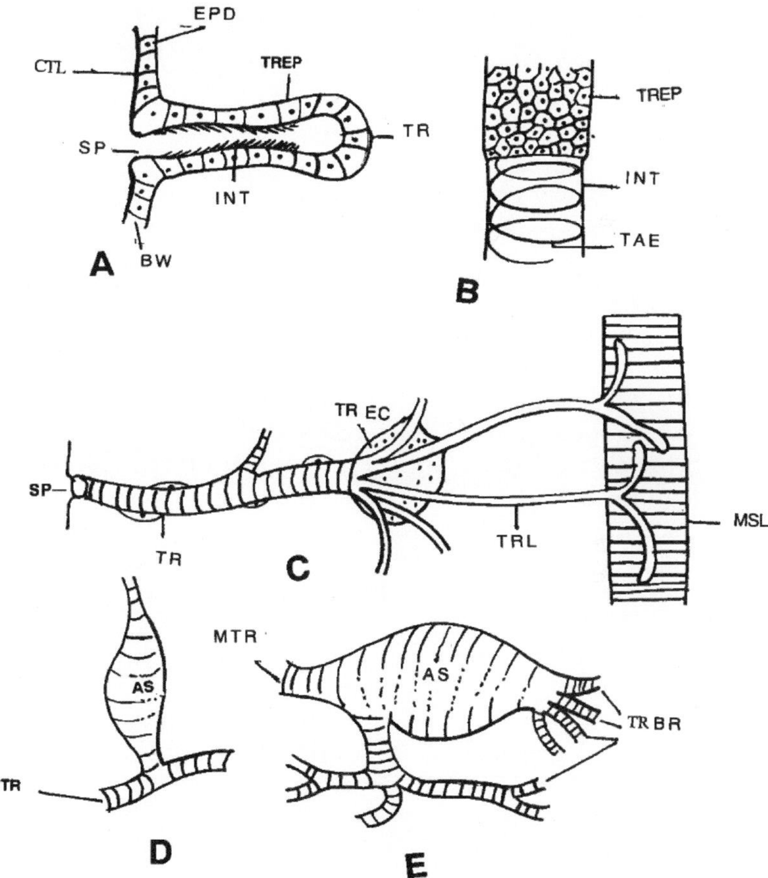

Fig. 33 A. Diagram showing origin of trachea, **B**. Diagram showing the structure of a trachea, **C**. Diagram showing open trachea of a generalized insect, **D**. Diagram showing vessel like air sac, **E**. Diagram showing bag like air sac.

Abbreviations : AS- air sac, BW- body wall, CTL- cuticle, EPD- epidermis, EPTH- epithelium, INT- intima, MSL- muscles, MTR- main trachea, SP-spiracle, TR-trachea, TREC-tracheal end cell, TAE-taenidium, TRL-tracheole, TREP- tracheal epithelium, TRBR - tracheal branches.

Tracheoles (Fig. 33C) : The trachea finally breaks up into a number of minute branches known as tracheoles which penetrate the cells of the body tissues. There is no sharp distinction between the tracheae and the tracheoles but the tracheoles are always intracellular and retain their cuticular lining at moulting which is not usually true of tracheae.

(B) GENERAL PLAN OF TRACHEATION

The general plan of body tracheation (Fig. 34A) in insects is as follows :

From each spiracle on either side is given off a short tracheal branch known as the spiracular trachea which joins the main tracheal trunk running along the full length of the body. The two main tracheal trunks are known as the lateral longitudinal trunks and they form usually the largest trachea in insects. Spiracular trachea or the lateral longitudinal trunk of each side gives off three branches viz. (1) a dorsal trachea entering the dorsal musculature of the body wall and to the dorsal blood vessel, (2) a ventral trachea supplying the ventral musculature and the ventral nerve cord and sending a branch to the leg in the leg bearing segment and (3) a median visceral trachea having its principal ramifications on the walls of the alimentary canal with branches to the fat body and in appropriate segments to the reproductive organs. In some insects, the dorsal, ventral and visceral trachea of successive segment become continuous to form the respective longitudinal trunks. Often the dorsal trachea of one side of a segment is connected by a commissure transversely to its fellow of the opposite side in the same segment. This is the dorsal tracheal commissure. There is a similar ventral tracheal commissure also in certain insects.

Modifications of the tracheal system

There are several insects which do not possess a fully equipped tracheal system. In many aquatic and parasitic species, the tracheal system is more or less reduced or rudimentary, the spiracles being shut or closed. This condition has developed as a result of secondary degeneration of the tracheae. In certain forms e.g. *Machilis*, groups of segmental tracheae with no longitudinal trunks are present. Again in insects like *Campodea*, trachea are present only in the thoracic region; the modification in the tracheal system have gone one step further in dipterous larvae and pupae where the dorsal longitudinal trunks become the principal respiratory passages and communicate to the exterior through dorsal spiracles.

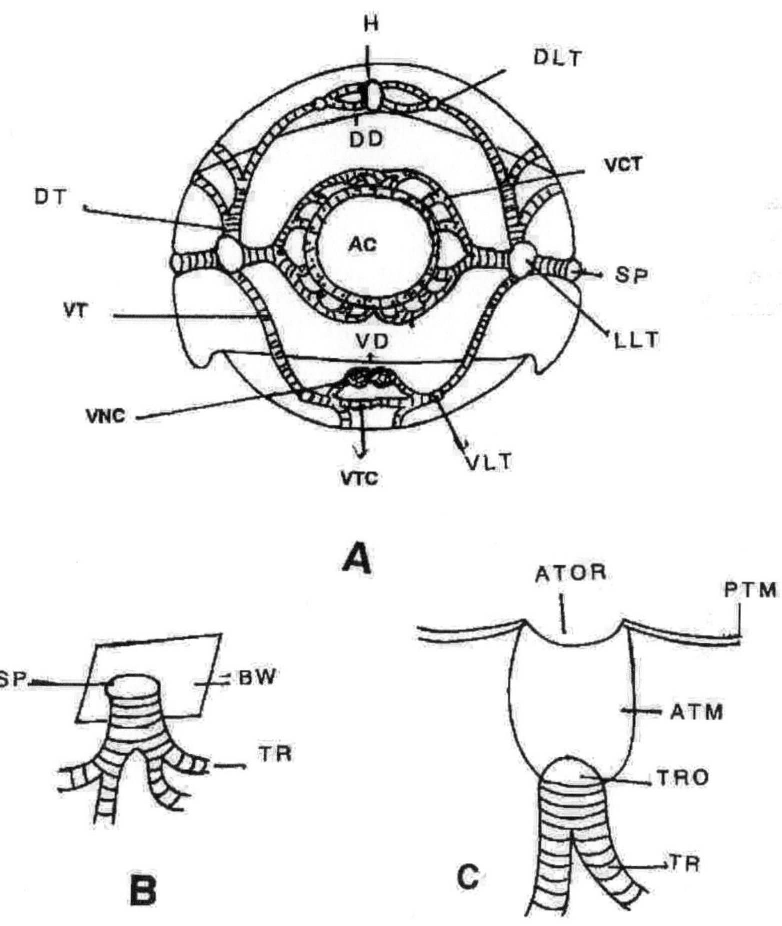

Fig. 34 A. Diagrammatic cross section of abdomen showing main tracheae, transverse trunks, **B**. Diagram showing simple spiracle, **C**. Diagram showing typical spiracle with atrium.

Abbreviations : ATM- atrium, ATOR- atrial orifice, BW- body wall, DD- dorsal diaphragm, DT- dorsal trachea, DLT- dorso- lateral trachea, H- heart, LLT- lateral longitudinal trachea, SP- spiracle, TRO- tracheal orifice, TR- trachea, VD- ventral diaphragm, VT- ventral trachea, VCT- visceral trachea, VNC- ventral nerve cord, VTC- ventral tracheal commissure, VLT- ventro lateral trachea, PTM- peritreme.

(C) TYPES OF RESPIRATORY SYSTEM

Except Diplura, the largest number of functional spiracles are ten pairs viz. two thoracic and eight abdominal. Accordingly, on the basis of the number and distribution of the functional spiracles, the respiratory system has been classified into 3 main types viz. (1) Holopneustic, (2) Hemipneustic and (3) Apneustic.

(1) Holopneustic - In this type the number of functional spiracles are ten pairs, two pairs of thoracic and eight pairs of abdominal, e.g. nymphs and adults of many orders.

(2) Hemipneustic : Here either one or two of spiracles are non-functional. This type has been further divided into (a) Peripneustic, (b) Amphipneustic, (c) Propneustic and (d) Metapneustic.

(a) Peripneustic - In this case the metathoracic spiracle is closed and the prothoracic and abdominal remain open e.g. Neuroptera, Mecoptera, Lepidoptera and Coleoptera.

(b) Amphipneustic - Here only prothoracic and posterior abdominal spiracles are open e.g. larval Diptera.

(c) Propneustic - In this case only the prothoracic spiracles remain open e.g. pupae of Diptera.

(d) Metapneustic - Here only the last pair of abdominal spiracles are open e.g. some larvae of Diptera.

(3) Apneustic : In this case functional spiracles are totally absent e.g. Collembola, Protura etc. Air enters the tracheal system by diffusion through the general body surface or gills or branchiae.

(D) TYPES OF SPIRACLES

(1) Simple spiracles (Fig. 34B): As already stated, spiracles are the external openings of the tracheae situated at the point of invagination of the body wall during the development of trachea. In their simplest form they are merely openings from the integument into the tracheae, representing the primitive apertures of the tracheal invaginations. Such spiracles have no provision for regulating the size of the tracheal aperture e.g. many Aptergyotes and Plecoptera.

(2) Typical spiracle (Fig. 34C) : In many insects the spiracles are more or less sunk into the secondary depressions of the integument, thus the external part of each spiracle becomes a pit-like or tubular chamber which is known as the spiracular atrium. This atrium opens on one hand to the outside by the secondary atrial orifice and on the other into the main trachea, through the primitive tracheal orifice. The atrial orifice is usually situated in a small sclerotic plate of the body wall and is called peritreme.

(3) Biforous spiracles (Fig. 35 A.B) : Biforous spiracle commonly, met within the larvae of certain families of Coleoptera. The characteristic feature of these spiracles is that the functional openings are secondary formations, since the primary atrial orifice is closed except during moulting.

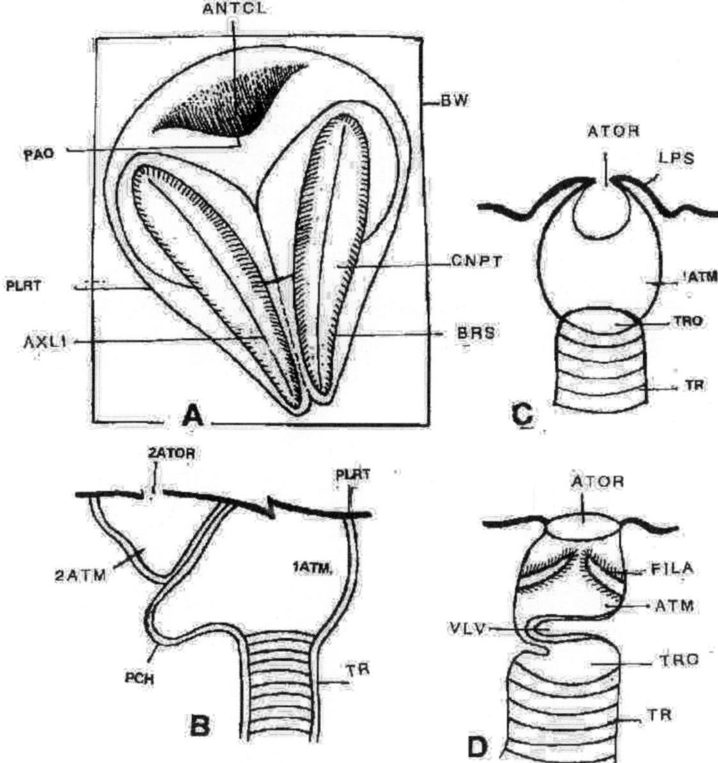

Fig. 35 A. External view of a biforous spiracle, **B**. Inner view of a biforous spiracle, **C**. Spiracle with external closing apparatus, **D**. Spiracle with internal closing apparatus.

Abbreviations : ANTCL- anterior sclerite, ATM- atrium, 2 ATOR- 2nd atrial orifice, 1-ATM- Ist atrium, 2-ATM- 2nd atrium, AXLI- axial line, BW- body wall, BRS- bristles, CNPT- convergent plate, FILA- filter apparatus, PAO- primary atrial orifice, PLRT- pleurite, PCH- pouch, TRO- tracheal orifice, TR- trachea, VLV- valve, LPS- lips.

External character of such a biforous spiracle shows an ovate peritremal area having anteriorly a dark sclerotic thickening and posteriorly two elongate convergent plates. This thickening marks the site of the closed primary aperture of the spiracle (original atrial orifice). Each convergent plate has a clear median area traversed by an axial line.

Internally, the spiracle consists of a closed atrial chamber leading posteriorly into a wide membranous pouch. External to the pouch and projecting beyond it posteriorly are two shallow cuticle-lined chambers situated beneath the convergent external plates of the spiracle. These chambers communicate to the exterior through openings formed along the axial line of the convergent plate. These are spiracular openings which establish communication of the trachea to the exterior in a round about way i.e. through the posterior chamber and the atrium (Fig. 35B).

Course of air inside a biforous spiracle

Air from outside enters the posterior chamber i.e. 2nd atrial chamber (2 ATM) through the openings of the convergent plates and then diffuses through the walls of the chamber into the 1st atrium (1 ATM) from where it enters the trachea.

(E) MECHANISM OF OPENING AND CLOSING OF SPIRACLES

There are essentially two types, by which the insect spiracles can be opened or closed in order to regulate the passage of air to and from the spiracular trachea. In the first type, the closing or opening of the atrial aperture is effected by outer lips of atrium and such spiracles are known as "spiracles with external closing apparatus" (Fig. 35C) The spiracles with external closing mechanism are of common occurrence on the thorax in most groups of insects.

The second type is the mechanism by which the tracheal orifice is regulated at the inner end of the atrium. Such spiracles are termed as "spiracles with internal closing apparatus" (Fig. 35D). Majority of the insect spiracles come under this category. The closing or opening of the tracheal aperture is brought about either by a pair of sclerotic bars in opposite wall of the atrium at the mouth of the trachea or by a valvular structure which arises as a fold at the inner end of the wall of the atrium. The spiracles of this type i.e. with internal closing apparatus are characteristics of the abdominal spiracles.

It may be mentioned here that in the spiracles with internal closing apparatus, certain accessory structures are usually present in the atrium. Such structures commonly have the form of opposing rows of tapering processes of the atrial wall thickly clothed with inter-lacing hairs. The whole structure is known as filter apparatus. The function of this apparatus is to freely permit the passage of air, but to prevent the entrance of foreign particles or water into the trachea. Sometimes the filter apparatus is present outside, guarding the atrial orifice itself, in such cases the foreign particles or water is effectively prevented from entering the atrium itself.

(F) PHYSIOLOGY OF RESPIRATION

The mechanism of respiration in insects is accomplished in two stages viz. (1) ventilation of the tracheal system by means of the movement of the body wall and (2) diffusion of gases inside the trachea.

(1) Respiration by tracheal ventilation

This ventilation is provided by means of the movement of the body wall produced by the somatic muscles and elasticity of the body wall, the movement generally being confined to the abdomen. They may consist of dorsoventrally flattering movement e.g. Orthoptera, Coleoptera or longitudinal telescopic movement e.g. Diptera, Hymenoptera. As a result of such movements most of the part of the tracheal system is kept filled with a gas which is approximately similar to the outside air. The exchange of gases between the trachea and various tissues is ultimately by diffusion. It is during tracheal ventilation that the air sacs play their prominent part. They function as "storage chambers" of air and like the `lungs' of the vertebrates they respond to pressure changes in the air. They are compressed with increasing pressure around them and inflated with reduction of pressure.

(2) Respiration by gas diffusion

Krogh (1920) has been the most successful worker to demonstrate the diffusion of gases within the trachea of an insect. He pointed out that the rate of diffusion varies directly with the diameter of tracheae and inversely to their mean length. According to this theory both carbon dioxide and oxygen pass from and to the tissues respectively through the permeable tracheal walls. It must be however noted that the amount of carbon dioxide produced and released through the tracheae is less than the amount of oxygen taken in by insects. Nevertheless, most of the carbon dioxide diffuses equally through the cuticle of the body wall.

(G) RESPIRATORY ADAPTATIONS IN AQUATIC INSECTS

Aquatic insects obtain oxygen directly from the air or from air dissolved in water hence they exhibit a variety of respiratory adaptations. They possess certain special structural modifications to carry out the function of respiration efficiently.

1. Air stores or physical gills

Some aquatic insects carry air bubbles or films of air which lie beneath the elytra (as in *Dytiscus*) or have a fine pile of erect hydrofuge hairs on other parts of the body which hold a layer of air between them and appear like a velvet when immersed in water as in *Notonecta*. These stores of oxygen, in the form of air films, air bubbles and hydrofuge hair-piles, containing air serve as physical gills.

The atmospheric oxygen is trapped in several ways by the aquatic insects. The spiracles of *Notonecta* are encircled by a fringe or crown of long hairs, or along the antennae as in *Hydrophilus*. On reaching the water surface a fringe of hydrofuge hairs spreads out by surface tension and the air space within is put into communication with atmosphere. Various forms of physical gills are found in aquatic Coleoptera and Heteroptera e.g. *Dytiscus, Hydrophilus,* (Coleoptera). *Notonecta, Corixa, Nepa, Naucoris* (Heteroptera) etc.

These external air stores have several functions viz. (1) they have an important hydrostatic function in enabling the insect to reach the water surface in the correct position to take in oxygen (2) they provide a store of oxygen which is gradually used up by the submerged insects and (3) they provide a mechanism for obtaining dissolved oxygen from the water and so function as a kind of physical gills.

(2) Plastron or air film (Fig. 36 C)

In aquatic Coleoptera like *Haemonia, Elmis* and *Coxelmis*, the air film is so held by the hydrofuge hairs that it cannot be replaced by water. These insects never come out on the surface i.e. the insects become independent of the atmospheric air e.g. *Elmis* can live for months below the surface if the water is well aerated. This very thin firmly held layer of gas is termed as plastron. The spiracles are situated just below the plastron but the air itself does not enter the tracheae.

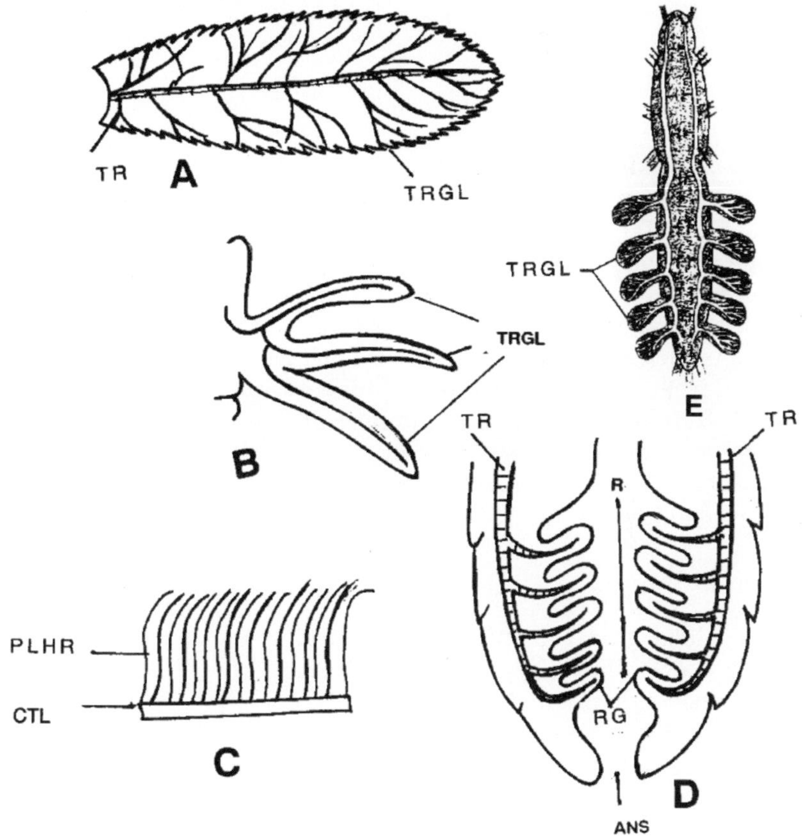

Fig. 36. A, B, E. Diagrams showing Tracheal gills, **C**. Plastron **D**. Rectal gills *Abbreviations:* ANS - anus, CTL- cuticle, PLHR-plastron hair, TR - trachea, TRGL-tracheal gills, RG-rectal gills, R-rectum.

In another group i.e. aquatic Heteroptera, the plastron is very well developed e.g. the Naucorid bug, *Aphelocheirus*. In this bug most of the body surface is covered with a very fine plastron held in position by an epicuticular hair pile. The spiracles are protected by a screen of long cuticular filament beset with fine hairs, the spiracles open into a chamber from which radiating channels run outwards in the form of a rosette. These channels communicate with the exterior i.e. with the plastron by minute openings along their course. The whole structure being lined by extremely fine plastron hairs.

The plastron can act as a permanent physical gill provided there is adequate oxygen dissolved in water which needs no renewal. The oxygen in water enters the plastron and then diffuses in the body of the insects.

3. Spiracular gills

These are variously termed as `tube gills' or `cuticular gills' but more commonly called `spiracular gills'. The spiracular gills arise as filamentous outgrowths of the ectoderm and are covered with cuticle. The lumen of the filament looses all connections with the body cavity. The epithelium more or less degenerates but cuticular wall becomes quite thick and is excavated to form air spaces which establish a connection with the tracheal system. Thus the air in the tracheae is separated from the water only by a delicate wall of cuticle about 0.1 μ thick. They are adapted for both aquatic and aerial respiration, since they do not collapse on drying. They are found in Coleoptera (family Psephenidae and Diptera (Family Tipulidae, Blepharoceridae, Deuterophlebiidae, Simuliidae etc.).

4. Blood gills

The blood gills are thin walled, tubular or digitiform and are some times eversible. They derive their name from the fact that they contain blood but are devoid of trachea or elsewhere are provided with only a few small branches. The function of blood gills is the absorption of water and inorganic ions rather than respiration. They are found in larvae of Trichoptera and Diptera (Culicidae and Chironomidae).

5. Tracheal gills (Fig. 36 A, B, E)

These are hollow, thin-walled, filiform or plate-like, sac-like or lamellate evaginations of the integument that are well supplied with tracheae and tracheoles. They are usually borne on the abdomen and are less frequently present on the thorax. They are found in the majority of the aquatic larvae and in some aquatic pupae of the order Ephemeroptera, Neuroptera, Trichoptera, Plecoptera and Odonata. In

Coleopteran larvae, tracheal gills are present in a few families (Gyrinidae, Hydrophilidae).

6. Anal and Rectal gills (Fig. 36 D)

In dipterous larvae (Culicidae), anal gills are found either in the form of papillae or lamillae. They are thin walled, elongate structures richly supplied with tracheoles. The respiratory siphon is reduced and the papillae are held in the current of water created by the mouth parts. In the Odonata (dragon fly nymphs) rectal gills are present in the form of a network of tracheae inside the rectum (Fig. 36D). The rectal gills may be papilla like or lamellate i.e. leaf like.

7. Integument

In a number of aquatic insects the gaseous exchange is effected through the body wall by means of simple diffusion. Cutaneous respiration is very common in the insects which remain submerged under water. During early stage of the larval life of *Chironomus, Simulium* and *Acentropus*, the tracheal system remains closed or filled with fluid and then respiration takes place by a simple diffusion of gases through skin.

(H) RESPIRATORY ADAPTATIONS IN PARASITIC INSECTS

These adaptations exhibited by many internal parasites leading a semiaquatic life within the blood and tissues of their host are almost similar to the respiratory adaptations shown by aquatic larvae. In the young stages of many of them the tracheae do not contain air, and the oxygen simply diffuses from blood of the host into the blood of parasite. In many forms there are often outgrowths from the body surface. In certain dipterous parasites e.g. Tachinidae the metapneustic type of respiration is seen. In such forms the parasite pierces the body wall on the large tracheal trunks of the host and breath the atmospheric air.

11

THE NERVOUS SYSTEM

The nervous system transmits electrochemical impulses from one part of the body to another, and functions as a connecting link between the sense organs and the effector organs, thus acting as a coordinating centre.

NERVE CELLS

The nervous system is composed of nerve cells or neurons which are ectodermal in origin. The nerve cell consists of a cell body containing the nucleus and long cytoplasmic projections or axons which extend to make contact with other neurons. Frequently, the axons have branches, and collaterals and end in a terminal arborization. Nerve impulses are transmitted from one neuron to the next via axons. Part of each nerve cell is specialized for the reception of stimuli which initiate conduction in the axon. This is known as dendrite and it may arise directly from the cell body or represent the distal endings of an axon in which case there is no anatomical differentiation between axon and dendrite.

The nerve cells are of three types viz., monopolar, bipolar and multipolar (Fig. 37 A, B, C). Generally the nerve cells of the insects are monopolar having only one axon. The peripheral sense cells are bipolar with a short distal dendrite receiving stimuli from the environment and a proximal axon extending to the central ganglia. The hypocerebral ganglion and frontal ganglion have some cells with more than two axons and represent the multipolar cells.

The nervous system can be distinguished into three distinct parts viz., (1) central nervous system (2) sympathetic nervous system and (3) peripheral nervous system and all the three parts are connected with each other.

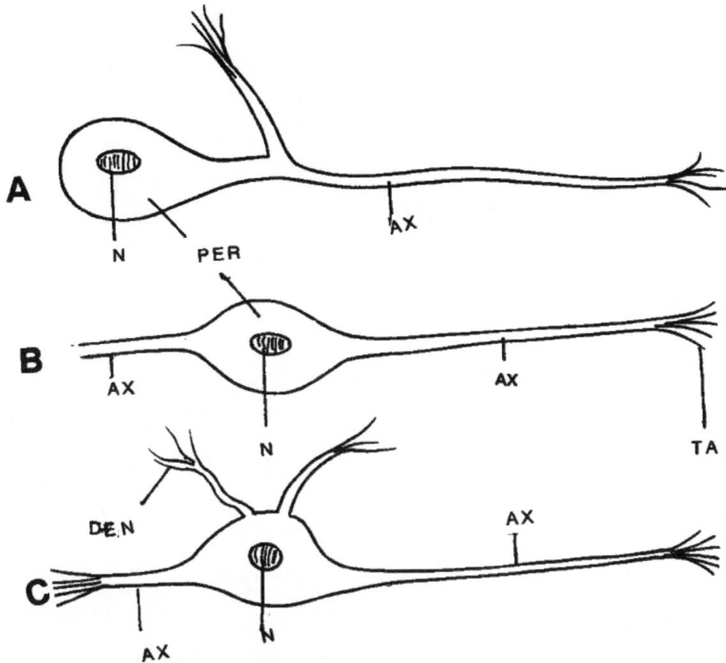

Fig. 37 Different types of neurons **A**. - Monopolar, **B**. - Bipolar, **C**. Multipolar.
Abbreviations : AX- axon, DEN- dendrite, PER- perikaryon, TA- terminal arborizations, N- nucleus.

1. CENTRAL NERVOUS SYSTEM (Fig. 38)

It consists of an anterior median brain, located in the head above the alimentary canal, sub-oesophageal ganglion below the oesophagus and the ventral ganglionated nerve cord situated in the mid ventral part of the body, in which the ganglia are united lengthwise by interganglionic connections.

Brain : It is situated in the head, just above the oesophagus region and therefore it is also known as supraoesophageal ganglion. It is always differentiated into three parts viz. (1) protocerebrum, (2) deutocerebrum and (3) tritocerebrum (Fig. 38).

Each of these consists of a pair of ganglia; the principal features to be distinguished within the brain are special group of cells, fibre tracts and certain compact bodies (Fig. 39A).

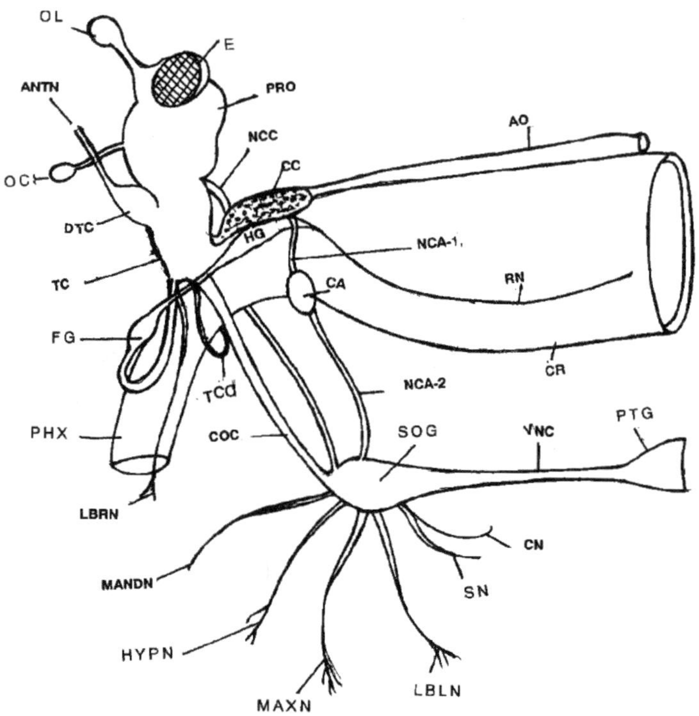

Fig. 38, Nervous system of a generalized insect.

Abbreviations : ANTN- antennal nerve, CC- corpus cardiacum, CN- cervical nerve, COC- circum oesophageal connective, DTC- deuto cerebrum, HYPN- hypopharyngeal nerve, LBRN- labral nerve, LBLN- labial nerve, MANDN- mandibular nerve, MAXN- maxillary nerve, OC- ocellus, OL- optic lobe, PRO- protocerebrum, SN- salivary nerve, TC- tritocerebrum, TCC- tritocerebral commissure, VNC- ventral nerve cord, E- eye, CA- corpus allatum, AO- aorta, NCC- nervi corporis cardiaci, NCA-1 - NCA-2- nervi corporis allati 1 and 2, SOG-suboesophageal ganglion, PTG- prothoracic ganglion, PHX-pharynx, FG- frontal ganglion, RN- recurrent nerve, CR- crop, HG- hypocerebral ganglion.

(a) *Protocerebrum* : The protocerebrum is the largest part of the brain and it represents a fused pair of preantennal ganglia. Normally, the optic lobe is not a part of the brain but in the adult stage it gets associated with the protocerebrum. The important fibrous structures present in the protocerebrum are pons cerebralis, corpus pedunculatum and corpus centrale. In addition to these fibre tracts, the two lobes are connected with the protocerebral commissural tract.

Pons cerebralis : It is also known as protocerebral bridge. It lies in the dorsal and posterior part of the pars intercerebralis. It is composed of fibrous mass or neuropile mass. It has been regarded as "posterior dorsal commissure" by C.B. Thompson (1913) but according to Snodgrass (1935) it is an association centre, since fibres enter it from many parts of the brain. Associated dorsally with pons cerebralis are two masses of dorsal globuli cells, the neurite of the globuli form the body of pons.

Corpus pedunculatum : The corpus pedunculatum is situated in the dorsal part of the brain on either side of the pars intercerebralis. These are also known as mushroom bodies due to their appearance like a mushroom with stalked cap. The cap or calyx is composed of globuli cells whereas the stalk is made up of nerve fibres. The fibres of stalk divide posteriorly into inner and outer lobes. The corpora pedunculata are the largest and the most highly developed association centres in the brain of Pterygote insects and are one of the most conspicuous features of the internal cerebral structure. It is regarded as the centre of all activity i.e. the cyclic centre of the brain.

Between these two stalked bodies, there are a number of small lobes which are connected with the ocelli and are called the ocellar lobes. They give out the ocellar nerves. The ventral bodies are situated in the ventrolateral parts of the brain just above the deutocerebrum. They are united with each other by means of a commissural tract. These ventral bodies are not universally present though they appear to be of large size in Coleoptera and Lepidoptera.

Corpus centrale : The corpus centrale or central body lies in the centre of anterior portion and is made up of a compact mass of nerve fibres which collectively form an oval mass. Although it is not directly connected with the neurons but it has been regarded as another important centre of association between the terminals of fibres from all other parts of the brain.

Optic lobes : These are the largest nervous bodies associated with the brain. Originally they are separate from the protocerebrum. In adult insects they get so closely associated with the protocerebral lobes that they may be regarded as the part of the protocerebrum. They are very complex in structure and contain the optic ganglion. According to the distribution of the nerve tracts, optic lobe is divided into three distinct region viz., (1) periopticon or ganglionic layer, (2) epiopticon or medulla externa and (3) opticon or medulla interna (Fig. 39A).

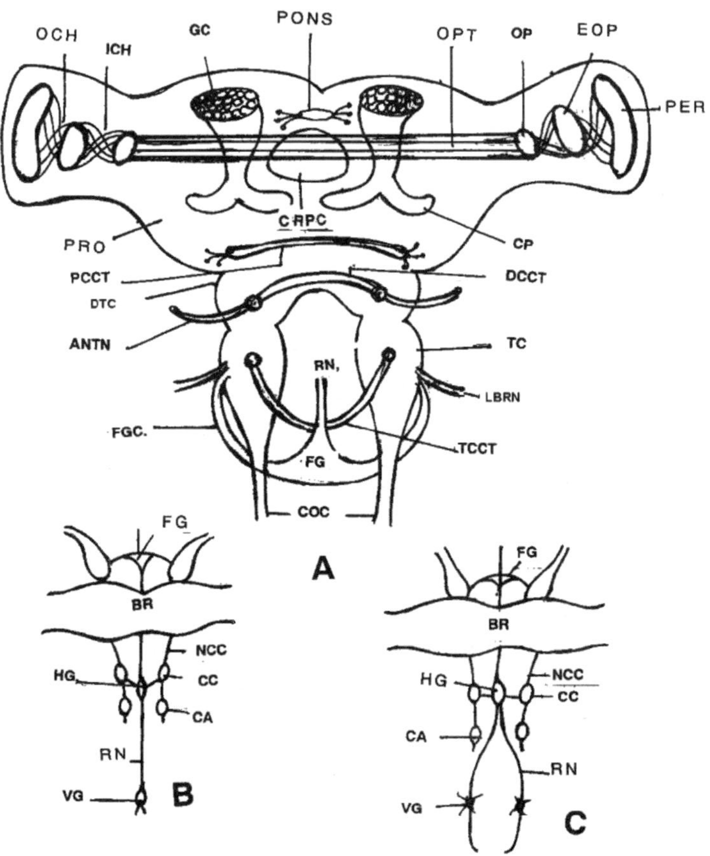

Fig. 39 A. Diagram of the brain showing internal fibrous bodies and fibre tracts. **B & C**. Diagrams of the visceral nervous system. *Abbreviations* : ANTN- antennal nerve, CP- corpus pedunculatum, CRPC- corpus centrale, COC- circum oesophageal connective, CA- corpora allata, CC- corpora cardiaca, DTC- deutocerebrum, DCCT- deutocerebral commissural tract, EOP- epiopticon, FG- frontal ganglion, FGC- frontal ganglion connective, GC- globuli cells, HG- hypocerebral ganglion, ICH- inner chiasma, LBRN - labral nerve, OPT- optic tract, OP- opticon, OCH- outer chiasma, PER- periopticon, PCCT- protocerebral commissural tract, PR- protocerebrum, RN- recurrent nerve, TCCT- tritocerebral commissural tract, TC- tritocerebrum, VG- visceral ganglion, NCC- nervus corporis cardiaci.

(1) Periopticon : This is the ganglionic plate which is nearest to the layer of ommatidia. Periopticon is connected with epiopticon by crossing fibres. It is these crossing fibres which form the outer chiasma.

(2) Epiopticon : It forms an external medullary mass. Sometimes it is also called as medulla externa. It is connected with periopticon by means of the outer chiasma.

(3) Opticon : It is subdivided into 2 fibrous masses which are connected to the fibres that cross with those from epiopticon to form the inner chiasma. Within its substance there are 4-layers of stratified fibres. It has an elaborate fibre connections through the optic tract with various parts of the brain.

(b) *Deutocerebrum* : These are situated as prominent swellings on the ventral aspect of the protocerebrum. The deutocerebrum consists of fused paired lobes derived from the antennary segment containing the centres of the antennal nerves. The two deutocerebral lobes are joined together by means of a deutocerebral commissural tract. Antennal lobe is connected with the corpus pedunculatum of its side by corpus centrale, forming optico-olfactory-chiasma.

(c) *Tritocerebrum* : This is the smallest part of the brain and is divided into two separate lobes. These two lobes are joined to each other by means of a tritocerebral commissure. Embryologically the tritocerebral lobes of the brain represent the first paired ganglia of the primitive ventral nerve cord; the union with the brain is secondary. An important feature of the tritocerebral ganglion is that it is connected with stomodaeal nervous system. The main nerves arising from tritocerebrum are the frontal ganglion connectives and the labral nerves.

Sub-oesophageal ganglion

The suboesophageal ganglion lies below the oesophagus. It is formed by the fusion of ganglia of the primitive gnathal segments. It innervates the mouth parts viz. mandibles, hypopharynx, maxillae, labium, salivary duct and the neck muscles. The nerve bundles contain both the types of nerve fibres, motor and sensory. The longitudinal nerve tract associated with the suboesophageal ganglion contains the connecting fibres between the sensory centres of the head and the motor centres of the body parts.

Ventral nerve cord : It consists of a series of ganglia which are connected together by means of a pair of connectives thus forming a longitudinal trunk or chain. It starts from the posterior border of the suboesophageal ganglion and lies on the ventral part of the floor of the thorax and abdomen. The first three ganglia are thoracic lying in the thorax and the rest are abdominal lying in the abdomen. From each thoracic ganglion two pairs of main nerves are given out, of which one supplies the general musculature and the other to the muscles of the leg. An additional pair of nerve is found in the ptero-thorax (meso- and metathorax) which controls the wing movement.

The number of abdominal ganglia varies, from order to order. Generally the first abdominal ganglion fuses with metathoracic and the terminal ganglion is always a composite structure formed by the fusion of atleast 3-primitive ganglia. A pair of nerves supplying to the muscles of its segment is generally given out from each abdominal ganglion.

Structure of a ganglion : Histologically all the ganglia of the ventral nerve cord are similar. Each ganglion is invested in a sheath - the neurilemma. The neurilemma is composed of two layers, (1) the outer non-cellular neural lamella and (2) the inner cellular perineurium. The neural lamella is continuous over the connectives, the nerves and all the ganglia. The neural lamella is secreted by the perineurial cells. The neural lamella contains neutral mucopolysaccharides and collagen fibres and is primarily a supporting structure. The perineurium helps to form a physiological barrier between the ionic and other constituents of the haemolymph and to those of the nervous tissue.

The cortex of the ganglion is formed by a group of monopolar neurons which are of two types - (1) ordinary neurons and (2) neurosecretory neurons. The latter perform a variety of functions in different insects. The central part of the ganglion is occupied by the neuropile mass.

In each ganglion there are six principal groups of nerve elements viz. (1) the cell bodies and roots of the motor fibres of the lateral nerves; (2) the roots of the sensory fibres of the lateral nerves; (3) the cell bodies and fibres of the intraganglionic association neurons, (4) the cell bodies and collaterals of the inter ganglionic association neurons, (5) the cell bodies and roots of the motor fibres of the median nerve and (6) the roots of the sensory fibres of the median nerve.

2. SYMPATHETIC NERVOUS SYSTEM

It can be divided into (1) stomodaeal nervous system, (2) ventral sympathetic nervous system and (3) caudal sympathetic nervous system.

i. Stomodaeal nervous system (Fig. 39 B,C)

It is also known as stomatogastric nervous system or anterior sympathetic nervous system. It lies on the dorsal surface of the foregut and consists of frontal, hypocerebral and visceral or stomachic ganglia and recurrent nerve. It is connected with the brain directly through nerves and innervates the foregut, midgut, heart, aorta, corpora allata and certain other tissues of the regions.

Cazal (1948) has described various type of arrangement of structures in this system. Two types of arrangements are very common in the Pterygota (Fig. 39 B,C). The one constant feature of this system is the presence of a small triangular ganglion situated on the dorsal wall of the oesophagus, a short distance in front of the brain. This is the frontal ganglion. Anteriorly it gives off a frontal nerve which passes to the clypeus. It is also connected with the tritocerebral lobes by a pair of frontal ganglion connectives. Posteriorly the frontal ganglion gives off a recurrent nerve which runs along the mid dorsal wall of the oesophagus, and expands into a hypocerebral ganglion. The hypocerebral ganglion is either absent or only occasionally present in Lepidoptera.

The hypocerebral ganglion in a number of insects lies in between the corpora cardiaca and either it is closely associated or connected with nerves. From the hypocerebral ganglion, the recurrent nerve passes backward either as a median or as paired lateral nerves to the hinder region of the foregut. On the foregut it terminates in a ventricular or stomachic ganglion which innervates the posterior part of the foregut. Behind the brain a pair of oesophageal ganglia are present on the dorsolateral aspect of oesophagus. These ganglia are also known as corpora cardiaca and are connected with the hypocerebral ganglion and protocerebrum. The corpora cardiaca include both nervous and endocrine cells. The corpora cardiaca are also connected with non-nervous epithelioid glandular bodies, the corpora allata by nervi corporis allati. In the Cyclorrhaphan Diptera i.e. fleshfly, housefly, blowfly larvae, hypocerebral ganglion, corpus cardiacum, corpus allatum and prothoracic glands fuse together to form a composite ring-shaped structure, known as Weismann's ring encircling the aorta (See p. 145). This system is variable and is reduced in the Apterygota but a frontal ganglion is always present.

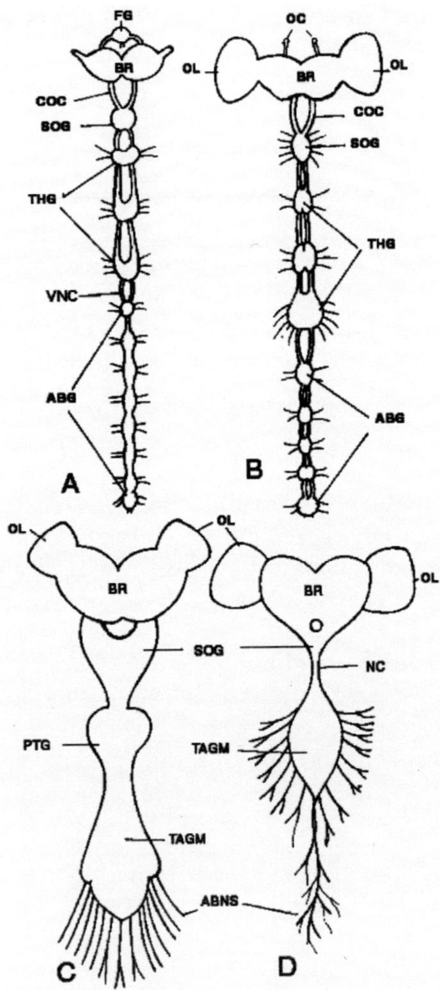

Fig. 40A. Nervous system of a caterpillar, **B**. Nervous system of a grasshopper, **C**. Nervous system of a plant bug, **D**. Nervous system of house fly. *Abbreviations :* ABG- abdominal ganglia, BR- brain, COC- circum oesophageal connective, FG- frontal ganglion, OL- optic lobe, OC- ocellus, THG- thoracic ganglia, SOG- suboesophageal ganglion, VNC- ventral nerve cord. PTG- prothoracic ganglion, TAGM- thoracico- abdominal ganglionic mass, ABNS- abdominal nerves, NC-nerve cord (Figs. A-D. Awasthi- unpublished).

ii. Ventral Sympathetic nervous system

Zawarzin (1924) has mentioned that in a typical condition this system consists of a pair of transverse nerves lying in association with each nerve cord ganglion. There is a median longitudinal nerve which connects the transverse nerve with the preceeding ganglion. These transverse nerves innervate the spiracles of their segment. The transverse nerves along their course give out one or more minute ganglionic swellings. These swellings, store and release neurosecretory material and are called perysympathetic neurohaemal organs.

iii. Caudal sympathetic nervous system

From the last ganglion of the nerve cord in the abdominal region, nerves are given to the reproductive organs, posterior portion of the alimentary canal, and Malpighian tubules which constitute the caudal sympathetic nervous system.

3. PERIPHERAL NERVOUS SYSTEM

It consists of all the nerves radiating from the ganglia of the central and sympathetic nervous systems. This system comprises of the distribution of all the nerves in the body. There are two types of sensory neurons in the peripheral system (1) bipolar neurons formed from the hypodermal cells and connected with the cuticular sensory hairs and (2) the multipolar neurons which are present just below the integument.

MODIFICATIONS OF THE NERVOUS SYSTEM

Various modifications have taken place in the nervous system of insects (Fig. 40 A-D). These modifications are correlated with the degree of development of the special instincts and activities e.g. there are structural differences in the brains of drone, worker and queen honey bees which are correlated with the activities of these three forms. Similarly the optic lobes are developed in proportion to the size of the eyes, and antennary lobes to the development of the antennal sense organs. Even in the same order there are differences of development of the brain.

Varying degrees of fusion of the thoracic and abdominal ganglia are seen by a comparative study of the nervous system of different insects. The most generalized type of ventral nerve cord is found in the Thysanura, many larvae and lower Pterygotes. In these forms there is (1) a suboesophageal ganglion (2) 3- thoracic ganglia and (3) 8-abdominal ganglia (Fig. 40A).

There is some concentration in Mecoptera, Trichoptera, Hymenoptera and most of the Orthoptra, where the metathoracic ganglion generally fuses with the first three abdominal ganglia and the 7th one along with the following ganglia forms a composite structure. In many Heteroptera, the suboesophageal and prothoracic ganglia are distinct but all other have fused together (Fig. 40C). There is only the suboesophageal ganglion and a single compound thoracico abdominal-ganglion in many sternorrhynchan-Homoptera and higher Diptera (Fig. 40D). Similarly in Coccoidea, Aphidoidea and some Coleopteran larvae, all the ventral ganglia including the suboesophageal, are united together forming a single ganglionic mass.

TRANSMISSION OF NERVE IMPULSE

In the following pages a brief account of nerve impulse transmission is given.

(1) Axonal transmisson

In the resting condition (polarized state) due to the presence of abundant sodium ions (Na^+) out side, the outer membrane is charged positively and the inner membrane is charged negatively. This is due to the movement of sodium ions from axoplasm to the out side of the axons. The outward flow of the sodium ions is known as sodium pump. On the otherhand the potassium ions are more abundant in the axoplasm (Fig. 41A).

Upon stimulation by an external stimulus the permeability of the membrane is changed and the flow of sodium ions stops. This causes the movement of sodium ions inside the axons and depolarization (Fig. 41 B,C) of the membrane takes place. Further more, on account of flow of sodium ions inside is very fast causing inner membrane positively charged and the outer part of membrane is charged negatively. Thus propagating a short electro chemical current or impulse along the axon(s). This is known as action potential.

The period of permeability to sodium-ions is short lived and is followed by a period of increased permeability to potassium ions. As a result of which the potassium ions flow out side the axon which negatively charges the inner membrane. This is called as falling phase of the action potential. In this way nerve impulses are propogated in the axons. As the impulse passes in the forward direction (Fig. 41D), the permeability of the membrane assumes original state i.e. polarized condition due to decreased flow of potassium ions out side the axons as a result of which repolarization takes place (See Fig. 41D).

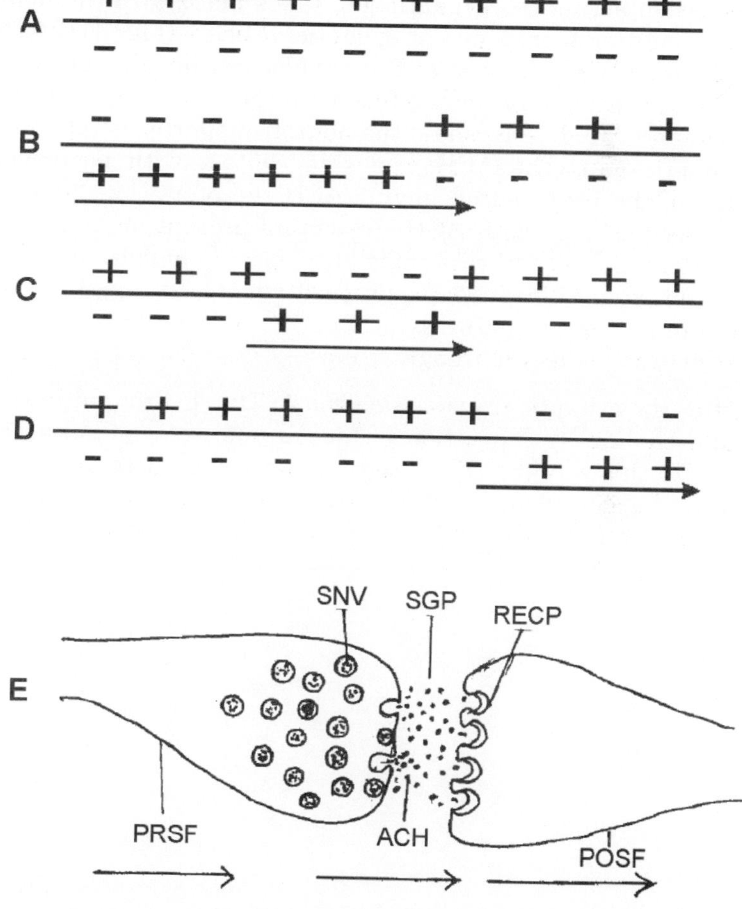

Fig. 41 A. Polarized condition, **B**. and **C**. Depolarized condition showing impulse conduction, **D**. Repolarization of membrane **E**. Diagram showing synaptic transmission.

Abbreviations: ACH-acetylcholine, SNV- synaptic vesicles, SGP- synaptic gap, RECP- receptors, PRSF- pre-synaptic fibre, POSF- post synaptic fibre.

(2) Synaptic transmission

The axon terminals of the neurons are not continuous but are contiguous with that of other terminals having a short gap. This gap is known as synaptic gap. When an inpulse has passed along an axon it must cross the synapses in order to stimulate another neuron. Transmission across the synaptic gap takes place with the help of a neurotransmitter stored in the synaptic vesicles (Fig. 41E). This is acetylcholine (ACH). As soon as the impulse reaches the terminal end of the axon, the synaptic vesicle fuse with the membrane of axon, the intermediate wall dissolves and the neurotransmitter is releazed into the synaptic gap. The ACH comes in contact with post synaptic terminal of the next neuron having ACH receptors (Fig. 41 E). Now ACH molecules bind with ACH - receptors present on post synaptic terminal. It changes the permeability of the membrane, causing depolarization and this initiates propogation of nerve impluse. In this way the impulses caused by external stimuli reach from one neuron to next neuron via axon through synapses.

After the synaptic transmission the ACH is hydrolized into acetic acid and choline by the action of an enzyme Acetyl Cholinesterase (ACHE). In this way the ACH - receptors become unoccupied in order to receive the second message. Thus, in the synaptic transmission the neurotransmitter (ACH) acts as stimulus to the receptor axons.

12

THE SENSE ORGANS

The sense organs are composed of various types of sensilla or sensory hairs and setae viz. (1) trichoid sensilla, (2) campaniform sensilla, (3) sensilla basiconica, (4) chordotonal organs, (5) sensilla placodea and (6) sensilla coeloconica etc. all these are formed by the modified epidermal cells.

(1) Trichoid sensilla (Fig. 42A)

These are sensory hairs. The trichoid sensilla covers the general body surface of the insect but are more concentrated on the antennae, tarsi and cerci. Each sensillum is provided with a trichogen cell, a tormogen cell, and a bipolar nerve cell. It is the nerve cell which acts as a sense cell and whose distal processes or dendrites receive stimulus and the proximal nerve process conveys message to the nervous system. These act as mechanoreceptor as well as chemoreceptor which depend on the place of their occurrence and thus perform a variety of functions.

(2) Campaniform sensilla (Fig. 42B)

In this case, the two epidermal cells and a bipolar spindle shaped nerve cell i.e. the sensory cell, is associated with each sensillum. This type of sensillum consists of a protuberance or dome-shaped cuticular structure. These are also distributed in the various parts of the body like head, thorax legs, cerci, wing bases, and mouth parts. They are found in groups or clusters and perform many functions like movement of wings and flight control.

(3) Sensilla basiconica (Fig. 42C)

These are cone-like setae and are composed of a trichogen cell, a tormogen cell and many bipolar nerve cells. These are sensitive to various chemical stimuli and they function as taste receptors or olfactory receptors, depending on their location in the body part.

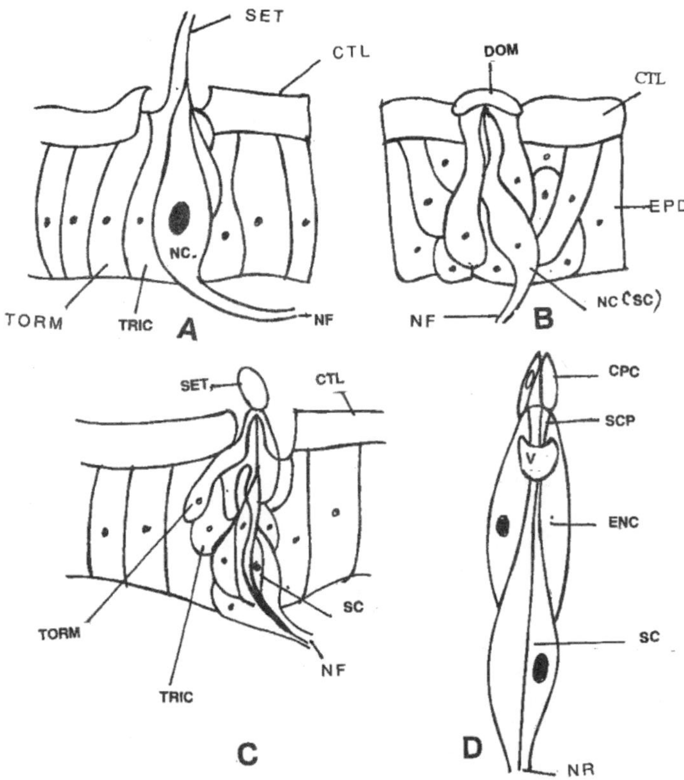

Fig. 42. Diagrams showing different types of sensilla, **A**. Trichoid sensillum, **B**. Companiform sensillum, **C**. Basiconic sensillum, **D**. A scolopophore. *Abbreviations* : CTL- cuticle, EPD- epidermis, DOM- dome, NF- nerve fibre, SC- sensory cell, SET- seta, TRIC- trichogen cell, TORM- tormogen cell, CPC- cap cell, ENC- envelop cell, NR- nerve, SCP- scolopale, V- vacuole.

(4) Chordotonal organs (Figs. 42D, 43B)

These are compound structures composed of a number of sensilla known as scolopophores or scolopidia, which contain a distinct rod-like structure, the scolopale. In the different parts of the body different types of scolopophores are found and each contains a sensory cell. Each chordotonal organ is composed of a bundle of spindle shaped scolopophores; usually the proximal and distal ends of the scolopophores are attached with the body wall but occasionally their distal ends are free. The two types may be distinguished respectively as integumental and subintegumental chordotonal organs (Fig. 43B). Each Scolopophore of the chordotonal organ is composed of a cap cell, a bipolar spindle-shaped nerve cell and a scolopale formed within the envelop cell. The cavity of the scolopale communicates basally with a liquid filled vacuole (Fig. 42 D). These are distributed in the legs, antennae, mouth parts and thorax, and are one of the most important organs of equilibrium or proprioceptors. In addition, they perform a variety of functions and are accordingly modified and named as subgenal organ, tympanal organ, Johnston's organ etc.

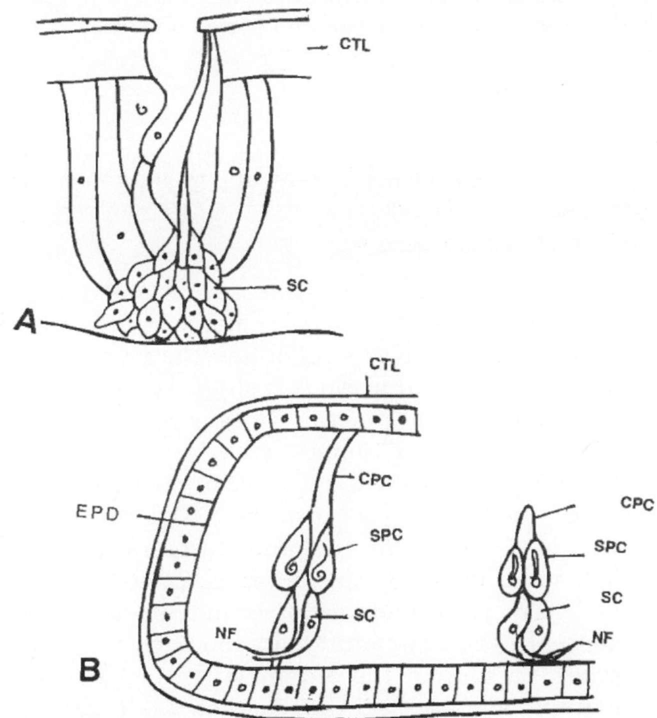

Fig. 43. Diagrams showing different types of sensilla. **A**. Placodeum. **B**. Showing two types of scolopophores viz. integumental and subintegumental of the chordotonal organ.
Abbreviations : CTL- cuticle, CPC- cap cell, EPD- epidermis, SC- sensory cell, SPC- scolopale cell, NF- nerve fibre.

(5) Johnston's organ

Johnston's organ is a highly specialized type of chordotonal organ. It is located in the second antennal segment of most of the pterygote insects and acts as a proprioceptor as well as sound receptor in different insects.

(6) Sensilla placodea (Fig. 43A)

In this case the circular or oval cuticle plate covering the sensilla is provided with large number of pores. The sensillum is composed of a trichogen, a tormogen and a bipolar nerve cell. The distal dendrites of the nerve fibres terminate in the cuticular plate of pores. These are found in the antennae of honey bee and act as olfactory receptors.

(7) Sensilla coeloconica

These are occasionally found in the antennae of some insects. These are also known as pit-peg organs due to sunken cuticular pegs. They have been described from the antennae of Coleoptera, Hymenoptera and Lepidoptera.

Essentially, all sensilla are composed of a trichogen cell, a tormogen cell and one or more bipolar nerve cells whose distal or terminal dendrite part receives stimuli and the proximal nerve fibre is connected with the central nervous system. The sensilla or sensory organs are distributed in the general body surface but more abundant in certain body parts like antennae, cerci, tarsi, tibiae, wing bases, legs, head, thorax, mouth parts etc. They perform different functions in different parts of the body as taste receptors (in mouth parts or tarsi), organs of equilibrium or proprioceptors (in legs, wings, antennae etc.), and control also the movement, twisting and vibration of the wings. They act as auditory or sound sensitive organs when present underneath the tympanum and olfactory receptors sensitive to odours, when present in the antennae. In fact one sense organ performs many functions e.g., the chordotonal organ of the 2nd antennal segment (Johnston's organ) acts as a gravity receptor, velocity receptor (rate of air flow), organ of equilibrium as well as auditory organ in different insects.

13

THE SOUND PRODUCING ORGANS

Insect sounds have diverse origin, kind, intensity and purpose. While they are chiefly caused by certain frictions and vibrations, they have a definite purpose in the life of the owner. Although their biological significance is not well understood even today, it is quite possible that these sounds play some important part in sexual attraction, sexual rivalry or defence and warning. Broadly speaking sound production can be classified into 5 types viz :

(1) Friction between part of the body and any external object

Here the sound production is either by tapping the head against the wall of the burrow or friction between the abdomen of the insects and the substrate or by knocking the head against the nest surface (e.g. termites and stone flies).

(2) Friction caused between any two parts of the insect body

A great majority of the sounds produced belong to this category. The organs producing the sound are known as stridulatory organs e.g. Orthoptera, Coleoptera and Hemiptera. In Orthoptera, particularly in the families Acrididae, Tettigoniidae and Gryllidae, the stridulation has been developed to a great extent. The sound producing organs vary greatly in these families e.g. it may be (a) in the form of a row of pegs situated on the inner side of each hind femur which would work against the outer surface of each tegmen (Acrididae), (b) a row of teeth situated on a tegminal vein which scrapes a sclerotized part of the margin of the opposite tegmen (Tettigoniidae and Gryllidae). Long-horned grasshoppers and crickets have a file on the large vein of one forewing (Fig. 43C) that is rubbed across the scraper at the edge of the other wing and is in the form of special ridges on the 2nd and 3rd abdominal segments which are rubbed by the posterior femora

(Gryllidae). The sounds produced by these insects vary in tone, melody and frequency.

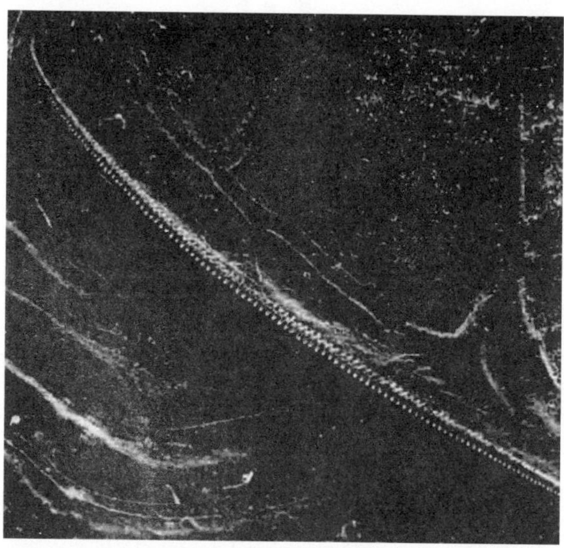

Fig. 43C. File on front wing of a male cricket (*Acheta assimilis*) that is rubbed across a scraper on the other forewing as the wings are moved. The result is vibration of the wings and a sound that attracts female cricket (From Elzinga, 1978)

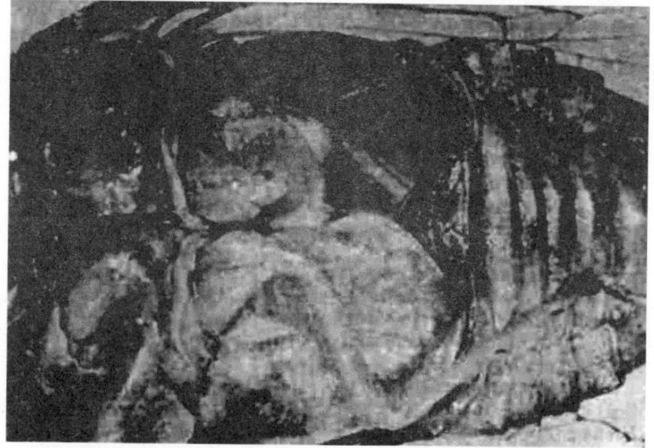

Fig. 43D. Tympanic organ in a male cicada (*Tibicen pruinosa*) which produces sounds that attract the female. The right metathoracic epimeron has been removed to expose the resonant chamber of the abdomen. Each species has its particular frequency and pattern of song which maintains specific isolation. (From Elzinga, 1978).

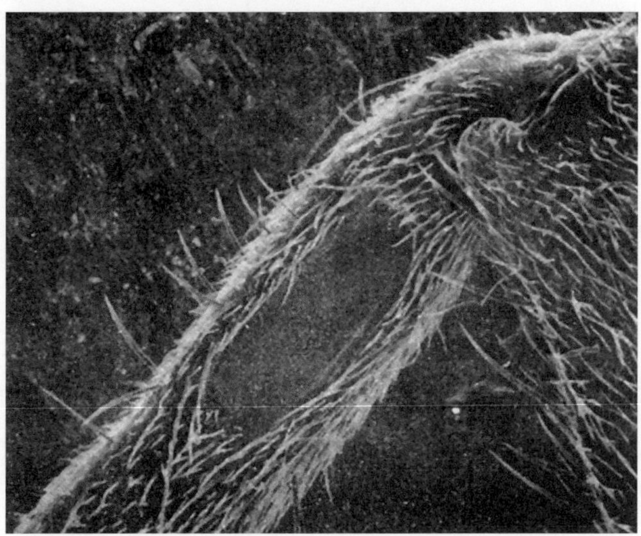

Fig. 43E. Tympanum on tibia of a female cricket (*Acheta assimillis*). (From Elzinga, 1978)

Among Coleoptera, stridulatory organs have been reported from various parts of the body in different species. Sometimes the exoskeleton forms a file like area which rubs the adjacent region and produces the sound. The purpose of sound production in Coleoptera might be both for sexual attraction and defence.

Sound production in Hemiptera is exhibited in both sexes or in male alone, as in the case of male corixid. The noise may be produced by the frictional contacts between the spines of the femur and the angular side margins of the face.

Stridulatory organs are also present in Lepidoptera and Hymenoptera. In the former group, the noises are produced either of a rustling or hissing type caused by (a) friction between fore and hind wings, (b) friction between legs and wings, (c) friction between adjacent parts of the thoracic wall and (d) friction between part of the third abdominal sternum and an infolding of the adjacent sterno-plural region. In the Hymenoptera, the organ consists of a file and scraper located on the mid dorsal region of the integument e.g. ants.

(3) By wing vibration

Sounds produced by wing vibration are in the form of humming or buzzing type which usually occur during the flight of insects.

Apparently the frequency of the emitted note corresponds with that of the wing beat e.g. bees, flies etc. Interestingly enough, the honey bees' wings beat 11,400 times a minute, and these wing movements make the bee's familiar buzz.

(4) By the vibration of special structures in the body

The organs responsible for sound production in this category are of a complex type, usually in the form of shell-like drums or tymbals situated at the base of the abdomen which vibrate by the action of powerful muscles. These organs are usually possessed by males and the noise produced is of high intensity. Normally, these organs are freely exposed but in some Cicadas they are covered by overlapping plates or opercula (Fig. 43D). Colonies of hundreds of male cicadas make loud noises to attract females. These sounds are as noisy as heavy traffic.

(5) Sounds of miscellaneous and unknown origin

The methods and organs responsible for sound production in this category are still uncertain. Some believe that sound is caused by the vibration either of a series of lamillae or a tongue like fold projecting into the lumen of the trachea close to the spiracles. According to certain authors there are special tympanal chambers behind the insertion of wings in close association with spiracles. Whatever may be the types, sound production is caused by one factor only i.e. by flow of air in this category e.g. certain Diptera and Coleoptera.

Sounds are received by the "ears" or *tympana* of the female; these tympana are located on the fore tibia (Fig. 43E) in crickets and long-horned grasshoppers and elsewhere in other insects. It is interesting to note that as sound production varies with temperature, receptor thresholds also respond correspondingly, thereby insuring success of the system.

Thus, sound production in insects can be popularly termed as the call and cry of the organisms for their mate location and also for the purpose of defence. While the cause and methods of sound production are very varied, the object and motive of the insects before the production of these particular noises are still a matter of some debate among Zoologist and Entomologists in particular.

14

THE PHOTORECEPTOR AND PHOTOGENIC ORGANS

(A) PHOTORECEPTOR ORGANS

The photoreceptors of insects are the simple eyes or ocelli and the compound eyes. In an ocellus, only one corneal lens is present whereas in the compound eye the cornea is divided into numerous separate facets. Thus, compound eyes are composed of a number of simple eyes or ommatidia, each ommatidium is the unit structure of the eye.

Ocelli : There are three simple eyes or ocelli which are located between the compound eyes and the frontal region of the head capsule. In each ocellus, there is a single cornea or lens which is formed by the transparent cuticle. It is secreted by the underlying corneagen cells. The shape of the cornea varies from species to species. Beneath the corneagen cells, lie the photo sensitive retina or retinulae. The retina is formed by the retinal cells. The apposed rhabdomeres of the retinal cells form a central or axial rhabdome. Slight variations are found in the ocelli from this basic pattern in different insect orders.

Compound eyes : There are two compound eyes, each lies on either side of the head capsule. Each compound eye (Fig. 44A) is composed of identical units, the ommatidia. The number of ommatidia varies considerably from species to species which may be from one to several thousands e.g. in the worker ant *Ponera,* each eye has only one ommatidium, in the housefly *Musca* each eye consists of about 4,000 ommatidia, and in the dragon flies each eye has some 10,000 to 28,000 ommatidia.

Structure of an ommatidium (Fig. 44B) : A single ommatidium consists of the following elements :

(1) **Cornea :** The cuticle covering the eye is transparent, colourless forming a biconvex corneal lens. The lenses in surface view form the facets of the compound eye.

(2) **Corneagen cells** : Each cornea or lens is secreted by two corneagen cells. These are modified epidermal cells.

(3) **Cone cells:** Beneath the corneagen cells or cornea lie four cone cells which produce a crystalline cone in eucone eyes. The crystalline cone and corneagen cells are encircled by two primary pigment cells, the iris cells.

(4) **Retina** : Just behind the crystalline cone are present 8-elongate photosensitive nerve cells known as retinulae.

(5) **Rhabdome** : The retinulae secrete a central refractive rod which is known rhabdome. In fact the inner part of the retinula is termed as rhabdomere which secretes rhabdome. In certain Diptera, each rhobdomere remains separate grouped around a central axis while in others they are contiguous and may fuse together. In the honeybee the rhabdomeres are fused in pairs whereas in some Orthoptera all the rhabdomeres are fused into a single unit.

(6) **The secondary pigment cells** : The retinular cells are surrounded by secondary pigment cells which isolate each ommatidium from its neighbours.

(7) **The inner ends of each retinular cells** is continued into an axon or a nerve fibre which passes through the basement membrane and connects with the brain through the optic lobe.

Types of compound eyes

On the basis of the nature of crystalline cone the compound eyes are classified into four types viz.

(1) **Eucone** : In this type each ommatidium contains a hard refractive crystalline conical body, secreted by the cone cells. Eucone eyes are found in Orthoptera, Lepidoptera, Coleoptera, Odonata etc.

(2) **Pseudocone** : In this type the cone is filled with a viscous or gelatinous liquid and the crystalline cone is absent e.g. some Diptera.

(3) **Acone** : In this case the refractive material or the crystalline cone is absent, there are elongated transparent cone cells e.g. Dermaptera, some Diptera and some Hemiptera.

(4) **Exocone** : In this type of ommatidium the crystalline cone is replaced by a cone like extension of the inner surface of the cornea or transparent cuticle and not from the cone cells e.g. some Coleoptera.

Formation of image

As remarked, a compound eye is composed of numerous ommatidia, each representing a visual unit. Thus, in each ommatidium only a fraction or part of the image is formed out of the entire visual field. The complete image is a collective one, or a mossaic image where all the ommatidia produce a series of spots of light of different intensities which together form a picture of the object.

The eyes are adapted to form two types of images : apposition and superposition; the former is formed in light and the latter is adapted to dark.

Apposition image (Fig. 44C)

In the day active or diurnal insects, in bright light apposition image is formed. In this case the screening pigment cells remain in the extended position and they isolate ommatidia. The light rays pass only through the central portion, hence the light rays entering from the other ommatidia are absorbed by the pigment.

Superposition image (Fig. 44D)

In dark adapted or nocturnal insects where the intensity of light is very low superposition image is formed. In nocturnal and crepuscular insects the pigment cells are contracted i.e. whereas in the light adapted insects the pigment extends inwards and isolates, in the dark adapted insects the pigment is condensed distally round the cornea. Thus it is obvious that in poor light the pigment is contracted and light can pass obliquely through several ommatidia, without being absorbed. As a result each rhabdome can receive light from several lenses and the image is formed in such a manner is a superposition image.

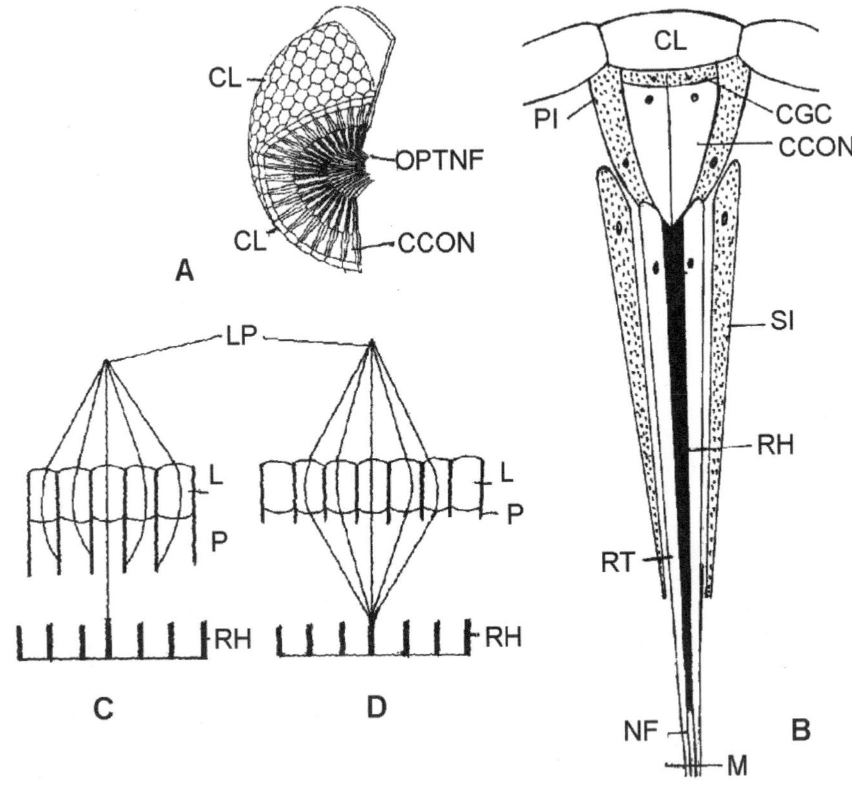

Fig. 44 A. Structure of a compound eye, **B**. Structure of an ommatidium. **C**. Apposition image, **D**. Superposition image, *Abbreviations* : CGC- corneagen cell, CCON- crystalline cone, CL- corneal lens, M- membrane, NF- nerve fibre, PI- primary iris, RH- rhabdome, RT- retina, SI- secondary iris. L - lens, LP- luminous point, OPTNF - optic nerve fibre, P- pigment.

(B) PHOTOGENIC OR BIOLUMINESCENT ORGANS

Bioluminescence is defined as a type of chemiluminescence in which light is emitted by living organisms as a result of cellular oxidation of some substrate in the presence of an enzyme. This form of luminescence occurs when the energy released during a biological oxidation forces (pushes) an electron of a molecule to a higher energy level. In this case, the oxidation energy serves the same purpose as the incident proton does in photoluminescence (McElroy, 1972).

There are some insects which are self luminous or light producing. The luminous properties possessed by these insects are because (a) some possess special photogenic organs, (b) some other possess this property due to the presence of light producing bacteria (c) in some cases due to ingestion of the luminous food.

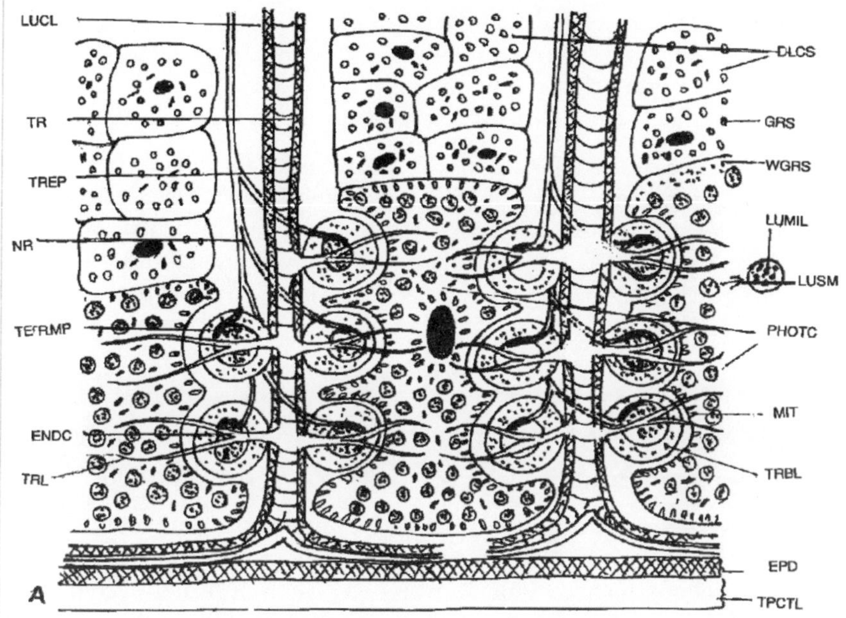

Fig 45A. A section through a part of the photogenic organ of firefly, *Photuris* *Abbreviations :* DLCS- dorsal layer cells, EPD- epidermis, ENDC-end cell, GRS- granules, LUSM- lumisome, LUML-luminelle, LUCL-lumen of cylinder, MIT- mitochondria, NR-nerve, PHOTC-photocyte, TR- trachea, TRL-tracheole, TERMP-terminal process, TRBL- tracheoblast, TREP- tracheal epithelium, TPCTL-transparent cuticle, WGR-waste granules.

These light producing organs are located beneath the areas of transparent cuticle. The histological structure shows various degrees of complicated nature. In case of *Phengodes*, the photogenic structures are large independent cells, just like the oenocytes. In other cases, it is a compact organ made up of masses of photogenic cells arranged in rosette-like fashion around tracheae from which short trunk runs into tracheal end cells (Fig. 45A). Large number of tracheoles are given out which penetrate in between the photogenic cells (Fig. 45A). The inner surface of the organ is covered by a layer of urate containing cells and thus form the so-called reflecting layer whose function is still uncertain.

Light is emitted when a substance known as "Luciferin" is oxidized by free oxygen in the presence of water and an enzyme Luciferase (Harvey, 1940). The amount of heat produced in the emission of light is very little and the energy for this change is got through preliminary reaction evolving ATP. The regulation of flashing in complex organ depends on a nervously controlled organ of a tracheal end cell which admits oxygen to the photogenic tissues.

In the fire flies, *Photinus* (Fig. 45E) and *Photuris* species the luciferin (LH_2) reacts with ATP and concomitantly, pyrophosphate (P-P) is liberated along with the conversion of ATP into adenosine monophosphate (AMP) and ultimately enzyme bound luciferyl adenylate (ENZ-LH_2-AMP) is formed. The latter compound i.e. ENZ-LH_2-AMP is oxidized to give a flash.

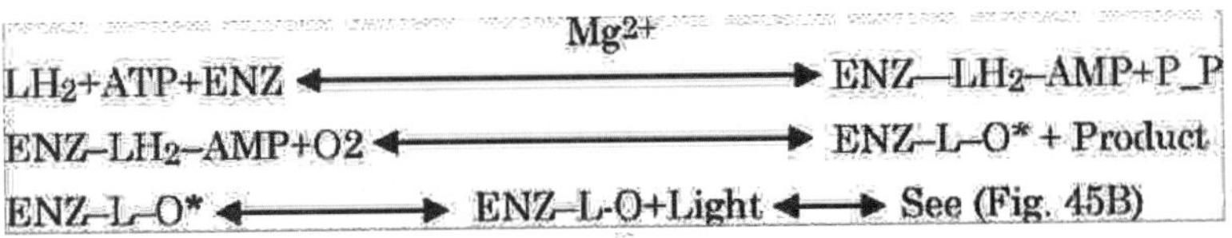

Ultrastructural studies of firefly photocyte suggest that a photocyte is composed of large membrane bound vesicles which are called photocyte particles or luminelles. Each luminelle comprises several smaller vesicles called lumisomes. All bioluminescent reactions occur in these lumisomes (Fig. 45 B,C) because the related proteins are enclosed within them. Upon stimulation of the insect the Ca^{2+} is supplied to lumisome which leads to a sequence of reactions resulting in the production of light as a single flash for a duration of 0.2 second (Fig. 45 B,C).

The exact method of control in simple organ which do not possess tracheal end cells is not clearly understood. The luminous insects are mostly confined to the order Coleoptera and various genera of the family Lampyridae (Fig. 45 D) and Cantheridae. The various genera of these families are : *Lampyris, Phengodes, Luciola* and some others. The Elatrid genera *Pyrophorus* and *Photophorus* are also luminous and so are the other species of Carabid, *Phesodera*. The larva of the Mycetophilids viz. *Arachnocampa* and *Ceroplatus* are also photogenic. In Lampyridae in some species the luminocity is found in all stages of

* Activated form

development and it is the character of the plasma. The luminous substance is diffused in the egg but it is localized in the photogenic organ in the postembryonic phases. These organs vary to a very great extent in shape, size and position in different species and according to sex and developmental stages e.g. there are 11 or 12 pairs of segmentally arranged organs in *Phengodes* and *Phrixothrix* whereas in many other species like *Lampyris* female, these are present mainly in the 6 or 7 abdominal sterna. They also occur on the pronotum and anterior abdominal sterna in *Pyrophorus*. These organs may be found in both the sexes or only in the females.

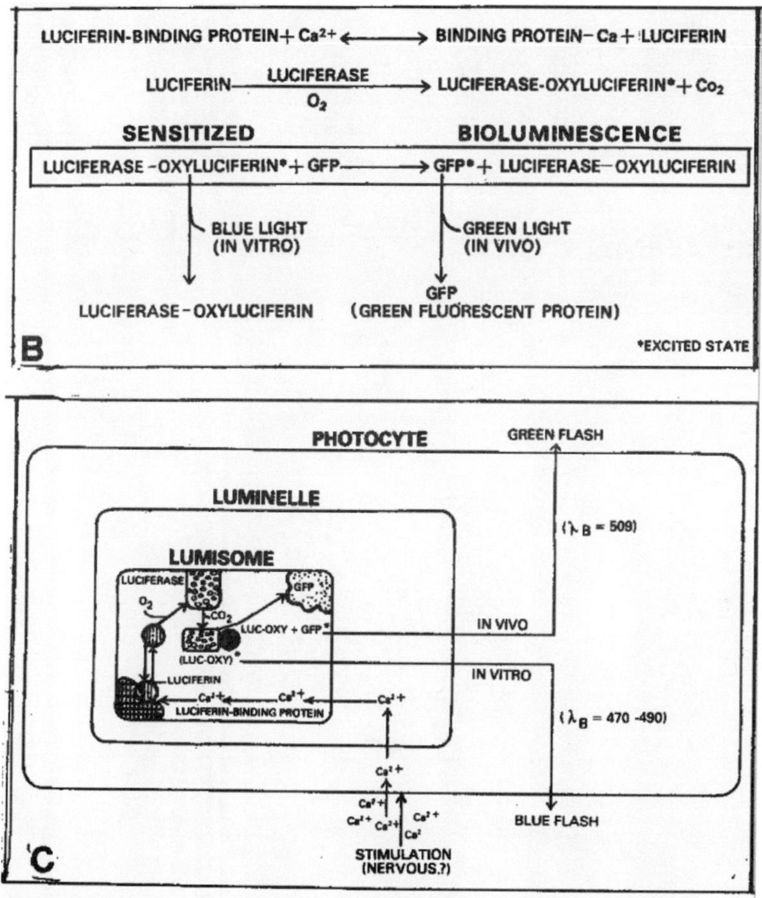

Fig. 45. B. Bioluminescent reactions in Firefly. **C.** Cellular mechanism of bioluminescence in Firefly. (Diagrammatic). (By courtesy of Shukla and Tripathi).

There are small number of insects which are luminous but they do not possess luminous organs, in many such cases the light is most probably associated with the presence of the bacteria e.g. some Ephemeroptera, certain Collembola, larval Diptera and Lepidoptera.

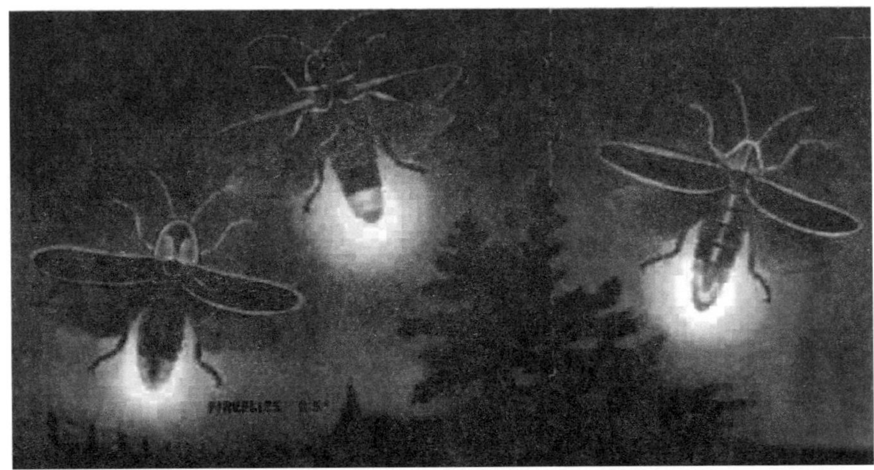

Fig. 45D. Fireflies (Lampyridae)

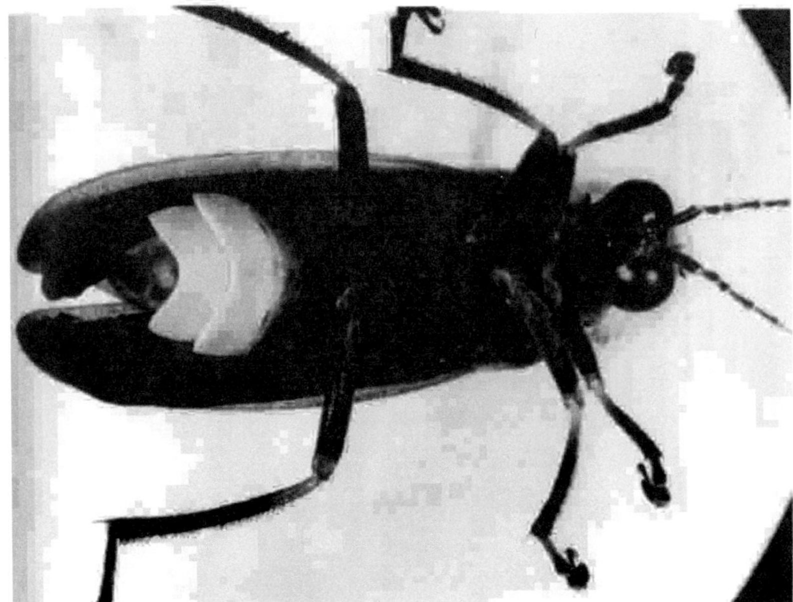

Fig. 45E. Ventral view of firefly (*Photinus pyralis*) showing abdominal luminescent organs (From Elzinga, 1978).

As stated, the light produced may either be a continuous glow or an intermittent or pulsating glow of variable intensity. It may be periodic short flashes. All this depends on the species concerned and the influence of the environmental factors. The light occupies the wave length over the range of 5000 to 6500 Å. It is yellowish-green in colour although in *Pyrophorus* the abdominal organ emits a reddish light during flight and the thoracic organ green light when the insect is not flying.

It is supposed that the light helps the insect in the attraction of sexes for mating. In *Photinus* (Fig. 45E) the female attracts the male by giving the flash at an interval of about 2 seconds.

15

THE ENDOCRINE ORGANS

The endocrine organs of insects are of two types: the neurosecretory cells found in the brain and various ganglia of the nervous system, and the endocrine glands like corpora cardiaca, corpora allata and ecdysial glands (Figs. 46, 47). In some insects the cephalic aorta is also associated with the endocrine organs.

1. THE NEUROSECRETORY CELLS

The neurosecretory cells (Fig. 48 A,D,I-K) are the glandular nerve cells which show cytological evidence of secretory activity. The product of these neurosecretory cells is called neurosecretory material (NSM) or neurosecretion. The presence of neurosecretory cells in the pars intercerebralis was first described by Weyer (1935) in honeybee, *Apis mellifica*, later their presence in the brain, the ganglia of the ventral nerve cord and other ganglia was demonstrated by numerous workers (Awasthi, 1985, Raabe 1989). The neurosecretory cells of the brain secrete a hormone known as brain hormone, which in fact is a mixture of several hormones secreted by different types of neurosecretory cells. The brain hormone activates other endocrine glands and also directly or indirectly controls almost all life processes. These cells are stained with specific neurosecretory stains like chrom alum haematoxylin phloxin (CAHP), aldehyde fuchsin (AF) paraldehyde fuchsin (PF), performic acid resorcin fuchsin (PARF), performic acid victoria blue (PAVB), performic acid alcian blue (PAAB) etc. The three techniques viz. CAHP, PF (AF) and PARF differentiate maximum types of cells and are used for the classification of the neurosecretory cells. Others stain only one or two types of cells. On the basis of staining properties two to four types of cells have been classified viz. A, B, C and D. Although in most of the cases the cells can be compared with similar types of cells of other

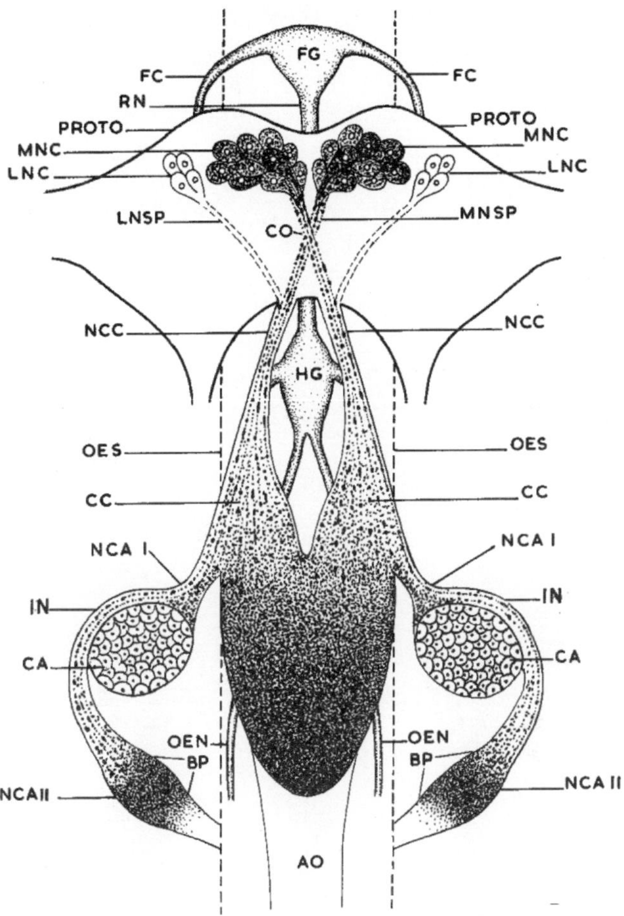

Fig. 46. Neurosecretory system of the house cricket, *Gryllodes sigillatus* *Abbreviations* - AO- aorta, BP- bulbous portion, CA- corpora allata, CC- corpora cardiaca, CO- crossing over, FG- frontal ganglion, FGC- frontal ganglion connective, HG- hypocerebral ganglion, IN- innervation, LNC- lateral neurosecretory cells, LNSP- lateral neurosecretory pathway, MNC- medial neurosecretory cells, MNSP- medial neurosecretory pathway, NCC- nervi corporis cardiaci, NCA-I- nervi corporis allati- 1, NCA-II- nervi corporis allati-II, OEN-oesophageal nerve, PROTO- protocerebrum, RN- recurrent nerve (From Awasthi, 1969).

insects but they may differ also due to their chemical nature. Hence it is not always possible to compare the work done on other insects. With CAHP the A-cells stain blue black and the B-cells pink whereas with PF, AF or PARF the A- cells stain purple, purple red or brick red and the B-cells green. Likewise C and D cells are also distinguished. Two types of cells viz. A and B are found in most of the insects whereas the C and D cells are found only in a few insect orders. It is probable that the visible secretion is only a carrier protein (neurophysin) to which the hormone molecule is attached. Thus carrier substance which is the stainable part acts as a marker for the hormone. When hormone is released it becomes separated from the carrier and is then free to enter the haemolymph.

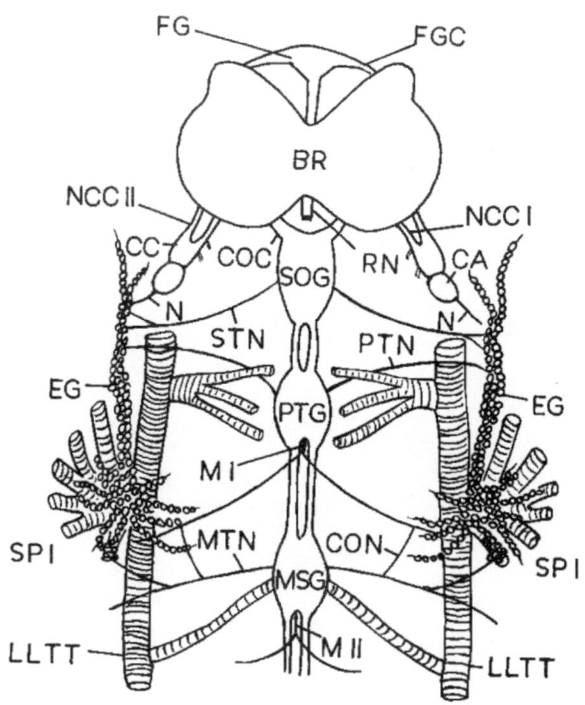

Fig. 47 Endocrine organs of *Amsacta collaris*.
Abbreviations: BR- brain, CC- corpora cardiaca, CA- corpora allata, CON- connective, EG- ecdysial gland, FG- frontal ganglion, FGC- frontal ganglion connective, LLTT- lateral longitudinal tracheal trunk, M-I- medial-I, M-II- medial-II, MTN- metathoracic transverse nerve, MSG- mesothoracic ganglion, N- nerve, NCC-I- nervi corporis cardiaci- I, NCC-II- nervi corporis cardiaci-II, PTG- prothoracic ganglion, PTN- prothoracic transverse nerve, RN- recurrent nerve, SP- spiracle, STN- suboesophageal transverse nerve, SOG- suboesophageal ganglion. (From Awasthi and Singh, 1983).

Under the electron microscope also several types of neurosecretory cells have been distinguished on the basis of the size of granules and electron density. The A-cells always contain membrane bound neurosecretory granules 1000 to 4000 Å in diameter and the matrix is electron dense. The granules in B-cells and other types of cells are much smaller in diameter and the matrix may be less electron dense, electron transparent or electron lucent.

It may be added that the amount of NSM in a cell at any one time is the result of a balance between its rate of synthesis and the rate of discharge. Some workers regarded a cell full of granular material as actively secreting while others regard such cells as an inactive one showing accumulated material. It is possible that both the views may be correct under different circumstances.

The neurosecretory cells are present in almost all the ganglia and in the following pages a brief description thereof, is given (For details see Awasthi, 1985, 2005).

1. The neurosecretory cells of the brain (Figs. 48 A,D, I-K and 50 A)

In the pars intercerebralis medialis of the protocerebrum, on either side of the medial line are found a cluster of neurosecretory cells in most of the insects. These form the medial groups of neurosecretory cells. Each medial group of neurosecretory cells is composed of either A or A and B or more types of cells. In some insects only A-cells are present, while in others A and B both e.g. in *Nezara viridula* the medial groups have only the A-cells, in *Gryllodes sigillatus, Lygaeus pandurus, Utetheisa pulchella, Bombyx mori,* A and B cells; in *Philosamia ricini, Achaea janata, Amsacta collaris, Orthetrum chrysis* A, B and C cells are present (Awasthi, 1985). Similarly 4-types of cells are found in *Schistocerca gregaria.* Thus, one to four types of cells may be present in each medial group depending on the insect order and species. The number of principal types of cells e.g. A-cells varies considerably from insect to insect e.g. in *Nezara viridula* each medial group has 5-A cells, *Dysdercus koenigii* 9-A cells, *Lygaeus pandurus* 5-A cells, *Sarcophaga ruficornis* 13-A cells, *Metochus uniguttatus* 9-A-cells, *Labidura riparia* 8 to 10A-cells, *Euborellia annulipes* 10-14 A-cells, *Gryllodes sigillatus* 150 A cells, and *Schistocerca gregaria* about 600 A cells (Awasthi, 2005). The A, B and C cells in several insects have been further classified into subtypes A-1, A-2, B-1, B-2, C-1, C-2 cells etc.

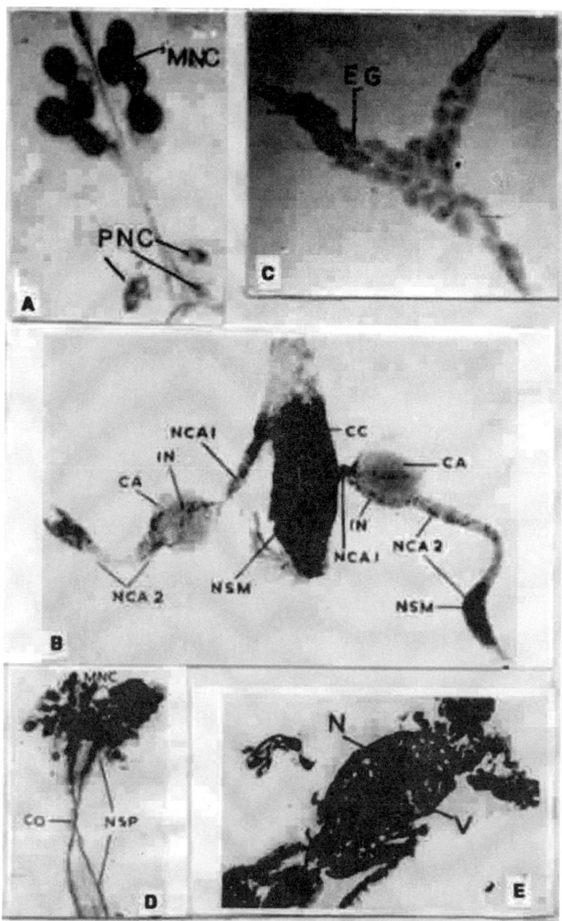

Fig. 48 A. Whole mount of the brain of *Amsacta collaris* showing medial and posterior groups of neurosecretory cells. **B**. Retrocerebral complex of the house cricket, *Gryllodes sigillatus*. **C**. Ecdysial glands of *Achaea janata*, **D**. Medial neurosecretory cells of the house cricket *Gryllodes sigillatus* showing crossing over of path ways at CO. **E**. L.S. of ecdysial gland of *Amsacta collaris*.

Abbreviations : CA-corpora allata, CC-corpora cardiaca, CO-crossing over, EG - ecdysial gland, IN-innervation, MNC- medial neurosecretory cells, N-nucleus, NCA-1-nervi corporis allati-1, NCA-2-nervi corporis allati-2, NSM-neurosecretory material, NSP-neurosecretory pathway, PNC-posterior neurosecretory cells, V-vacuoles. (A - Awasthi and Singh, 1981, B. - Awasthi, 1968, C - Singh and Awasthi, 1981, D - Awasthi, 1969, E - Awasthi and Singh, 1983).

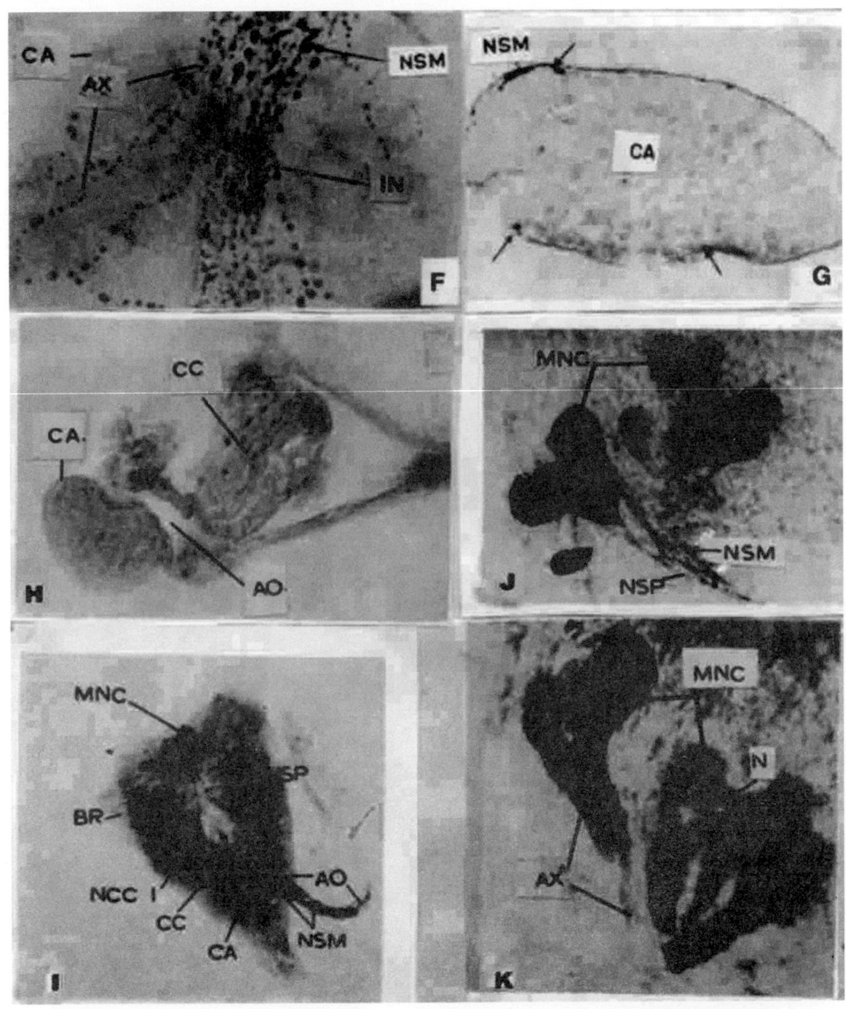

Fig. 48 F.G. Corpora allata of *Gryllodes sigillatus* showing NSM on the outer parts, **H.** T.S. of ring gland of adult *Sarcophaga ruficornis*. **I.** The neurosecretory system of *Nezara viridula*. **J**. Brain of *Metochus uniguttatus* showing one group of medial neurosecretory cells. **K.** Brain of *Lygaeus pandurus* showing one group of medial neurosecretory cells. *Abbreviations* : AO-aorta, AX- axon, BR- brain, CA-corpora allata, CC-corpora cardiaca, IN-innervation, MNC- medial neurosecretory cells, N-nucleus, NCC-I-nervus corporis cardiaci-I, NSM- neurosecretory material, NSP-neurosecretory pathway, (F. - Awasthi, 1968, G - Awasthi, 1969, H - Awasthi, 1972, I- Awasthi, 1972, J-Awasthi, 1973, K-Awasthi, 1980).

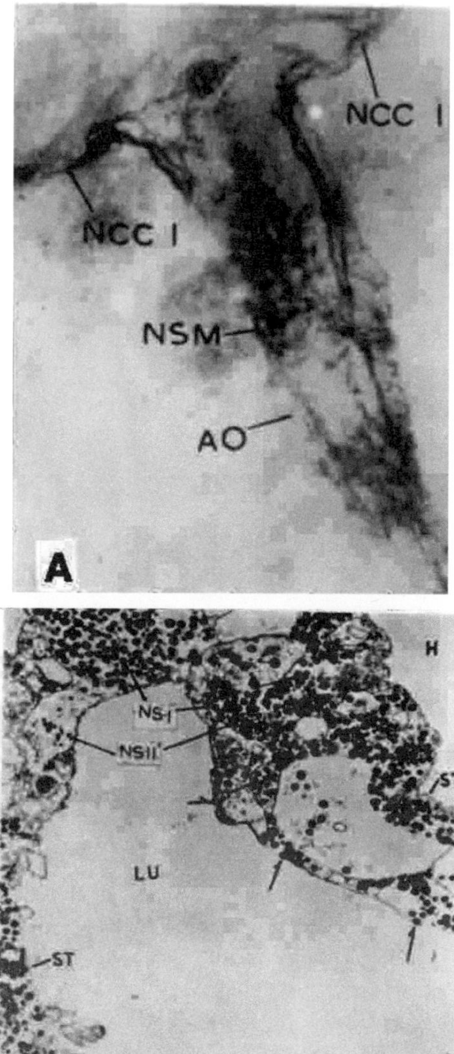

Fig.49 A. Whole mount of cephalic aorta of a Hemipteran insect, *Nezara viridula*, **B**. Electron micrograph of the cephalic aorta of the Earwig. *Euborellia annulipes*

Abbreviations : AO-aorta, NCCI- nervi corporis cardiaci-1 NSM- neurosecretory material, NS-I- neurosecretory granules of type 1, NS- II- neurosecretory granules of type II, LU- lumen of aorta, ST- stroma, H- haemocoel. (A- Awasthi, 1972, B. Awasthi, 1976).

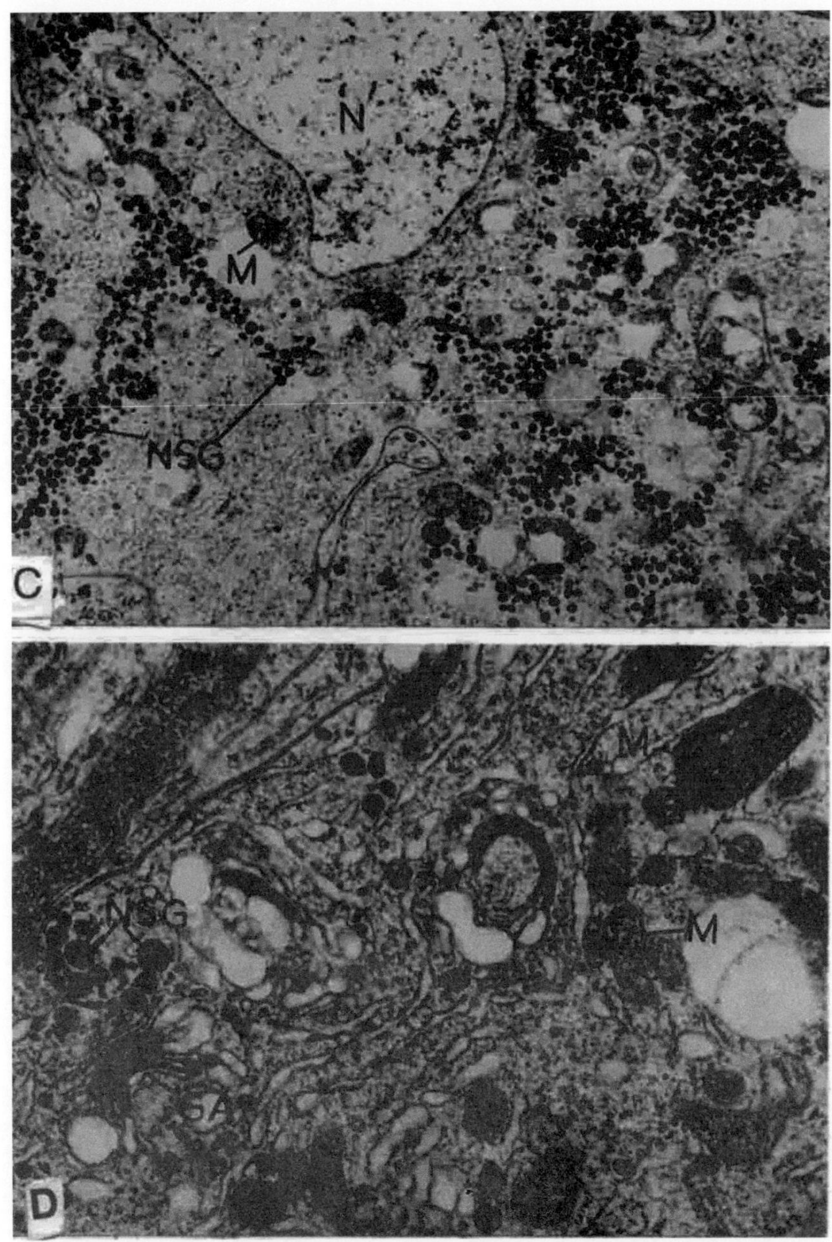

Fig. 49. C.D - Electron micrographs of neurosecretory cells of *Pieris rapae crucivora* showing NS- granules, N- nucleus, M- mitochondria and GA- Golgi apparatus. (C,D By Courtesy of Prof. Y. Kono).

The 2nd paired group of neurosecretory cells is present in the lateral part of the protocerebrum and constitutes the lateral group of neurosecretory cells. The number and position of cells varies considerably in this group. The lateral group also consists of either only A cells, or A and B both the types of cells. In some insects the lateral group is absent. In addition to the medial and lateral groups, in certain insects other groups viz. optic, ventral, posterior have also been reported. (Awasthi and Singh, 1981).

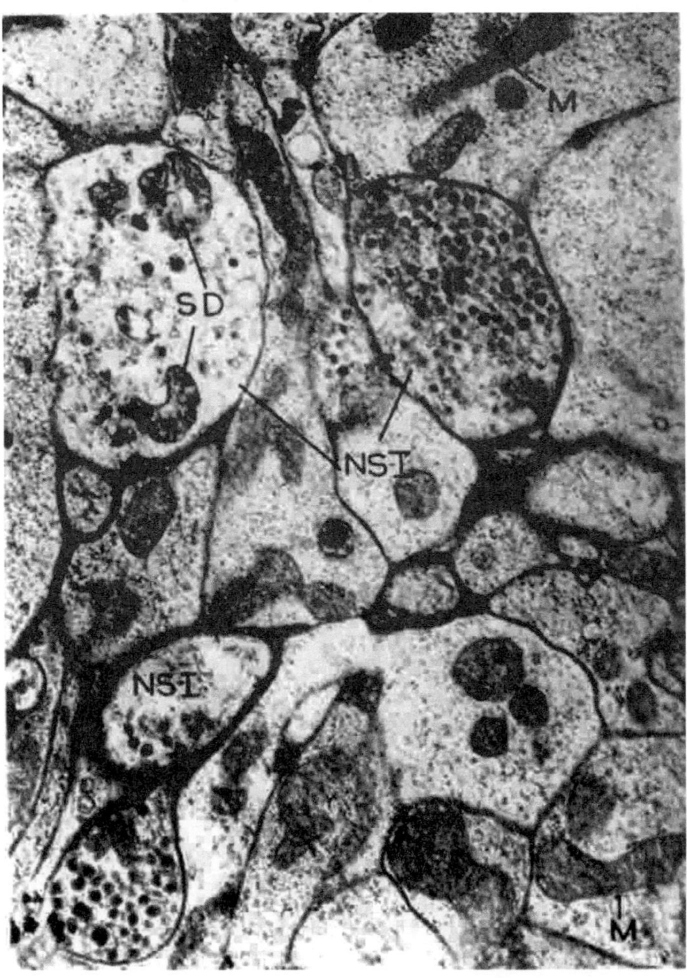

Fig. 49E. Electron micrograph of a part of corpus cardiacum of *Gryllodes sigillatus* showing several axons containing neurosecretory granules type I (500-1400 Å) (Awasthi, 1977)

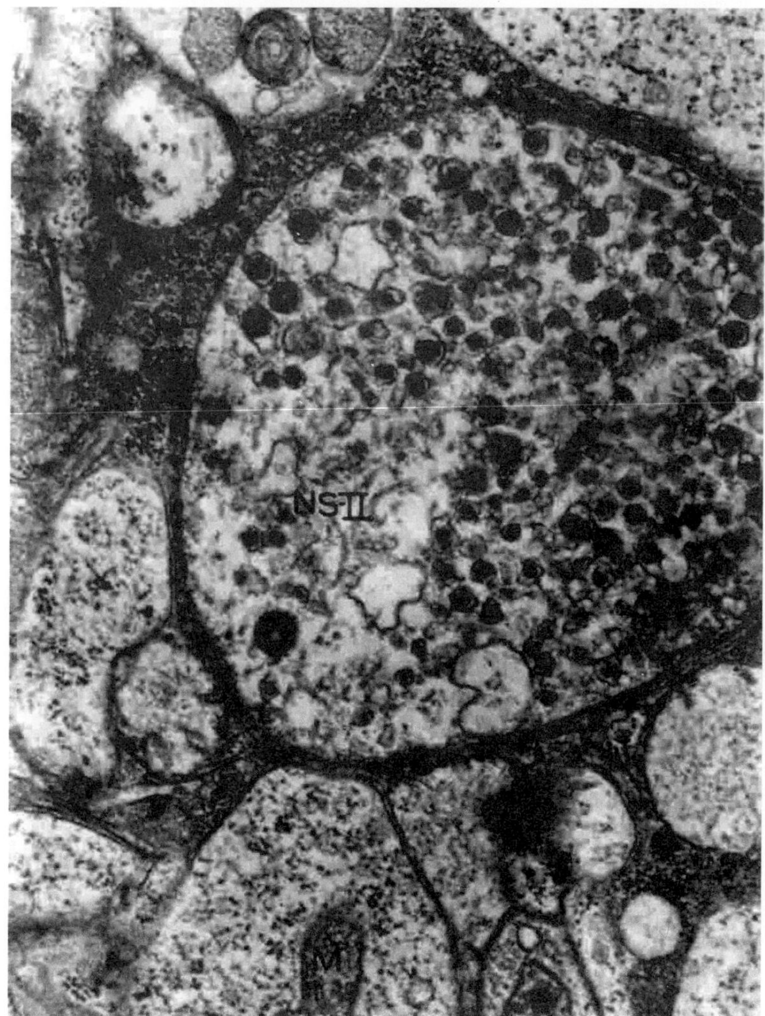

Fig. 49F. Electron micrograph of a part of corpus cardiacum of *Gryllodes sigillatus* showing axon containing neurosecretory granules type II (900-1700Å) with electron lucent space between the boundary membrane and the matrix. The axon is covered by glial sheath (GL). In addition, ordinary axons having mitochondria (M) are also visible (Awasthi, 1977).

Neurosecretory cells in the suboesophageal ganglion

The suboesophageal ganglion consists of numerous different types of neurosecretory cells. There are two prominent neurosecretory cells (Fig. 50C) of A type, which are located in the middle or posterioventral part of the ganglion in Hemiptera. In addition to the two conspicuous A-cells, there are numerous B type of cells present in the suboesophageal ganglion. The A and B type of cells in staining properties are similar to the A and B cells of the brain.

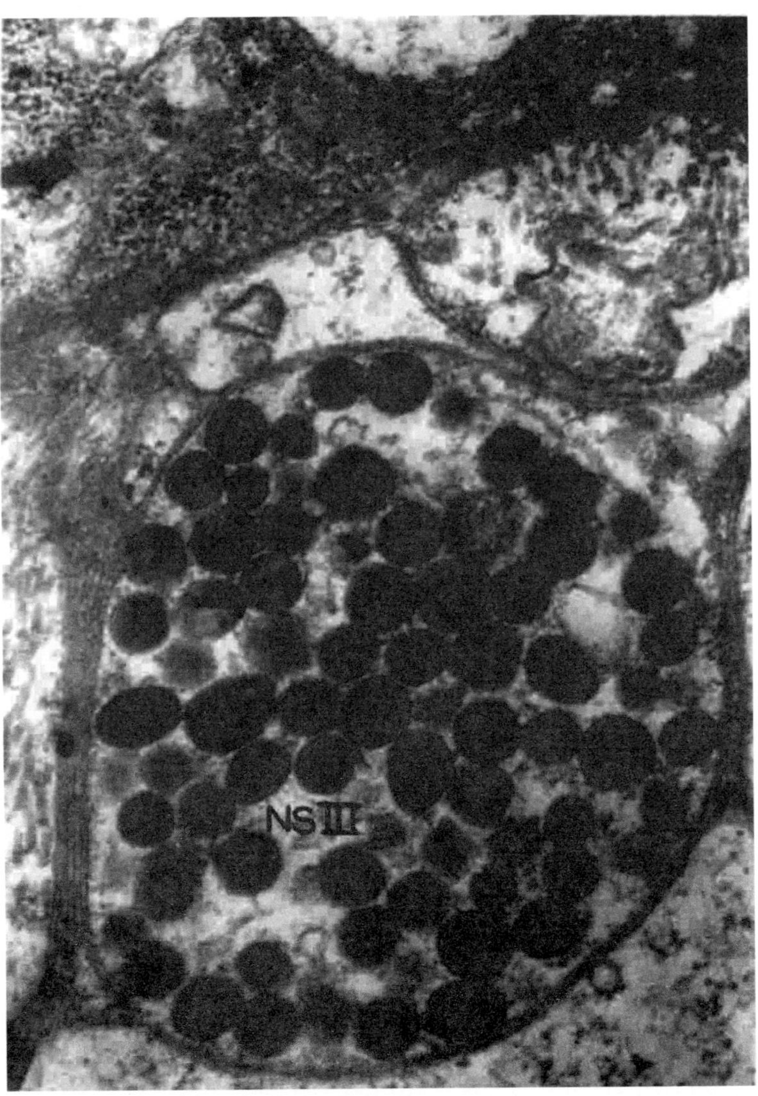

Fig. 49G. Electron micrograph of a part of corpus cardiacum of *Gryllodes sigillatus* showing the axon containing electron dense, oval or spherical neurosecretory granules type III (1100 to 2300 Å) (Awasthi, 1977)

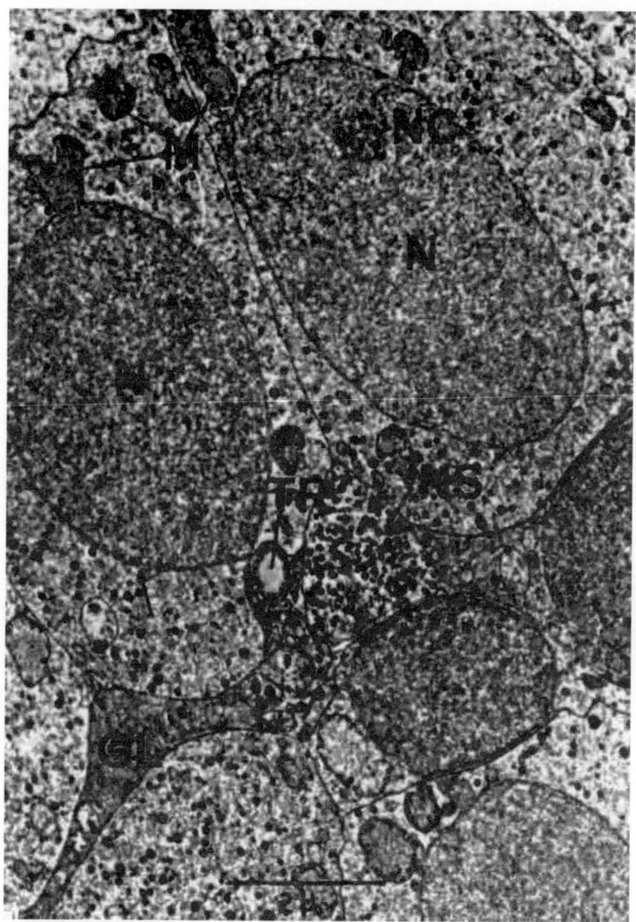

Fig. 49H. Electron micrograph of a part of corpus allatum of *Euborellia annulipes* showing axon containing electron dense neurosecretory granules (NS) tracheole (TR), mitochondria (M) glial cell process (GL), and oval nuclei (N) are also visible (Awasthi, 1976).

Neurosecretory cells in other ganglia

All the ganglia of the ventral nerve cord consist of A and B types of neurosecretory cells. The cells are arranged in several clusters. These cells also secrete specific hormones which are stored in the perysympathetic neurohaemal organs. The neurosecretory cells have also been reported from the frontal and hypocerebral ganglia in Dictyoptera and Lepidoptera (Fig. 50 B, D).

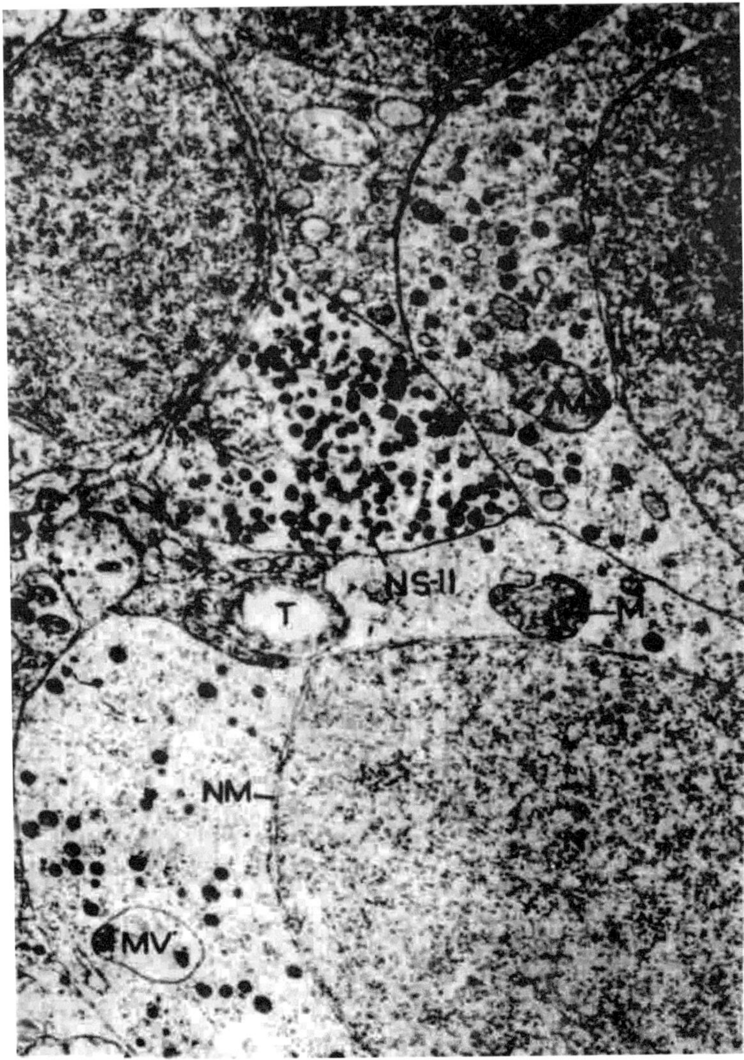

Fig. 49 I. A magnified part of the corpus allatum of *Euborellia annulipes* showing the neurosecretory axon (NSII) associated with the parenchymal cells. A branch of tracheole (T) is also seen. Other structures are vesicle clusters and mitochondria (Awasthi, 1979).

Synthesis of neurosecretory material

The raw material for the synthesis of NSM is synthesized in the granular endoplasmic reticulum. From here it is transported to the

Golgi cisternae (Fig. 49 D). In the Golgi bodies elaboration of NSM takes place through the mediation of alkaline phosphates (Tewari and Awasthi, 1968) and finally the membrane bound vesicles are budded off in the form of neurosecretory granules. These are stored in the perikarya (Fig. 49C) and transported to the neurohaemal organs via axons.

Transport of neurosecretory material

It may be pointed out here that all the axons of the medial neurosecretory cells of one side converge at one point and form a medial neurosecretory pathway. The medial neurosecretory pathways formed in such a manner, run on the dorsal side of the protocerebrum and decussate (Fig. 48D). After the decussation, the pathways diverge and run on the dorsum of the protocerebrum of the opposite side. In some insects they emerge from the posterior part of the protocerebrum while in others they traverse the deutocerebrum and emerge from the posterior part of the tritocerebrum as nervi corporis cardiaci-I. In a similar manner lateral neurosecretory pathways are formed but their axons do not cross over each other i.e. there is no chiasma formation in the lateral neurosecretory pathways. These emerge from the brain as nervi corporis cardiaci-II. In some insects both NCC-I and NCC-II fuse in the hinder part of the brain before emergence and they come out in the form of a common nervus corporis cardiaci (NCC). In some insects a third nerve emerges from the brain and is termed as nervi corporis cardiaci-III. (NCC-III). The NCC-I, NCC-II or the common NCC as the case may be enters either in the corpora cardiaca or cephalic aorta (the neurohaemal organs) as the case may be.

The neurohaemal or storage release organs

The neurohaemal organs are those which store the neurohormones secreted by the neurosecretory cells of the brain and other ganglia and released them into the haemolymph. The NSM is stored for a longer or shorter duration in the neurohaemal organs. The neurohormone may be released as such or prior to release it may be modified in the neurohaemal organ by the intrinsic secretions. Two principal types and numerous other sub types of neurohaemal organs have been recognized in insects (see : Awasthi, 1974, 1976).

The corpora cardiaca

In most of the insect orders the corpora cardiaca (Figs. 46, 47, 48B,H), perform the function of principal neurohaemal organ. The corpora cardiaca are a pair of oval, spherical, spindle-shaped, elongated or U-shaped neuroglandular bodies, lying behind the brain on the dorsal part of the foregut and are associated with the cephalic aorta.In some insects they are found on the lateral part of the foregut in between the dorsal and ventral dilator muscles. They are absent in Collembola. The corpora cardiaca receive the neurosecretory axons of the neurosecretory cells of the brain via neurosecretory pathways and nervi corporis cardiaci. The cardiac bodies are fused in some insects and separate in others. In some insects the corpora cardiaca are differentiated into two distinct regions, one region is composed of neurosecretory axons and axon terminals of cerebral neurosecretory cells and the other glandular part containing intrinsic secretory cells, but in many insects no such distinction is possible. Thus, the corpora cardiaca are composed of extrinsic and intrinsic materials (see Fig. 49 E-G). The extrinsic elements are the neurosecretory axons and their terminals of cerebral origin containing neurosecretory granules of diverse nature. The intrinsic elements of the corpora cardiaca are the chromophil or intrinsic neurosecretory cells and chromophob or glial cells. The chromophil or intrinsic neurosecretory cells are similar to the neurosecretory cells of the brain. It is believed that the intrinsic cells secrete a hormone which regulates heart beat. The corpora cardiaca function as a principal neurohaemal organ in Phasmida, Dictyoptera, Orthoptera, Neuroptera, Coleoptera, Diptera, Isoptera, Homoptera, Hymenoptera, Plecoptera, Ephemeroptera etc. but not in Dermaptera, Heteroptera and Lepidoptera (Awasthi, 1976, 2005).

The cephalic aorta (Fig. 49 A,B)

The cephalic aorta is the second principal neurohaemal organ. This is also situated behind the brain and dorsal to the oesophagus and has a wider anterior part which receives the axons of the nervi corporis cardiaci-I (Fig. 49A). It is very conspicuous in the living condition in Dermaptera and Heteroptera, where it appears under dark field illumination as a silvery-white organ due to a large quantity of NSM. The axons after entering the aorta, ramify profusely and form a network on the entire aortic wall. In mature insects the aorta is provided with a copious amount of NSM. In Dermaptera and Heteroptera the aortic walls in the anterior region are thicker than the remainder. Since NSM occurs in axons as well as in the intercellular spaces, the thick wall with loose tissues probably facilitates the storage of NSM in this region. Ultrastructural studies have also confirmed the role of aorta as a storage release organ for the product of medial neurosecretory cells of the brain (Fig. 49B). The corpora cardiaca in Dermaptera, and Heteroptera function as a neurohaemal organ only for the products of the lateral neurosecretory cells of the brain.

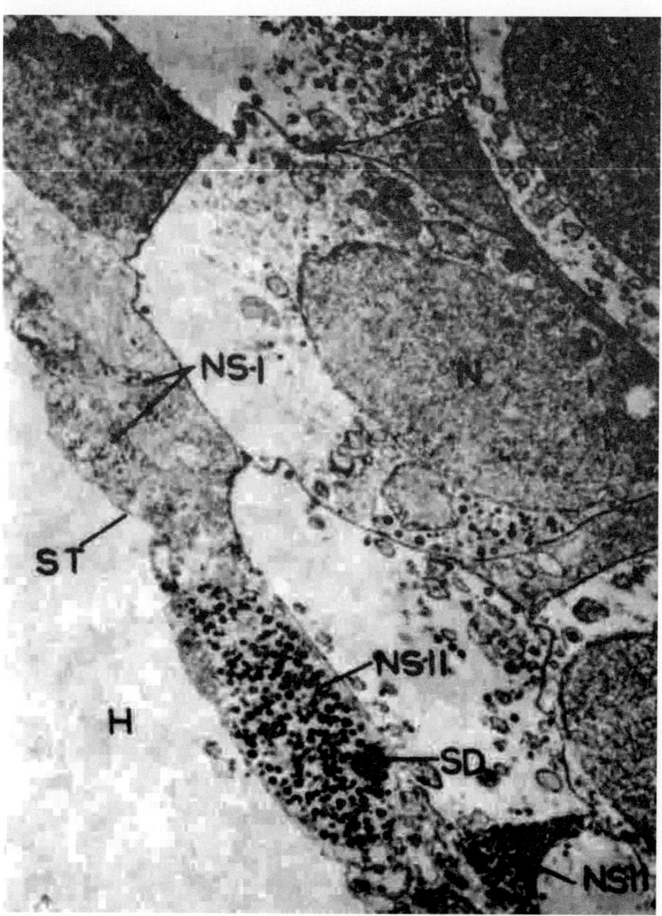

Fig. 49J. Electron micrograph of a part of corpus allatum of *Euborellia annulipes*. Note the distribution of two types of axons containing neurosecretory granules in the peripheral part and other portion of the gland (NS-I and NS-II). The stromal layer (ST) at the axon terminal that bath the haemocoel (H) is extremely thin. A large secretion droplet (SD), probably formed by the fusion of neurosecretory granule is also seen (Awasthi, 1979).

Other neurohaemal organs

There are numerous other neurohaemal organs found in insects e.g. the hypocerebral ganglion and oesophageal nerve in *Epilampra*, nervi corporis allati-2 in *Gryllodes*, neuropilar neurohaemal organs within the brain and corpora allata in Lepidoptera (moths) (Fig. 50 E,F). The perisympathetic neurohaemal organs along the ganglia of the ventral nerve cord are found in a large number of insects which store and release the product of the neurosecretory cells of the ventral ganglia.

Corpora allata

The corpora allata (Figs. 46, 47, 48B,F-H) are non-nervous, epithelioid, oval, spherical or leaf like glandular bodies which lie on either side of the corpora cardiaca. They are paired structures in most of the insects but in some insects e.g. Dermaptera and Gymnoceratan Heteroptera they are unpaired medial bodies. Each is connected with the corpus cardiacum of the same side by the nervi corporis allati-I which carries fibres from the neurosecretory cells of the brain. The gland consists of parenchymal cells and is innervated by neurosecretory axon-terminals and pre terminals containing different types of neurosecsetory granules (Fig. 49 H-J).

Four types of corpora allata have been recognized in insects viz.

(1) *The vesicular type* : In Thysanura and Phasmida the corpora allata are hollow balls of cells, with gland cells forming the walls.

(2) *The pseudolymphoid type* :These are formed by small cells with reduced, deeply staining, cytoplasm resembling that of the lymphoid tissue in vertebrates and are found in Odonata and Ephemeroptera.

(3) *The microcellular type* : The corpora allata of this type are most common, found in Orthoptera, Dictyoptera, Coleoptera, Hemiptera etc. They are composed of numerous small cells.

(4) *The macrocellular type* : Found in Hymenoptera, Lepidoptera, Diptera and Trichoptera. The gland consists of a few large cells with large, often lobular nuclei.

The corpora allata show cycles of secretory activity associated with which are the changes in their size. The active glands have large cells with a lot of dense cytoplasm with scattered nuclei. The inactive gland is small, with poor cytoplasm and aggregated nuclei. The corpora allata produce juvenile hormone which controls metamorphosis and yolk deposition in the eggs. A number of other functions like tanning of the cuticle, green-brown polymorphism in locust, mating behaviour in social insects, etc. are performed by the juvenile hormone. Since in young stages it maintains larval characters and inhibits adult differentiation, it is also called as neotenin or juvenile hormone.

Ecdysial glands

The ecdysial glands (Figs. 47, 48 C,E) are also known as prothoracic glands or thoracic glands. The first name, ecdysial glands relates to the function i.e. ecdysis or moulting and the 2nd name relates to the location or position of the glands in the insect body. They are paired, glandular structures, located in the ventrolateral areas of the prothorax extending into the mesothorax in some species and sometimes partially fused in the median line. They are often closely associated with the lateral longitudinal tracheal trunks near the thoracic spiracles (Fig. 47). In some insects e.g. *Bombyx mori* and *Achaea janata* they appear as compact bodies, in others they may be diffusely distributed in the prothorax and mesothorax as in *Philosamia ricini* and *Hyalophora cecropia.* Like corpora allata, they are also non-nervous epithelioid bodies. In the living condition, the gland cells appear transparent. In each larval instar there is a distinct cyclicity in the secretory activity. The greatest size and secretory activity of the ecdysial glands is reached in the last larval instar. At rest the nuclei are small and oval but in the active gland they become enlarged and lobulated, the basophilic cytoplasm is deeply stained. The ecdysial glands degenerate in majority of the insects after the final moult but they remain preserved in some cases like *Locusta* and *Amsacta.* These glands produce a hormone which is known as ecdysone or moulting hormone.

Weismann's ring : It is also known as ring gland. It is found in the larvae of Cyclorrhaphous Diptera like house fly, flesh fly and blow fly larvae. The ring gland is formed by the corpus cardiacum, hypocerebral ganglion, corpus allatum and prothoracic glands. The basal part of the ring is composed of fused corpus cardiacum and hypocerebral ganglion, the sides are formed by the cells of the ecdysial glands or prothoracic glands and the upper part is formed by the corpus allatum. The central part of the ring is occupied by the cephalic aorta. Thus all the components of the retrocerebral endocrine glands participate in the formation of the ring gland. It is connected to the brain by a pair of nerves and is also richly tracheated.

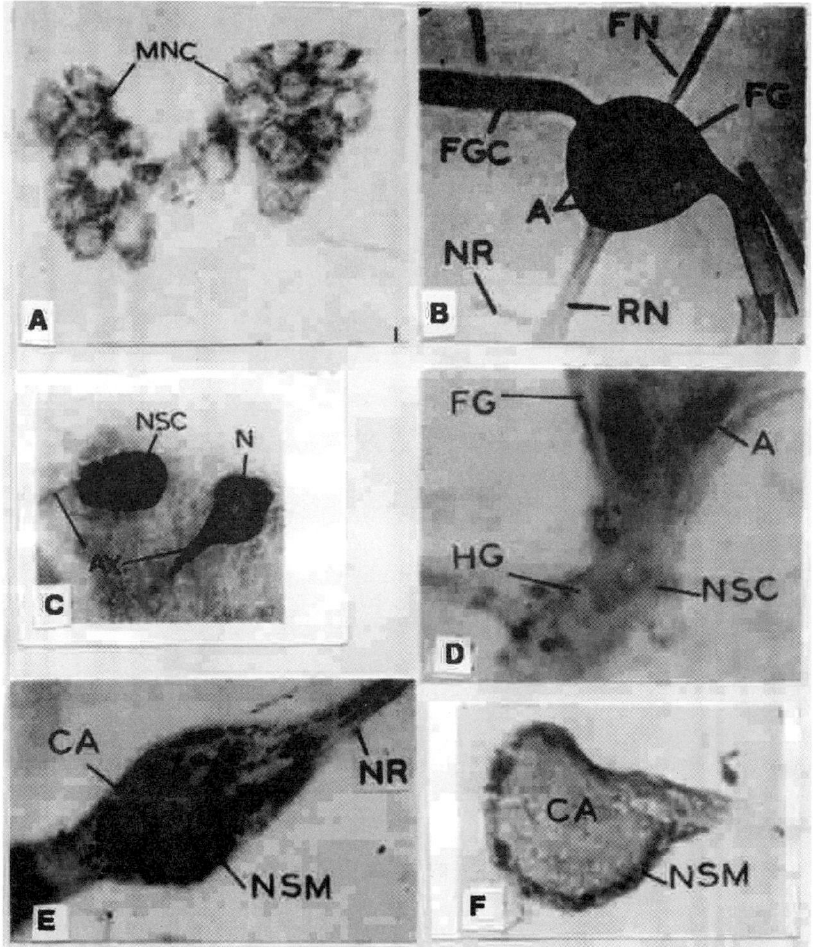

Figs. 50 A - Brain of flesh fly, *Sarcophaga ruficornis* showing 2-fused groups of medial neurosecretory cells. **B**. Frontal ganglion of *Achaea janata* showing 2-A cells of mature larva. **C**. Suboesophageal ganglion of *Metochus uniguttatus* showing 2-A cells with axons. **D**. The neurosecretory A cells in the frontal and hypocerebral ganglion of *Amsacta collaris*, **E.F**. - Showing storage of neurosecretory material on the outer part of the corpora allata in *Amsacta collaris*.

Abbreviatins : A- A cell, Ax - axon, CA-corpora allata; FG- frontal ganglion; FGC- frontal ganglion connective, FN - frontal nerve, HG-hypocerebral ganglion, MNC-medial neurosecretory cells, NSC-neurosecretory cells, NSM- neurosecretory material, N - nucleus, NR-nerve (A - Awasthi, 1972, B-Awasthi, 1980, C-Awasthi, 1973, D, E, F, Singh and Awasthi, 1981).

Release of the hormones : The neurohormones produced in the neurosecretory cells of the brain, reach their target organs in several ways. A direct transfer along the axons probably occurs in aphids. Similarly, neurosecretory axons pass from the corpora cardiaca of *Calliphora, Sarcophaga* and other insects to the pericardial cells, aorta, corpora allata and alimentary canal.

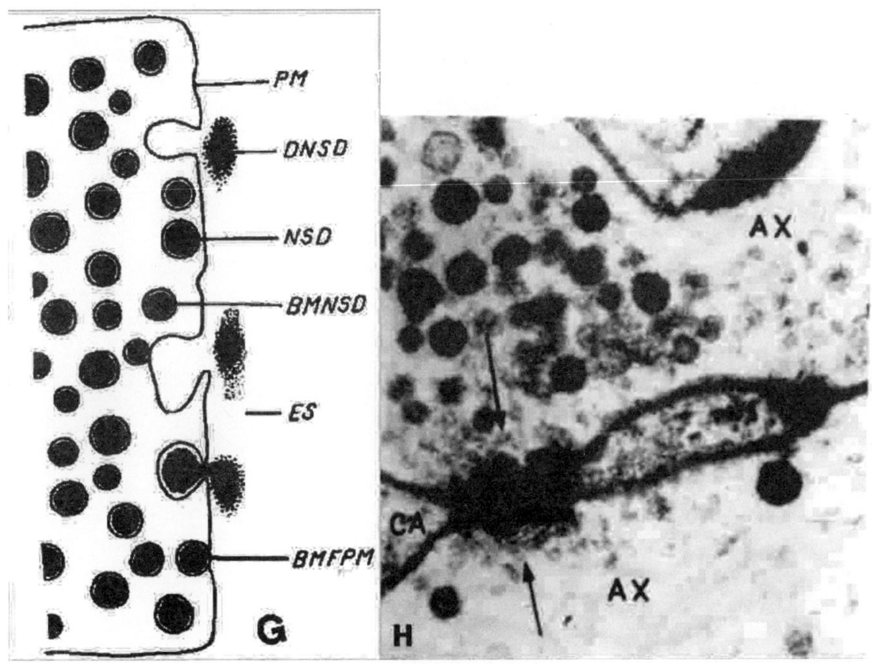

Figs. 50G. Diagram showing release of NSM through exocytosis, **H**. Diagram showing the release of neurosecretory material by disintegration of granules and diffusion of material.
Abbreviations : AX- axon, BMNSD- boundary membrane of neurosecretory droplet, BMFPM- boundary membrane fused with plasma membrane, ES- extracellular space, CA-corpus allatum, DNSD- discharged neurosecretory droplet, PM-plasma membrane, NSD- neurosecretory droplet [G Based on Smith and Smith 1966; H-Johnson, 1966].

Majority of the workers are of the opinion that the hormones are released into the blood from where they reach the target organs. The hormones are released in two ways viz., (1) by exocytosis and (2) diffusion.

By Exocytosis : It is a well known fact that the neurosecretory granules are encased in a delicate membrane contributed by Golgi complex. This membrane fuses with the plasma membrane of the cells, the wall in between dissolves and the contents of the droplet are released in the extracellular or extra axonal spaces as the case may be e.g. *Carausius morosus* and *Calliphora erythrocephala* (Fig. 50G).

By Diffusion: In this case the boundary membrane of the neuro-secretory granules disintegrates and the contents of the granules are released in the adjacent area from where they pass out through diffusion e.g. *Calliphora stygia* (Fig. 50H).

16

SOME HORMONALLY REGULATED PHENOMENA

1. HORMONAL CONTROL OF METAMORPHOSIS

Insect metamorphosis is controlled by a variety of hormones of which the most important are (1) brain hormone, (2) ecdysone and (3) juvenile hormone. Most of the work with insect hormones has been carried out by legaturing the larvae or pupae at various levels between anterior and posterior ends, by making grafts or by the injection of extracts of endocrine glands or synthetic hormones.

(1) BRAIN HORMONE

This is secreted by the neurosecretory cells of the brain which are usually 4-clusters a pair of medial and a pair of lateral cluster. The hormone secreted by these cells pass down the nerve axons of these cells and accumulates in the corpus cardiacum. Release of the hormone into the blood occurs at the corpus cardiacum level through several ways i.e. either by disintegration of granules and their diffusion into the blood or by means of exocytosis.

Experimental investigations in *Cecropia* silkworm show that both the median and lateral neurosecretory cells are necessary for the formation of this hormone. In fact the brain hormone is a mixture of several hormones secreted by different types of neurosecretory cells present in the brain.

Role of brain hormone in metamorphosis : This hormone has an indirect effect on metamorphosis. The chief function of this hormone is to control the secretion of the prothoracic glands. Consequently, the removal of the brain has the same result as the removal of the prothoracic glands. In both cases the secretion of ecdysone is stopped and the insect enters into a state of diapause. Thus, diapause is due to the failure of the endocrine activity.

(2) ECDYSONE

This is also known as moulting hormone. Fraenkel (1935) found that when mature fly larvae were legatured, only the front end was able to form a puparium while the hinder part remained unchanged. Blood transfusion from the front to the hind end even in the legatured fly larvae immediately caused a change resulting in the pupation of the posterior part. From these observations, Fraenkel suggested that the growth of the larva into the pupa was dependent on a hormone secreted somewhere in the anterior end. Fukuda (1940) clearly demonstrated that it is secreted by the prothoracic glands or ecdysial glands. Finally it was Karlson (1954) who isolated the hormone in a highly purified stage and named it as "Ecdysone". Chemical study of ecdysone has clearly revealed that it is composed of two closely related molecules viz. α-ecdysone and β-ecdysone and it is steroidal in nature.

Effect of ecdysone on metamorphosis

Ecdysone is the only hormone which has a central control on the developmental events at the cellular level in insects. When the prothoracic glands stop the secretion of ecdysone the result is developmental stand still. This is what happens during diapause. It must be mentioned here however that the development of muscles during metamorphosis is not governed by ecdysone alone. There is another factor called neurosecretory factor or a neurosecretory control (hormone) which when produced during the same period when ecdysone is released in the blood causes muscle development. This observation clearly points that there is a close integration of the prothoracic secretory process with the activity of the neurosecretory system. Further integration is through the neurosecretory control of the brain over the prothoracic gland.

(3) JUVENILE HORMONE

This is secreted by corpora allata, a paired or unpaired small cephalic endocrine glands. The hormone is known as juvenile hormone (JH), corpus allatum hormone or neotenin which prevents the real change of stage in the developing insect. Wigglesworth (1934) was the first to demonstrate this function of the juvenile hormone in his experiments with *Rhodnius prolixus*. However other workers such as Bounhiol (1938) and Fukuda (1944) and several others have confirmed the observations of Wigglesworth by their experiments on caterpillars. The JH is terpenoid in chemical nature.

Role of Juvenile hormone in metamorphosis

Its chief function is to maintain the immature stages i.e. larvae or nymphs of insects without causing pupation or the development of adult stage i.e. the actual metamorphosis is blocked. In a normal metamorphosing insect the corpora allata decline its secretory activity and thus a sudden change from an immature young stage to adult happens. Removal of corpora allata in the juvenile period of growth in insects have clearly shown a precocious metamorphosis in the life history.

In short the process of metamorphosis takes place as given below:

The brain hormone secreted by the neurosecretory cells stimulates the prothoracic glands to secrete ecdysone or moulting hormone, but moulting occurs only when the juvenile hormone is either absent or present in very small amounts. In the last larval instar of Holometabola the corpora allata becomes almost inactive, hence a low concentration of JH is present causing larva to moult into a pupa. Similarly during later part of the pupa or late pupa, again the corpora allata become inactive and in total absence of JH the pupa moults into an adult insect. In Hemimetabolous insects also the same phenomena is repeated i.e. the corpora allata become inactive in the last instar nymphs and in the complete absence of JH imaginal ecdysis takes place. Thus metamorphosis in insects is controlled by a delicate balance in the timing of secretion and concentration of two hormones viz. JH and MH secreted by the corpora allata and ecdysial or prothoracic glands. The total absence of JH is the most significant factor for imaginal ecdysis in both, Hemimetabolous and Holometabolous insects.

If much JH is present then larva will moult into a larva, if small amount of JH is present then pupation takes place and in the total absence of JH adult emergence occurs. And if the balance of hormones is disturbed then monsters, intermediate between larva and pupa or between pupa and adult are produced and the phenomena is termed as prothetely or metathetely.

2. HORMONAL CONTROL OF HEART-BEAT

The activity of the heart in insects is controlled by a variety of factors among which endocrine or hormonal control of the heart beat seems to be one. The corpus cardiacum plays an important role in controlling the beating of insect heart. Cardioacceleratory substances have also been extracted from the ganglia of the ventral nerve cord. Likewise metameric neurohaemal organs also contain substances which stimulate heart beat in insects. Mordue and Goldsworthy (1969) reported that the corpus cardiacum extract accelerated the heart beat. These workers found that the glandular lobes of the corpora cardiaca of locusts produce 53% and storage part 37% increase in the heart beat rates.

Unger (1957) separated two physiologically different substances from the brain, suboesophageal ganglion, corpora cardiaca as well as from the ventral nerve cord of *Periplaneta americana* and termed them as neurohormone-C and neurohormone-D. The neurohormone-D stimulated the frequency of the heart beat and increased its amplitude but use of high concentration of neurohormone-D prevented the heart beat. The neurohormone-C increased the frequency of heart beat but decreased its amplitude. At higher concentration neurohormone-C stopped the heart beat causing systolic collapse.

Traina *et al.* (1976) separated four heart acceleratory factors from the corpora cardiaca of *Periplaneta americana* and named them 1a, 1b, 2a, 2b. All the four factors in chemical nature are peptides.

A variety of chemicals have been extracted from the corpora cardiaca of insects. Barton-Brown and others (1961) have suggested that the active substance which increase the heart beat in chemical nature is similar to adrenalin. Cameron (1953) isolated O-diphenol from the intrinsic cells of the corpora cardiaca of *Periplaneta americana* which was similar to adrenalin. Davey (1961, 1962) and Ralph (1962) demonstrated that the substance increasing the heart beat is a peptide, which acts on pericardial cells. The substance extracted from the pericardial cells was indole alkylamine. The indole alkylamine is secreted by the pericardial cells when corpus cardiacum extract is injected. Colhoun (1963) and Gersch *et al.* (1961) have shown presence of serotonin in the cerebral neurosecretory cells and corpora cardiaca of *Periplaneta americana.*

Thus, O-diphenol, serotonin and an adrenalin like substance, were extracted from the corpora cardiaca and indole alkylamine from the pericardial cells upon stimulation by the corpus cardiacum extract. In fact which one is responsible for the heart beat control cannot be said with certainty. However, it is clear from experimental evidences in hand, that the corpus cardiacum directly or indirectly controls the heart beat. The detailed process and the exact chemicals involved in the process are yet not clearly understood in insects in general (For details see Awasthi, 1983).

3. HORMONAL CONTROL OF DIAPAUSE

The period of suspended development at any stage of the life cycle, under adverse condition(s) accompanied by greatly decreased metabolism is called diapause. This phenomenon is governed by a variety of external and internal factors. This is an adaptation to tide over unfavourable environmental conditions like cold, draught, extremes of temperatures, humidity, photoperiod, nutritional, vitamin and other deficiencies. The other factor(s) which determine the diapause and are equally responsible are the endocrine glands.

The hormonal regulation of diapause was earlier proposed by Wigglesworth (1934) and was confirmed by the experiments of Williams (1946) on *Hyalophora cecropia* which normally enters diapause just after pupation and may be induced to resume development by several months of chilling at 3-5°C. In Saturniids, the pupal diapause is caused by the absence of ecdysones, which results from inactivity of the neurosecretory cells of the pars intercerebralis. Diapause has been broken in a number of species by the injection of α or β-ecdysone, with the onset of adult development. A permanent diapause may be produced even in related non-diapausing species by brain removal. This response will vary from species to species.

Thus, the neurosecretory cells of the brain are the controlling elements in the initiation and termination of hormonally regulated diapause. However, developmental arrest is not necessarily the result of a lack of ecdysone in all cases. Adult diapause may be brought on by corpora allata inactivity under the control of neurosecretory cells of the protocerebrum.In this case also as in the case of ecdysone regulated diapause, reactivation of the neurosecretory cells in the brain is required to terminate the conditions and restore reproductive activity.

Three principal types of diapause can be distinguished on the basis of their hormonal mechanism :

(1) Diapause caused by a lack of brain hormone and moulting hormone. According to Novak (1975), to this category belong all cases of larval and pupal diapause known so far and doubtless also, embryonic diapause.

(2) Diapause caused by lack of brain hormone and juvenile hormone, this mechanism is concerned with imaginal diapause.

(3) Diapause caused by the action of neurosecretory factor produced by the suboesophageal ganglion of the female and affecting the development of the eggs. This category comprises early embryonic diapause.

The diapause caused by a deficiency of the brain hormone and moulting hormone is the most common type of diapause and the type so far most thoroughly investigated. It consists of an arrest of growth and development due primarily to the absence of moulting hormone itself is a consequence of an inhibition of brain hormone production. It is likely that the majority of both the pupal and the larval diapause results from the failure of the cerebral neurosecretory cells to produce thoracotropic hormone.

Adult Diapause

The adult diapause is the result of a deficiency in the brain hormone and juvenile hormone i.e. inactivity of the neurosecretory cells of the pars intercerebralis and corpora allata. Morphologically, the adult diapause is only seen in the interruption of growth of the ovaries and in the suppression of the functions of the accessory glands of both the male and female individuals. In the fly *Calliphora*, the implantation of corpora allata can induce the ripening of some of the eggs but complete termination of diapause is possible only by the simultaneous action of juvenile hormone and brain hormone. Apart from this, the presence of BH seems indispensable for the activity of CA. Evidences for CA (JH) regulation of adult diapause comes from a number of other species e.g. *Pyrrhocoris apterus, Hypera postica, Necrophorus fossor, Galeruca tenaceti, Pterostichus nigrita, Leptinotarsa decemlineata, Aedes aegypti, Culex pipiens, Calliphora* and *Gryllodes sigillatus* etc. The experimental evidences clearly suggest that the basic intrinsic factor of imaginal diapause lies in the neurosecretory cells (NSCs) of the brain which appears to be part of the mechanism of action of external stimuli.

Pupal Diapause

The principal cause of pupal diapause is the failure of endocrine activity i.e. activity of the NSCs of the brain and the ecdysial glands. Thus, to break the pupal diapause, presence of active ecdysial glands (EGs) is necessary in addition to BH. Implantation of active EGs also breaks the diapause. It is important to note that implantation of one or more active brains into an isolated abdomen in *Hyalophora* has no effect whereas the implantation of active EGs breaks the diapause and development continues. Injection of ecdysone has the same effect as the implantation of EGs. Pupal diapause can take place at any point during the pupal stage. Its occurrence in a given species is bound to be in the same phase of the pupal period and in most species it is genetically fixed. The common example of pupal diapausing insects are *Amsacta moorei, Amsacta collaris, Achaea janata* and others.

Larval Diapause

It has been found that in the diapausing penultimate instar nymphs of *Gryllus campestris* ecdysial glands show reduced secretory activity and the brain appears to be responsible for this. Both the brain and the EGs have also been shown to control diapause in larvae of *Sialis lutaria*. It has been remarked that the brain controls the activity of the EGs and CA, and is the principal control centre of diapausing larvae.

Early Embryonic Diapause

The early embryonic diapause is associated with the diapausing hormone (DH) secreted by the suboesophageal ganglion. Removal of the suboesophageal ganglion from diapause producing pupae shortly after pupation completely eliminates diapause from the eggs eventually produced; when the operation is delayed, a progressively larger proportion of eggs enter diapause. The incidence of the diapause in the eggs is conditioned by endocrine activity of the neurosecretory cells of the suboesophageal ganglion of females. The extirpation of suboesophageal ganglion at the beginning of the pupal stage causes eggs which were determined for diapause through their mother (e.g. female *Bombyx mori),* to develop without diapause. In the same way univoltine females can be caused to lay nondiapausing eggs, a thing which they never do under normal condition, conversely if upto a critical period a female pupa which has been conditioned to lay non-diapausing eggs receives a suboesophageal ganglion from female which would have laid diapausing eggs, it also will lay diapausing eggs. The same effect can be produced by the implantation of the suboesophageal ganglion of either females or males of a univoltine race. That the relevant DH is non-specific for species, genus or family was shown by implantations of the suboesophageal ganglion from other Lepidoptera. But in this case also the neurosecretory cells of the brain play a key role. In diapause producers, the brain stimulates the production and release of DH from the suboesophageal ganglion, in nondiapause producers the brain inhibits hormone production and release from suboesophageal ganglion.

4. HORMONAL CONTROL OF REPRODUCTION

Wigglesworth (1936) for the first time clearly pointed out that corpora allata secrete a gonadotropic hormone in *Rhodnius prolixus* . Now it is well established that in many insects the egg maturation is controlled by the corpora allata. In several insects viz. *Pieris brassicae, Galleria mellonella, Leucania separata* and *Carausius morosus* removal of the corpora allata in the larvae reduced the total egg production of the adult (Engelmann, 1968). In the cockroach, *Leucophaea*, the implantation of active corpora allata into normally fed pregnant whose own corpora allata were inactive, caused egg maturation in about 80% of the females but only 45% starved females when received active corpora allata matured eggs. Thus, nutritional factor is also involved in yolk deposition in the maturing eggs.

During the time of egg maturation i.e. yolk deposition in the oocytes, the size of the corpora allata increase markedly in majority of the insects showing enhanced activity (Engelmann, 1970). Cyclic changes in the corpus allatum volume which are correlated with egg maturation cycles have been reported for many insects.

In *Calliphora erythrocephala* the neurosecretory cells are essential in reproduction. Removal of the neurosecretory cells eliminated egg maturation completely. Reimplantation of neurosecretory cells into females previously deprived of neurosecretory cells resulted in resumption of egg maturation in some flies. The mosquito,

Aedes taeniorhynchus failed to mature eggs in the ovaries after destruction of the pars intercerebralis but after reimplantation of pars intercerebralis, maturation resumes. Unlike Calliphora, in *Aedes* taeniorhynchus corpus allatum is indeed essential for egg maturation. Allatectomy is followed by complete cessation of egg maturation (Lea, 1967). In these cases both the neurosecretory cells of the brain and corpus allatum furnish hormones that are needed for egg maturation. In *Sarcophaga bullata* also two hormones, one secreted by the neurosecretory cells of the brain and the other, by corpus allatum is needed for egg maturation. In *Sarcophaga bullata* egg maturation was abolished after the destruction of neurosecretory cells of the brain. In the case of *Sarcophaga bullata* it was suggested that corpus allatum promotes protein synthesis essential for egg maturation and the neurosecretory cells of the brain act as a true gonadotrophic hormone, facilitating the incorporation of yolk in the oocytes.

In *Locusta migratoria* also, both corpora allata and neurosecretory cells of the brain are essential for egg maturation i.e. both the hormones viz. corpus allatum hormone and brain hormone are needed for egg maturation. Removal of the neurosecretory cells of brain completely prevented egg maturation in several insects e.g., *Schistocerca gregaria, Schistocerca paranensis, Anacridium, Gomphocrus, Aedes aegypti,* and *Tenebrio* etc. which clearly indicates that only neurosecretory cells of the brain are essential in these insects for egg maturation.

The enhanced activity of the neurosecretory cells of the brain was observed during egg maturation. In *Schistocerca,* the rate of protein synthesis in the fat body as well as concentration of protein in the haemolymph was reduced in the pars intercerebralis cauterized females. Failure to mature eggs in the pars intercerebralis cauterized/ operated insects was the consequence of low protein metabolism, which did not allow the activation of the corpora allata. It was suggested that the corpora allata facilitate the deposition of available proteinaceous yolk into the growing oocytes but not involved in protein synthesis.

Thus both the hormones are needed for maturation of eggs. The principal function of the hormone secreted by the corpora allata or neurosecretory cells of the brain or both, as the case may be, is to promote the incorporation of the proteinaceous and lipid yolk in the growing oocytes. It was found that the blood protein level decreased with the approach of complete egg growth and was parallelled by a decreased corpus allatum activity. In addition to hormonal factors, several other factors like nutritional, mating, environmental etc. are essential indirectly with egg maturation process.

It can be said that the neurosecretory cells stimulate and activate the corpora allata and the secretion of corpora allata, facilitate yolk synthesis and yolk deposition in the eggs i.e. maturation of eggs.

5. HORMONAL CONTROL OF METABOLISM

Recent studies of a large number of investigators have shown that almost all vital events, including metabolism of protein, lipid, carbohydrate, respiratory and water are controlled by the insect hormones. The work has been reviewed by Awasthi (1979).

(1) Protein metabolism

It has been demonstrated that in *Calliphora* deprived of their medial neurosecretory cells, the activity of protease was greatly decreased. Reimplantation of the neurosecretory cells caused normal release of proteases. Production of proteolytic enzyme is regulated by eating of meat and presence of medial neurosecretory cells. In *Calliphora, Nauphaeta, Glossina, Rhodnius, Periplaneta,* midgut protease synthesis is controlled by medial neurosecretory cells but in *Leucophaea, Dytiscus, Sarcophaga* and *Blatta* secretion of intestinal proteases is independent of medial neurosecretory cells; it may be controlled by neural stimuli or secretogogues.

In the insects which are deprived of their medial neurosecretory cells and corpora allata, an accumulation of protein in the haemolymph was observed. In some insects only corpora allata regulate protein synthesis. In *Schistocerca,* neurosecretory cells control the haemolymph protein concentration i.e. protein availability to the oocytes. There are two views, according to one corpora allata produce a gonadotropin but the protein metabolism associated with egg production is under the control of neurosecretion. The other view is that corpus allatum hormone is involved in the regulation of protein metabolism in *Sarcophaga bullata.* Thus, in some insects protein metabolism during egg maturation depends on either corpora allata or neurosecretory material or on the secretions of the both.

The female specific protein-synthesis is controlled by the corpora allata or corpus allatum hormone. In *Leucophaea* female specific protein or "vitellogenin" which is induced by juvenile hormone analogue, makes up about 90% of the total yolk protein. The female specific protein is essential for egg maturation. In *Drosophila* yolk protein is synthesized in the fat bodies outside the ovaries and is stored in the haemolymph where it appears as female specific protein and from here it is taken up by the growing oocytes.

(2) Carbohydrate metabolism

The corpora cardiaca regulate the carbohydrate metabolism in insects in various ways. The action of corpus cardiacum extract depends on the lowering of the amount of glycogen and increasing the haemolymph trehalose concentration. The corpora cardiaca of *Calliphora* and *Periplaneta* contain a hyperglycaemic hormone which when liberated into the blood increases the trehalose contents. In *Periplaneta* two hormones are present in the corpora cardiaca which are responsible for increasing the haemolymph sugar concentration. After injection of corpus cardiacum and brain extracts an increase in the trehalose content in the haemolymph in *Periplaneta* was observed and it was remarked by Ralph and McCarthy (1964) that the active factor of the corpus cardiacum is produced by the neurosecretory cells of the brain. But Steele (1961) is of the view that the hyperglycaemic hormone is secreted by the intrinsic cells of the corpora cardiaca which stimulate the production of blood trehalose from glycogen stored in the fat body cells. Highnam and Goldsworthy (1971) are also of the opinion that the hyperglycaemic factor extractable from the corpus cardiacum is the product of neurosecretory cells of the brain. It appears that the hyperglycaemic factor is synthesized in the brain and stored in the corpus cardiacum from where it is released into the blood.

(3) Lipid metabolism

In the mosquitoes, the neurosecretory system causes a decrease in glycogen synthesis while it stimulate the synthesis of triglyceride. Cauterisation of medial neurosecretory cells from the brain in the female *Calliphora*, soon after the final moult, caused reduction in the quantity of lipid in the fat body. According to Verman *et al.* (1965) the allatectomized females of *Periplaneta* store 68% more lipid than is done by the unoperated females, the turn over rate of both triglycerides and phospolipids being slowed down in operated females. Gilbert (1967) has pointed out that the uptake of lipids by the ovaries in *Leucophaea* was stimulated by corpus allatum hormone. It is thus clear that the corpora allata are involved in the lipid metabolism and facilitating egg maturation.

Allatectomy of one-day old female desert locusts caused an accumulation of lipid in fat body which was due to the continuation of lipid deposition in the fat body after the expiry period of somatic growth. Lipid accumulation in allatectomized locusts is more likely to be related directly to the lack of juvenile hormone. Implantation of active corpora allata into one day old adult female locusts resulted in a premature development of oocytes and a decrease in fat body lipid accumulation. The juvenile hormone may have a dual effect on the fat body metabolism, that of suppressing lipid synthesis and of stimulating vitellogenic protein synthesis. Increased synthesis of lipid by the fat body would then account for the accumulation of lipid in the fat body after allatectomy.

(4) Respiratory metabolism

Removal of corpora cardiaca depresses the respiratory metabolism in the adult male cockroach, *Blaberus*. In the cardiacectomized-allatectomized cockroaches the rate of respiratory metabolism was very low. The corpora cardiaca extract produces an effect on respiratory metabolism at the level of mitochondria. Brain hormone has a stimulative effect on oxygen consumption of the whole body or on isolated body parts. The fat body, Malpighian tubules, pericardial cells, intestine and corpora allata show increased respiratory metabolism in response to brain hormone. Removal of neurosecretory cells of the brain led to a 20% reduction of the normal oxygen consumption and considerably reduced nutritional uptake (Keely, 1967).

The corpora allata also play an important part in respiratory metabolism. Decrease in oxygen consumption was caused by allatectomy in *Calliphora* and *Carausius* whereas an increase in oxygen consumption was observed following implantation of corpora allata. In *Tenebrio* the juvenile hormone analogue influences the respiratory metabolism in the same way as the implantation of corpora allata i.e. an increase in oxygen consumption and protein synthesis was observed following corpus allatum implantation. Similarly an increase in the corpus allatum hormone concentration in the haemolymph of adult insects is followed by an increase in respiratory metabolism. However, in certain insects like *Gryllus bimaculatus, Pyrrhocoris* and *Leucophaea* removal or implantation of corpora allata or corpus allatum hormone have no pronounced effect on the respiratory metabolism.

(5) Water metabolism

The various parts of the neuroendocrine organs probably contain diuretic and antidiuretic factors or a factor is released by one of the same parts of the endocrine system which performs different functions in different species (Raabe, 1989).

In addition to brain and corpora cardiaca, the corpora allata also play an important role in regulating water excretion in insects. Delphin (1963) reported that the cells in the ventral ganglia of *Schistocerca gregaria* produce an antidiuretic factor. In *Rhodnius*, the diuresis is controlled by a hormone produced by small paired neurosecretory cells present in the ventral ganglionic mass. The diuretic hormone is released by the axon terminals of the neurosecretory cells of thoracico-abdominal ganglionic mass in *Rhodnius*. An antidiuretic hormone in the thoracic ganglia and a diuretic hormone in the abdominal ganglia of *Chaoborus* was demonstrated by Gersch (1969). Mordue (1963) demonstrated both the diuretic and antidiuretic factors in *Locusta*. In *Periplaneta* the neurosecretory cells produce an antidiuretic hormone causing water retention. Highnam *et al.* (1965) are of the opinion that the neurosecretory cells of the brain in *Schistocerca* produce a diuretic factor. According to Nicolson (1976) the diuretic hormone is probably produced in the brain and released from the corpora cardiaca. Both diuretic and antidiuretic effects of neuroendocrine system depend on the (1) age, (2) the water availability to the insects for maintaining the water balance as well as on (3) several internal and external factors.

6. HORMONAL CONTROL OF PIGMENTATION OF THE CUTICLE

Hormonal control of tanning of the cuticle is now an established fact. Fraenkel and Hsiao (1962) have demonstrated that the pars intercerebralis controls the tanning in *Calliphora*. They are of the view that the neurosecretory cells of the brain contain a melanogenic hormone. Likewise, Girardie (1964) has shown that colouration of the cuticle in *Locusta* is controlled by the cells of the pars intercerebralis. According to Roussel (1966) implantation of a pair of corpora allata of an adult male or female into nymphs of the last but one or the last instar induces at the subsequent moult an overall clearing of the cuticle due to the dispersion or disappearance of the black pigment.

Extirpation of corpora allata in the last but one or the last instar induces at the subsequent moult a complete black pigmentation in all the operated *Gryllus bimaculatus*. According to Roussel (1967) the corpora allata strictly control pigmentation, their absence always induces black pigmentation. The control by corpora allata is a direct one and acts on the integument.

7. HORMONAL CONTROL OF GREEN/BROWN POLYMORPHISM IN LOCUSTS AND GRASSHOPPERS

Rowell (1967) performed a number of experiments to find out the factor(s) involved in the colour changes in the locusts/ grasshoppers. He observed that the nymphs of *Acanthacris ruficornis, Humbe tenuicornis* and *Gastrimargus africanus* were implanted with a single corpus allatum from a young adult donor of the same species and in all the 3-species of the post-operated nymphal instars included significantly more green forms than the controls and in *A. ruficornis* and *H. tenuicornis* significantly more green adults. It is concluded that corpus allatum control green/brown polymorphism in Acridoidea in general. Pener (1967) reached to the same conclusion that corpus allatum directly or indirectly controls colour changes in *Schistocerca gregaria*. Thus it is clear from ablation and implantation of corpus allatum experiments performed by numerous workers that colouration in the locusts/grasshoppers is controlled by endocrine glands, specially the corpora allata.

17

THE PHEROMONES

The social attraction of an insect colony appears to be maintained primarily through the utilization of a series of communication chemicals which function to transmit specific bits of information to the members of a single population. These chemicals are termed as "pheromones". The term pheromone was coined by Karlson and Butenandt (1959). According to Karlson and Butenandt (1959) the pheromones are chemicals which are secreted to the outside by animals, in this case insects, and which if passed to another individual of the same species, cause it to respond in a particular manner. These are produced in several specialized tissues or glands called exocrine glands, which evacuate their contents into the environment in response to specific stimuli. Social insects such as hymenopterans and isopterans, possess many types of exocrine glands. It is well known that the pheromones modifies the insect behaviour in a variety of ways and now several hundred pheromones are known whose chemical properties etc. have been fully elucidated (see : Beroza, 1970 and Jacobson, 1972). Many aspects of insect biology and behaviour are controlled by these chemicals - the pheromones. The pheromones which have been identified are sex pheromones, trail marking pheromones, swarming pheromones, alarm pheromones, aggregation pheromones, territorial-marking pheromones and spacing pheromones etc. Account of some of these pheromones is given in the following pages.

The insect pheromones have been reported from Orthoptera, Heteroptera, Homoptera, Diptera, Isoptera, Neuroptera, Siphonaptera, Coleoptera, Hymenoptera and Lepidoptera. In some of the cases, they have been studied in great detail e.g. Hymenoptera, Orthoptera, Diptera, Coleoptera and Lepidoptera while in other groups comparatively less information is available. The pheromones coordinate and control a variety of biological and behavioural physiology in insects.

SEX PHEROMONES

Male and female conspecific insects often communicate with chemical *sex pheromones.* Mate location and courtship may involve chemicals in two stages, with *sex attraction pheromones* acting at a distance followed by closeup *courtship pheromones* used prior to mating. The sex pheromones involved in attraction often differ from those used in courtship. Production and release of sex attractant pheromones, tend to be restricted to one sex, mostly the female e.g. honey bees (Fig. 51A) although there are several lepidopterans and mecopterans in which males are the releasers of distance attractants that excite females. The producer releases volatile pheromones that stimulate characteristic behaviour in those members of the opposite sex within the range of the odorous plume. An stimulated recipient raises the antennae, orientates towards the source and walks or flies upwind often in a zig-zag track until the source is reached. Each successive action appears to depend upon an increase in concentration of this air borne pheromone. As the insect approaches the source, cues such as sound and vision may be involved in close-up courtship behaviour (Gullan and Cranston, 1994).

It has been observed that the courtship which involves coordination of the two sexes may require close-up chemical stimulation of the partner with courtship pheromones. This pheromone may be simply a high concentration of the attractant pheromone but *aphrodisiac* chemicals exist as seen in the butterfly *Danaus gilippus.* The males of this species as with several other lepidopterans have *extrusible abdominal hair pencils*, which produce a pheromone that is dusted directly on to the antennae of the female while both are in flight. The effect of this pheromone is to placate the natural escape reaction of the female who descends, folds her wings, and allows copulation. In *D. gilippus* this male courtship pheromone a pyrrolizidine alkaloid is called "danaidone". The scent pencils are also present in Heliconian butterfly that secrete and disperse attractant or aphrodisiac pheromone (Fig. 51B). It is of profound importance to note that the butterfly cannot synthesize it without acquiring the chemical precursor by feeding on selected plants as an adult. Likewise in several butterflies and moths, the precursor is acquired through feeding the specific plant species. As remarked earlier, sex pheromones are mostly species specific and they are the mixture of several chemicals. Therefore, a single component of the mixture may elicit only one part of the sex attraction behaviour or a partial or complete mixture may be needed. Often the mixture produces a greater response than any individual component, a synergism which is widespread in insects that produce pheromone mixtures.

Fig. 51A. This worker honey bee is exposing scent glands on her abdomen and fanning her own scent into the air with her wings (From Blaney, 1977)

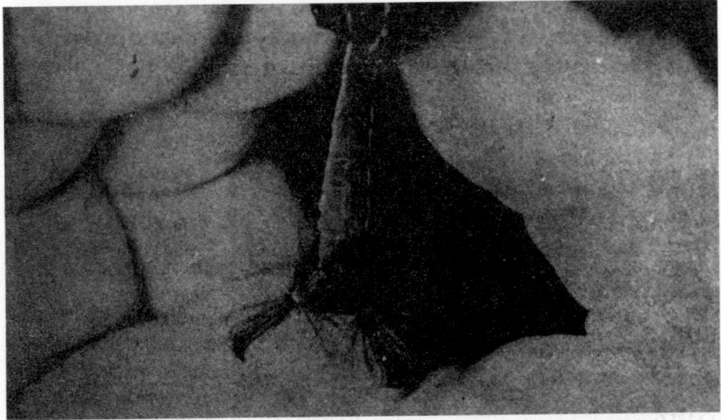

Fig. 51B. Scent pencils on male Heliconian butterfly from Ecuador (From Elzinga 1978)

Fig. 51 C, D. Pectinate antennae of giant silkworm moth (*Antheraea polyphemus*) (C) lateral view; (D) SEM micrograph of filaments showing sensory setae used in detecting female sex pheromones (From Elzinga, 1978).

In certain moths, the virgin female assumes a calling position and emits from glands near the tip of the abdomen, specific *sex attractant pheromones* that diffuse out into the environment. These pheromones are usually specific and are detected by antennal receptors located on the male antennae (Fig. 51 C,D). If a minimal threshold of molecules (10^{-12} µg for the gypsy moth) come in contact with the antennal receptors, the male becomes "excited" and flies upwind, the most likely direction of pheromone source. Olfactory attractants bring the male close to the female, where other sensory stimuli such as: visual, tactile, or wing beat clues become important. The distance of initial attraction varies. Male gypsy moths have been reported to fly nearly 4 km to a female, but effective attraction is probably less than 100 m (Elzinga, 1978). It was also found that the males of the Indian moon moth *Actias selene* (Fig. 51E) liberated at a distance of 11 km., were able to return to the females (Wigglesworth, 1965).

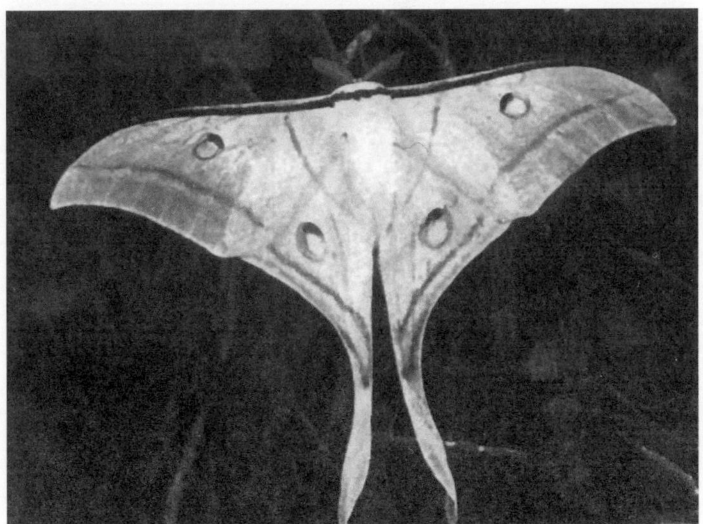

Fig. 51E. Indian moon moth, *Actias selene*

TRAIL MARKING PHEROMONES

Many social insects use pheromones to mark their trails particularly to food and nest. Most of the trail marking pheromones are volatile and short lived chemicals that evaporate within days unless reinforced. As stated above, the trail pheromones are found in insects of the order Hymenoptera and Isoptera.

Ant Trail Pheromones

The ant trail pheromones originate in several different glandular tissues like hindgut, mandibular gland, poison gland, Dufour's gland, Pavan's gland and metathoracic legs of formicidae (Hymenoptera). Odour trail pheromones apparently originate in the hindgut in three sub-families of ants but the exact place of origin of the hindgut trail pheromone is not clearly understood. The Dolichoderine ants have developed a novel social organ, the Pavan's gland in which the trail pheromones probably are synthesized (Wilson and Pavan, 1959). Since this exocrine structure is found in a large number of ants, it appears that they play a major role in the social biology of the ants (Fig. 51F).

Fig. 51F. Leafcutter ants (*Atta* sp.) following a chemical trail across a piece of string back to the colony (From Elzinga, 1978).

It is only in the ant of Myrmicinae that the glands associated with the sting have been utilized as the source of the trail pheromone. According to Wilson (1959) the contents of Dufour's gland of the fire ant, *Solenopsis saevissima* are evacuated through the sting in order to generate odour trails. On the other hand, several myrmicine genera lay odour trails with the poison gland which are used for defensive purpose (Blum *et al.,* 1964). In general, the trail pheromones appear to be relatively stable compounds which should be quite amenable to isolation and structural elucidation. Sheneirla and Brown (1950) have demonstrated that in nature *Eciton* trails retain their activities for several weeks. Similarly Blum *et al.* (1964) have shown that artificial trails prepared from attine poison glands retain their activities for several days at room temperatures. Beroza (1970) has demonstrated that n-hexane solutions of the trail pheromone are stable for at least 2 years when stored at 4°C. Moser and Silverstein (1967) have demonstrated that the trail pheromone of *Atta texana* was composed of at least two substances one of which was relatively non-volatile. The volatile fraction could not be detected by the ants after 60 min., whereas the non-volatile fraction was active for at least 6 days. It is presumed that the non-volatile fraction is oxidatively inactivated since artificial trails were highly active 4 months after preparation, provided that they were held under a vacuum of 0.5 mm of mercury.

Termite Trail Pheromones

In primitive as well as higher termites, odour trails are laid from a sternal gland in the ventral part of the fifth abdominal segment (Stuart, 1961). In the primitive genus *Zootermopsis* trails are laid to breaks in the nest structure and workers are thus recruited rapidly to a specific point in the nest. Alarm behaviour is also exhibited by *Zootermopsis* workers when they have discovered a nest break and the subsequent laying of an odour trail is part of a defensive reaction which results in building behaviour (Stuart, 1967). The number of workers recruited to the break is related to the intensity of the stimulus, the odour trail. In higher termites, food sources produce low-level alarm reactions when they are discovered and trails are laid from the food finds back to the nest. Thus, in both lower and higher termites, alarm stimuli release trail-laying behaviour (Beroza, 1970).

Bee Trail Pheromones

Lindauer and Kerr (1958) have demonstrated that the workers of many species of stingless bees, *Melipona* spp. (Meliponinae) lay trails from food finds to their nests. The trail is laid as a series of droplets which are placed by the scout bee at specific intervals on either vegetation or the ground as the bee moves from the food source to the nest. The distance between the scent marks varies with each species of bee but in all cases so far examined the marking pheromones originate in mandibular gland secretions (Lindauer and Kerr, 1960). Recruited bees orient to an aerial trail which results from the volatilization of the deposited mandibular gland droplets. These trails are rapidly reinforced by returning recruits and thus an effective concentration of trail pheromone is maintained during the critical period, the initial recruitment phase. The efficiency of recruiting in the higher Meliponini is as great as that achieved by the honeybee *Apis mellifera.* In addition, stingless bee trails can be laid in the vertical component, whereas the honeybee is able to communicate only horizontal distances (Lindauer and Kerr, 1958). Therefore, stingless bees can recruit workers to flower concentrations in tall tropical trees, but honeybees must necessarily restrict their recruiting activities to the flora near the ground (Beroza, 1970).

ALARM PHEROMONES

Many species of social insects can rapidly communicate colonial disturbance by secreting pheromones from exocrine glands. Muschwitz (1964) has demonstrated that all major groups of hymenopterous insects employ alarm pheromones and Moore (1968) has shown that these excitatory substances are secreted by certain species of termites (Fig. 51G). The widespread distribution of alarm pheromones emphasizes the probable importance of these chemical releasers in the biology of social insects. Alarm pheromones appear to be least specific of the classes of chemical releasers of social behaviour.

Fig. 51G. Termite nasutes (*Nasutitermes* sp.) attracted by alarm pheromones to a hole in the nest (From Elzinga, 1978).

Although alarm pheromones are far from species specific, it is nevertheless very clear that ants have a degree of olfactory acuity which enables them in the presence of their own excitatory odorant, to release alarm behaviour at lower threshold values than is possible with related compounds.

Alarm is provoked by the presence of a predator or in many social insects by a threat to the nest. The alarm behaviour of many eusocial insects is most familiar to humans when nest disturbance induces many ants, bees or wasps to an aggressive defence. Alarm pheromones attract aggressive workers and these recruits attack the cause of the disturbance by biting, stinging or firing repellent chemicals. Emission of more alarm pheromone mobilizes further defenders.

Alarm pheromones are also utilized as releaser of several different types of behaviour. Wilson (1958) has reported that the alarm pheromone of *Pogonomyrmex badius* functions as an attractant at low concentrations and a releaser of alarm behaviour at higher concentration. High concentrations of 2-heptanone release digging behaviour in excited workers of *Conomyrma pyramica*, a reaction also observed with workers of *P. badius*, which are exposed to their alarm pheromone. The alarm pheromone used by the stingless bee *Lestrimelitta limao* has been adapted to function as a key element in the robbing behaviour of this species. Workers of *L. limao* rob colonies of stingless bees in the genera *Trigona* and *Melipona* and their alarm pheromone "citral" (Blum, 1966) subserves several crucial functions during the robbing act. Citral which is released by the initial robbers, attracts other workers of *L. limao* to the nest which is under attack. Blum *et al.,* (1969) have demonstrated that citral is highly effective in releasing escape behaviour in workers of stingless bee species normally raided by *L. limao*. A detailed account on this subject has been given by Beroza (1970).

THE QUEEN-SUBSTANCE

Among social insects the only sex attractant that has been identified is that of the queen honey bee *Apis mellifera*. The queen substance, 9-oxodecenoic acid is produced in the mandibular glands of the queen honey bee, which serves as a sex attractant to the drone during mating and keeps the workers in the hive informed of the continued presence of the queen. The secretion is obtained by beating the body of the queen and passed throughout the community by food exchange, so long as it is present the building of queen cell by the workers is inhibited.

Worker honey bees which are non-reproductive females do not produce queen substance in their mandibular glands. Butler and Simpson (1967) demonstrated that the keto-acid would attract swarming bees which had lost their queen. Further more, Butler *et al.,* (1967) demonstrated that extracts of queens of *Apis cerana* and *Apis florea* were as attractive to drones as those of the *Apis mellifera* and further that the mandibular glands of *A. cerana* and *A. florea* produce the same chemical as the *A. mellifera*. It has been reported that in conjunction with another scent from the queen, the queen substance suppresses the maturation of the ovaries of the workers and yet another secretion keeps the workers in the swarm in close contact with queen (Wigglesworth, 1965).

Thus, it can be concluded that the "queen substance" produced by the mandibular glands of the queen bee and transmitted by food sharing to the worker bees, on which it has two effects (1) it prevents them from building the large queen-rearing cells and (2) it suppresses ovary development. This substance does not act alone; other factors, perhaps some volatile odour contribute to ovary inhibition and removal of the mandibular glands reduces the inhibition of queen-cell building but does not eliminate it completely. The active material is 9-oxodecenoic acid of which 0.13 μg per bee will inhibit queen-cell production. It also serves as a sex attractant in mating. But the mandibular glands contain other substances; one of these, probably 9-hydroxy-decenoic acid has been recognized as the pheromone, which prevents the cluster of the bee swarm from breaking up (Wigglesworth, 1965).

TERRITORIAL PHEROMONES

It has been pointed out that the male bumble bees *Bombus* spp. use their mandibular gland secretions in order to mark territorial sites which may play an important role in their mating biology. While the scent marks produced by a male of one species do not attract males of other species and therefore function as territorial markers. The marks do attract both males and females of the same species thus bringing the reproductives of both the sexes together and increasing the probability that mating will occur. Probably the scent marks also physiologically initiate the males and females for subsequent copulatory activity and thus play the role of sexual stimulants (Beroza, 1970).

Bergstrom *et al.* (1968) have identified 2, 3 dihydro-6-trans farnesol as the main component in the mandibular gland secretion of *Bombus terrestris*. Several other chemicals have been isolated from the mandibular glands of different species of *Bombus* (see : Beroza, 1970).

AGGREGATION PHEROMONES

The release of an aggregation pheromone causes conspecific insects of both the sexes to crowd around the source of the pheromone. Aggregation may lead to increased likelihood of mating but in contrast to many sex pheromones both sexes may produce and respond to aggregation pheromones. The potential benefits provided by the response include security from predators, maximum utilization of a scarce food resource, overcoming of host resistance and attraction of social insects as well as to the chance to mate.

SPACING PHEROMONES

Gillott (1995) remarked that a relatively new discovery in pheromone research are those pheromones that stimulate insects to spread out so as to maintain an optimal population density. They may be produced by immature and adult insects and may be olfactory or tactile. Spacing pheromones appear to have been best studied in relation to oviposition. For example, after laying, the female apple maggot fly, *Rhagoletis pomonella* releases, a pheromone from her ovipositor that deters oviposition on the fruit by other females. Several Lepidoptera also produce oviposition - deterring pheromones which limit the number of eggs laid on a given plant. Further, the presence of feeding larvae of *Pieris brassicae* inhibits egg laying, suggesting that the larvae also produce a pheromone. Thus, the pheromones which control density of the insect population are called spacing pheromones or dispersion pheromones. Spacer pheromones of colony-specific odours may be used to ensure an even dispersal of colonies of conspecifics, as in African weaver ants, *Oecophylla longinoda* (Formicidae : Hymenoptera) (Gullan and Cranston, 1994).

CASTE REGULATING PHEROMONES

The regulation of caste differentiation in social insects is a morphogenetic phenomenon, just as are the changes from larva to larva, larva to pupa and pupa to adult. Such changes depend on the activity of corpora allata for their manifestation. It is not surprising therefore to learn that the pheromones that regulate caste differention including development of reproductives, exert their effect ultimately via corpora allata. In lower termites, e.g. soldier formation can be induced in experimental colonies by administration of juvenile hormone through feeding, topical application or as vapour. Thus, the soldier-inhibiting pheromone may act by inhibiting the corpora allata or by competing with juvenile hormone as its site of action (Luscher, 1972).

SEMIOCHEMICALS

Gullan and Cranston (1994) have remarked that the communication chemicals (semiochemicals) may function between individuals of the same species in which case they are called pheromones, or between different species in which case they are called `allelochemicals'. Interspecific semiochemicals may be grouped according to the benefits they provide to the producer and receiver like kairomones, allomones and synomones.

Kairomones

These are the communication chemicals that benefit the receiver and disadvantageous to the producer, for example myrcene, the terpene produced by a ponderosa pine when it is damaged by the western pine beetle, acts as a synergist with aggregation pheromones, to lure more beetles. Thus, myrcene and other terpenes produced by damaged conifer trees can be kairomones (i.e. the chemicals that disadvantage the producer), in this case by luring damaging timber beetles. A kairomone need not be a product of insect attack e.g. elm bark beetles - *Scolytus* spp. (Scolytidae) respond to α - cubebene, a product of the Dutch elm disease fungus, *Ceratocystis ulmi* that indicates a weakened or dead elm tree. As remarked by Gullan and Cranston (1994) the elm beetles will inoculate previously healthy elm trees with the fungus but pheromone - induced beetle aggregation form only when the kairomone (i.e. fungal α-cubebene) indicates suitability for colonization. Host plant detection by phytophagous insects also involves reception of plant chemicals that therefore act as kairomones.

Allomones

These are the chemicals that benefit the producer but have neutral effects on the recipient e.g. defensive or repellent chemicals are allomones that advertize distastefulness and protect the producer from lethal experiment by prosepctive predators. The effect on a potential predator is considered to be neutral since it is warned from wasting energy in seeking a distasteful meal.

Synomones

As stated, the terpenes produced by damaged pines are kairomones for pest beetles but if identical chemicals are used by beneficial parasitoids to locate and attack the bark beetles, the terpenes are acting as synomones i.e. benefiting both the producer and the receiver (Gullan and Cranston, 1994). Thus, according to the above authors, the α-pinene and myrcene produced by damaged pines are kairomones for species of *Dendroctonus* but synomones for the pteromalid hymenopterans that parasitize these timber beetles. In a similar manner α-cubebene produced by Dutch elm fungus is a synomone for the braconid-hymenopteran parasitoids of elm bark beetles (for which it is a kairomone) (Gullan and Cranston, 1994). It has been further remarked by Gullan and Cranston (1994) that if a damaged host plant produced a repellent odour then the chemical would act as :

1. An allomone that deters non-specialist phytophages.
2. A kairomone that attracts a specialist phytophage.
3. A synomone that lures the parasitoid of the phytophage.

DEFENSIVE SECRETIONS

Furthermore, many insects produce "defensive secretions" which are repellent to potential predators. These comprise a vast range of saturated and unsaturated hydrocarbons and their derived alcohols and aldehydes (ranging from hexenal and tridecene to pentadecane), fatty acids such as butyric and caproic acids, a great number of phenolic substances and quinones such as phenol, *m*-cresol, benzoquinone, toluquinone etc. and a variety of terpenoids such as geraniol, farnesol, citral, citronellal etc. These different substances perform different functions in different insect species.

Wigglesworth (1970) has summarized the work on pheromonal control of aggregation, dispersal, trail formation and social coherence. According to Wigglesworth (1970) defensive secretions as such would not fall within the definition of pheromones. But what are commonly regarded as repellent or defensive secretions against outsiders, can be utilized within the species for recognition, aggregation and trail formation. The trail pheromone of *Zootermopsis* contains many compounds, but one of the most active is caproic acid. The "Nassanoff glands" of the honey bee worker secrete a mixture of terpenoids (viz. citral, geraniol etc.) and this odour is used by foragers as an attractant to secure recruits for their expeditions to rich collecting grounds. In *Lasius niger* and Ponerine ants generally, the contents of the stomach and hind gut will serve as an effective scent trail; in Myrmecine ants the secretion of the poison gland is used.

Defensive secretions may also play the opposite role. They may serve to alert and disperse aggregations. That is seen in *Formica rufa* where formic acid is a potent means of defence and offence but serves also to alert the colony to a threatened attack and so to induce a general agitation throughout the nest. The citral produced by the mandibular glands of the leaf cutting ant *Atta* likewise, serves this same dual purpose of repellency and alarm and in the cotton stainer *Dysdercus,* which produces a mixture of hydrocarbons ranging from hexenal to pentadecane, this product can serve both as a repellent against predators and as a warning signal that causes dispersal of the aggregations which occur in this species.

A very large number of alerting pheromones have been isolated and identified in ants and termites, many are terpenoids, others are aliphatic ketones and esters. In the fire ant, *Solenopsis*, a whole series of complex acts of behaviour, mass foraging, trail following, colony emigration, alarm, grooming and clustering all are largely induced by chemical (i.e. pheromone) releasers of this kind. The mandibular glands of the queen honeybee contains a variety of organic acids, one of which is 9-hydroxy-decenoic acid, which serves as ``cohesion pheromone", which keeps the swarm of worker bees clustered around their queen. There is indeed evidence for the presence of widespread pheromones in other social insects notably ants and termites (vide supra) which are distributed throughout the colony by food sharing between adults and serve the purpose of colony cohesion but the chemical nature of these materials is not well known.

Among the social insects one of the best known chemical messengers (i.e. pheromones) of this kind is the queen substance, 9-oxodecenoic acid also secreted by the mandibular glands of the queen bee. This secretion is smeared over the integument of the queen during her grooming activities. The queen is usually surrounded by a circle of workers, all facing her and licking her and passing on the material in the course of food exchange with other workers. This probably is the chief means by which workers are kept informed of the presence of the queen. If the queen is removed, a change in behaviour among the workers becomes apparent and within 24 hours they begin building large "queen cells" to rear new queens. At the same time the ovaries of many of the workers ripen and they begin to lay eggs.

It may be that the absence of the same queen substance, in combination with a special queen scent is responsible for both these effects. The queen scent is a product of glandular cells in the mid dorsal area of the abdomen. There is no evidence that queen substance or odour act directly on the reproductive system, it seems more probable that they act upon the sense organs and the central nervous system and thus cause indirectly a change in hormone secretion. During the first few days after the removal of the queen the corpora allata of the worker bees increase in volume, their activity seems normally to be inhibited by the queen substance (See: Wigglesworth, 1970).

18

THE REPRODUCTIVE SYSTEM

The reproductive systems consist of paired gonads viz. the ovaries of the female and testes of the male, paired genital ducts into which the sexual products are discharged and a median duct viz.: the vagina in the female and the ejaculatory duct in the male. In conjunction with genital ducts are found specialized glands and organs for storage of spermatozoa. The gonads and genital duct are of mesodermal origin and the vagina and ejaculatory ducts are of ectodermal origin, lined with cuticle. In most of the insects, the sexes are separate and the sexual dimorphism is distinct i.e. the male and female individuals can be easily identified by examining the tips of the abdomen.

Snodgrass (1935) and Wigglesworth (1965) have given a detailed account of the female and the male reproductive organs, which has been followed in the text with some minor changes.

THE FEMALE REPRODUCTIVE ORGANS

The female reproductive organs (Figs. 52A, 53A, 54A) consist of a pair of ovaries, a pair of lateral oviducts and the median oviduct formed by the union of two lateral oviducts. The median oviduct opens posteriorly through a slit like opening the gonopore. Besides these, there is usually a saccular spermatheca for reception and storage of spermatozoa and a pair of accessory glands of diverse function and a copulatory pouch of the body wall. The latter is either an open genital chamber or a tubular exit passage from the median oviduct termed as vagina.

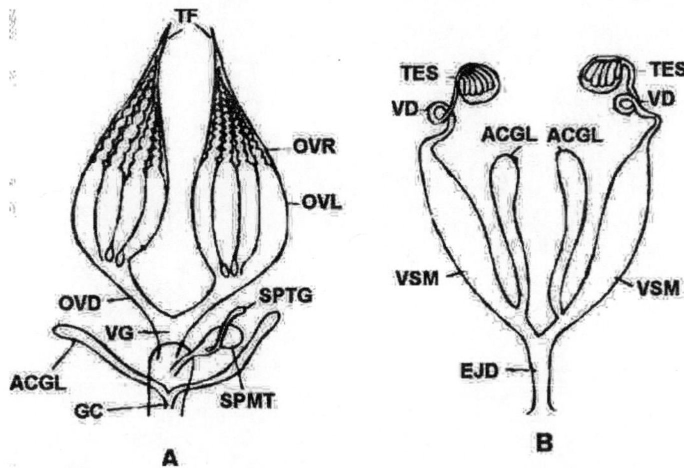

Fig. 52 A. Female reproductive system of a generalized insect. **B.** Male reproductive system of a generalized insect
Abbreviations - ACGL- accessory gland, EJD- ejaculatory duct, GC- genital chamber OVD- oviduct, OVR- ovary, OVL- ovariole, SPTG- spermathecal gland, SPMT- spermatheca, TES- testis VD- vas deferens, VSM- vesicula seminalis, VG- vagina, TF- terminal filament.

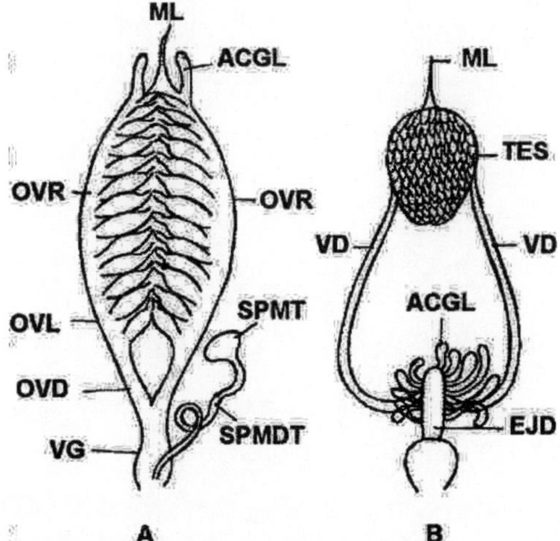

Fig. 53 A. Female reproductive system of a grasshopper, **B**. Male reproductive system of a grasshopper
Abbreviations : ML- medial ligament, TES- testis, VD- vas deferens, ACGL- accessory gland, EJD- ejaculatory duct. OVR- ovary, OVL- ovariole, SPMT- spermatheca, OVD- oviduct, VG- vagina, SPMDT- spermathecal duct.

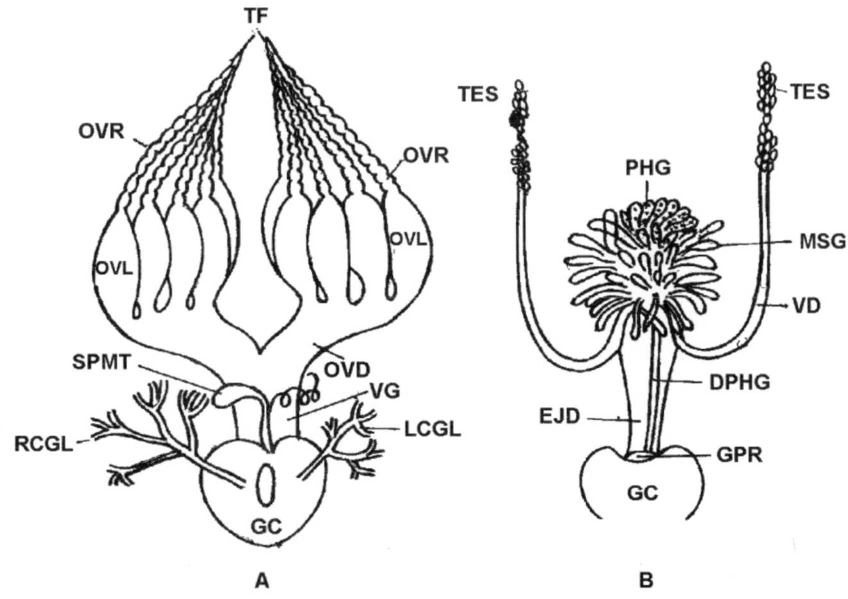

Fig. 54A. Female reproductive system of cockroach, **B**. Male reproductive system of cockroach

Abbreviations: TES- testis, PHG- phalic gland, MSG- mushroom gland, VD- vas deferens, DPHG- duct of phalic gland, EJD- ejaculatory duct, GPR- gonopore, GC- genital chamber, TF- terminal filament, OVL - ovariole, OVR- ovary, OVD- oviduct, VG- vagina, SPMT- spermatheca, RCGL- right colleterial gland, LCGL- left colleterial gland.

The ovaries

The ovaries consist of a group of cylindrical or tapering units known as ovarioles. The ovarioles converge at the anterior end to form filaments that ultimately unite and form a suspensory ligament. The ligament may end in the neighbouring fat tissue but generally it is attached either to the body wall or to the dorsal diaphragm. In certain instances, the ligaments from the two ovaries are combined to form a median ligament which is inserted in the ventral wall of the dorsal blood vessel. In young individuals, the ovary is enclosed in a peritoneal sheath of connective tissues which is absent in adults. In the dipterous insects, it is retained in the form of a membrane enveloping the ovarioles. Each ovary contains a variable number of ovarioles which varies from species to species and from group to group.

Structure of ovariole

Each ovariole (Fig. 54 C,D,E) consists of three distinct parts namely (1) a terminal filament, (2) an egg tube and (3) a supporting stalk or pedicel.

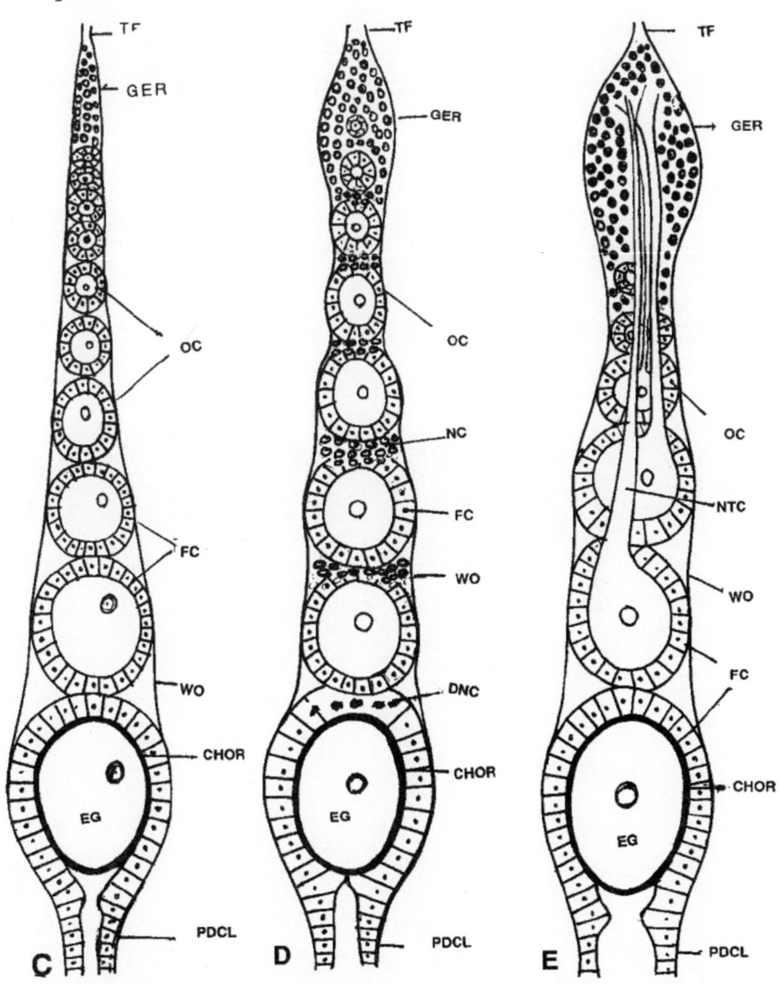

Fig. 54. Diagrams showing different types of ovarioles. **C**. Panoistic, **D**. Polytrophic, **E**. Acrotrophic

Abbreviations : CHOR- chorian, FC- follicle cell, GER- germarium, NC- nurse cell, NTC- nutritive cord, OC- oocyte, PDCL- pedicel, DNC- degenerated nurse cell, EG- egg, WO- wall of ovariole, TF- terminal filament.

The Terminal Filament

The slender, thread-like filament that forms the anterior part of an ovariole is called the terminal filament. The terminal filaments form the suspensory apparatus of the mature ovary.

The Egg Tube

Egg tube is the main part of the ovariole containing germ cells and their derivatives. In adults, it is an elongate sac filled with a mass of cells derived from germinal elements, some of which become the eggs or ova. The wall of the egg tube is formed by the tunica propria. The entire egg tube is lined inside the tunica propria by a layer of follicular epithelium.

Each egg tube has two main divisions : The anterior part has germ cells in an active stage of division and incipient differentiation. This part is the (i) germarium. Beyond the germarium is the region in which the egg cells grow and attain their mature size. This part of the egg tube is the (ii) zone of growth or vitellarium.

The Germarium

The germarium represents the primary egg tube of young ovary in which the germ cells are lodged. These germ cells develop into oogonia and then into oocytes or young egg cells which are generally accompanied by nurse cells. The germarium also contains smaller cells which develop into follicle cells in the lower part of the egg tube. The follicle cells are designated as cystocytes and the nutritive cells as trophocytes. Both are accompanied by developing eggs.

The Vitellarium

The vitellarium is formed during the development of the insect as an extension of the egg tube beyond the germarium. As the oocytes multiply, their mass and volume increase causing the vitellarium to enlarge. At the anterior end, the follicle cells have an apical position in the tube and begin to exhibit a definite epithelial arrangement while the nutritive cells assume an axial position. The growth of the oocytes distends the vitellarium into a series of egg chambers or follicles, which become successively larger towards the posterior end of the egg tube. Beyond the last egg chamber, a mass of follicle cells forms a plug that closes the posterior end of the egg tube.

The oocytes are produced continuously from the oogonia so that the first encysted oocyte becomes the lower most and is the first to mature. With successive addition of an oocyte to the series, the egg tube becomes lengthened between the last formed egg chamber and the germarium by a rapid multiplication of the follicle cells in this region. The last egg in each egg tube, when fully formed is usually abruptly larger than any of those preceding it and is enclosed in the chorion. However, this fully formed egg is still in the oocyte stage, because maturation does not take place until the egg has been deposited. Hence it is improper to speak of the full grown ovarial oocytes as `mature' or "ripe eggs". If an oocyte is fully formed in all the ovarioles at the same time, the insect may deposit at one laying as many eggs as there are ovarioles in the ovaries. However, some insects deposite single egg in a continuous succession. With a few insects, the fully grown oocytes are produced alternately first in one ovary and then in the other. Finally a large number of eggs may accumulate in the exit passage before they are discharged.

The Pedicel

The pedicels or stalks of the ovarioles are short ducts connecting the egg tubes with the lateral oviduct. They are formed from the strand of mesodermal cells along the lower margin of the embryonic gonad which is continuous posteriorly with the primitive oviduct. Each pedicel has a lumen that communicates with the end of the duct but the lumen is closed at the upper end beneath the epithelial plug of the egg tube. When the first egg is ready to be laid, the cells of the epithelial plug, and the cells of the retaining walls of the pedicel are dissolved to open a passage through the pedicel from the egg tube into the oviduct. The walls of the pedicels may be composed of a simple elastic epithelium or they may be covered by an extension of the muscular sheath of the oviduct.

When the oocytes are fully developed, a chorion similar to that of cuticle forms over the entire surface of the oocyte except for one or more small points through which the sperm from the spermatheca enters through micropyle opening in the chorion. Usually these eggs are laid out side the insect body, however, in certain instances they are retained within the body until they are ready to hatch, resulting in the production of viviparious larvae e.g. flesh fly, *Sarcophaga ruficornis*.

The Trophic function of the egg tubes

The growth of the eggs in the ovary which is mostly the

accumulation of yolk involves the utilization of much nutrient material. This material is supplied either from the daily food of the insect absorbed into the blood or from the food reserves stored in the body mainly in the fat tissue. In insects which do not feed in the adult stage, all the added egg material must be drawn from the fat tissues.

Since the eggs in the egg tubes do not have direct access to the nutrient elements of the blood, an important function of the ovary is that of an intermediate trophic organ between the blood and the eggs. Such a function of egg tube is performed in various ways.

The structure of the egg tubes differs according to the manner of the nourishment of the developing oocytes. In general two principal types of structure are distinguished by the presence or absence of special nutritive cells within the follicular tubes.

(1) Panoistic type of egg tube (Fig. 54C)

In this type of egg tube there are no nutritive cells differentiated from the egg cells. The food products elaborated by the follicular epithelium are absorbed directly by the oocytes. Panoistic ovarioles are found in the insects belonging to order Apterygota, Ephemeroptera, Odonata, Orthoptera, Siphonaptera, Isoptera etc.

(2) Meroistic type of egg tube

When nutritive cells are present in the egg tube then it is called meroistic type. In this case the nutritive cells as well as germ cells arise from the primitive sex cells and contribute to the nourishment of the oocytes in the early stages of their development. This is further divided into two subtypes: (a) polytrophic and (b) acrotrophic according to the position of the nurse cells or trophocytes in the egg tube.

(a) Polytrophic egg tube (Fig. 54D)

In this type, the egg tube contains an alternate succession of oocytes and nurse cells or trophocytes. The polytrophic type of egg tube occurs in Anoplura, Neuroptera, Coleoptera (Adephaga), Lepidoptera, Hymenoptera and Diptera.

(b) Acrotrophic or telotrophic egg tube (Fig. 54E)

In a few insects particularly Hemiptera and some Coleoptera (Polyphaga), the nutritive cells or trophocytes are confined to the apex of each ovariole and are connected to the developing oocytes in the early stages of their development by means of long protoplasmic nutritive cords. Hence, the germarium in acrotrophic type of egg tube also functions as an apical feeding chamber for the oocytes.

GENITAL DUCTS AND ACCESSORY STRUCTURES

Lateral oviducts

The lateral oviducts are long, tubular structures. The wall of oviducts is composed of glandular cells and muscles. The anterior end of each lateral oviduct is somewhat expanded forming the calyx into which open the pedicels of the ovarioles. The oviducts are generally simple tubes without accessory structures of any kind. In Acridoidea the anterior ends of wall are glandular. The cellular walls of the lateral oviducts are covered by a muscular sheath consisting of circular or longitudinal fibres or sheath.

Median oviduct

The two lateral oviducts join a median oviduct which is ectodermal in origin, hence it is lined with cuticle. The median oviduct is more muscular than the lateral oviducts with circular and longitudinal muscles. The posterior opening of the median oviduct is the female gonopore. The posteriormost part of the median oviduct is modified into a sac like structure, the genital chamber or the vagina. The vagina opens outside and the opening is termed vulva. Frequently, the vagina is well developed to form a pouch, the bursa copulatrix which receives the penis, while in viviparous Diptera the anterior part of the genital chamber is enlarged to form the uterus in which the larval development occurs. However, in most Lepidoptera, there are two openings, one on the segment 9th serves for the discharge of eggs and is known as the oviporus while the other on segment 8th is the copulatory opening, the vulva. The latter, i.e. vulva leads to the bursa copulatrix which is connected with the oviduct by a sperm duct. In addition to female Lepidoptera, two openings also occur in certain water beetles.

Spermatheca (Receptaculum seminis)

The spermatheca is ectodermal in origin and is lined with cuticle. The size, shape and structure of the usual single spermatheca is highly variable in different insects but generally the organ is sac like in form with a slender duct. Often a diverticulum of the duct forms a tubular spermathecal gland, producing secretions which probably provide nutrients for the sperms.

Since in the insects insemination of the eggs is not generally accomplished during the act of mating but takes place at a varying length of time afterwards, most female insects are provided with a sperm receptacle in which spermatozoa are stored and from which they can be ejected upon the eggs, the latter are extruded from the oviduct. This organ is called the receptaculum seminis or spermatheca.

Female accessory glands

The pair of glands associated with the exit apparatus of the female genital organs usually has some function associated with the laying of the eggs. These glands are therefore, known as accessory glands of the reproductive system. The opening of the accessory glands lies on the venter of the 9th abdominal segment, between the bases of the 2nd valvifer of the oviposter. They are ectodermal in origin.

In the majority of female insects the accessory glands produce an adhesive substance for attaching the eggs to a support or for gluing them together in a mass as they are laid, hence they are commonly termed as colleterial glands. The frothy secretions which form the egg pods of grasshoppers and the gelatinous sheath of *Chironomus* eggs are produced by the accessory glands. In *Hydrophilus* the accessory glands produce silk which forms the cocoon in which the eggs are laid. The accessory glands perform a variety of functions in female Hymenoptera. The poison used to paralyse the prey in Hymenoptera is derived from the accessory glands. In *Periplaneta* the function of accessory glands have most fully been investigated. In this insect the eggs are laid in an ootheca consisting of a tanned cuticle-like substance which is produced by the accessory glands.

THE MALE REPRODUCTIVE ORGANS

The male reproductive organs (Figs. 52B, 53B, 54B) consist of a pair of testes, a pair of sperm ducts viz. the vasa deferentia and the ductus ejaculatorious. In most insects there are also a number of accessory glands which open into the vasa deferentia or ejaculatory duct.

Testis and spermatogenesis

The testis is made up of a series of tubular follicles varying greatly in number and arrangement in different insects. Each follicle consists at first a layer of epithelium resting on a basement membrane. Later, the entire group of follicles is enclosed in a connective tissue sheath often pigmented. In Lepidoptera, the two testes are bound together in a single capsule. Each follicle contains a succession of zones in which the sex cells are in different stages of development (Fig. 52C).

(1) The Germarium or Zone of spermatogonia

These are distinguished as the germarium or zone of spermatogonia, containing densely packed primordial germ cells or spermatogonia. At the upper extremity of the follicles in Lepidoptera and certain other insects, large apical cells are present which serve as nurse cells in the early stages of spermatogenesis.

(2) Zone of spermatocytes

As each primitive spermatogonium moves backwards, it becomes covered with a mantle of somatic cells, at the same time it divides repeatedly and distends this sheath to form a cyst. In the begining the cysts are spherical but as they increase in size they become polyhedral from mutual pressure. Each contains the spermatocytes from 64 to 256 in number derived from a single spermatogonium

(3) Zone of maturation and reduction

In this region the spermatocytes divide twice, at first they form two spermatids and then into four spermatids. In majority of the insects the first division is the reduction division at which the number of chromosomes become haploid or "x".

(4) Zone of transformation

In this zone the rounded or spherical spermatids enclosed within the cysts are transformed into elongate, flagellated spermatozoa. The mature sperm of *Drosophila* measures about 1.76 mm in length, which is more than three times the length of the egg (0.5 mm).

Structure of mature spermatozoa

The mature sperms are filamentous in form (Fig. 53C). The head and tail of sperm are of approximately the same diameter. The greater part of the head region is occupied by the nucleus with only a thin coating of cytoplasm and in front of the nucleus is the acrosome. The acrosome is probably concerned with attachment of the sperm to the egg and possibly also with the lysis of the egg membrane, thus permitting sperm entry (For details see: Chapman, 1972).

Fig. 52C. Diagram of testis follicle showing the stages of development of the sperm (From Wigglesworth, 1965).

Abbreviations: a. apical cells surrounded by spermatogonia; b. zone of spermatogonia; c. zone of spermatocytes; d. cysts with mitoses of first maturation division, e. cysts with mitoses of second maturation division, f. zone of spermatids; g. zone of spermatozoa.

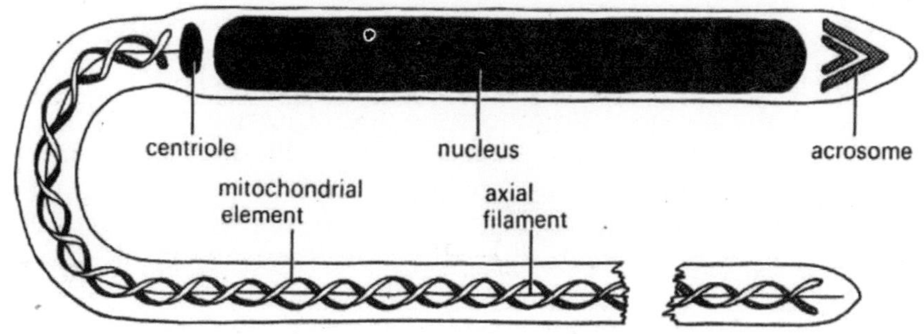

Fig. 53C. Diagram showing the possible structure of an insect spermatozoon (From Davey, 1965)

GENITAL DUCTS AND ACCESSORY STRUCTURES

Genital ducts

Vas deferens

The vasa deferentia receive the sperms or spermatophores from the testes and help in their transport to the median ejaculatory duct, which is formed by the two vasa deferentia. The vasa deferentia have a thick cellular epithelium coated with circular muscle fibres. The vasa deferentia are primarily of mesodermal origin but they may be extended posteriorly by ingrowths from the ectoderm or more or less replaced by the ectoderm. In some insects e.g. locusts and grasshoppers the posterior part of the vasa deferentia enlarges to form the seminal vesicle for the storage of sperms. Frequently, a part of each duct is much convoluted and the coils may form a definite epididymis for storing the sperms.

The ejaculatory duct

The vasa deferentia unite and form a common duct, the median ejaculatory duct of ectodermal origin. The ejaculatory duct is internally lined with cuticle and is provided with a powerful muscle coat made up of outer layer of circular muscles and inner layer of longitudinal muscles. The terminal part of the ejaculatory duct is often enclosed in an evagination of the body wall to form the intromittent penis or aedeagus.

Male accessory glands

The glandular structures associated with the male reproductive organs usually perform the function of secreting a mucous or viscid fluid, which is either discharged as liquid with the sperms or hardens about them to form a covering or capsule known as "spermatophores" or "sperm bundles". The accessory glands generally arise from the anterior end of the ejaculatory duct or from short divergent anterior branches of the duct. They are in the form of elongate sacs or tubes. The latter are greatly convoluted in nature. Usually there is one pair of accessory glands but in some insects a series of glands arises from each side of the ejaculatory duct, while in others there may be a compact mass of tubules about the terminus of the vas deferens. The walls of the accessory glands have a muscular sheath continuous with that of the ejaculatory duct.

MATING, SPERM TRANSFER AND FERTILIZATION

Mating

Different kinds of stimuli of diverse nature are associated with the mutual attraction of the male and female individuals at the time of mating. In many Diptera, Trichoptera, Plecoptera and others, the females are attracted to the males by their dancing in swarms, in crickets and grasshoppers by their chirping songs, and the males of Lampyrid beetles are attracted by the luminous organs of the female. But the most important stimuli are probably the scents. The female individuals of many insects have glandular scent producing organs in the vicinity of the sexual openings. These are well developed in the female Lepidoptera, in which the eggs of the female are ripe for laying at the time of emergence from the pupa. Male moths are provided with plumose antennae which bear olfactory sense organs and are specialized for perception of the female scent. By following the odour or the scent, they can locate the female from a distance of several miles and are readily attracted to other objects contaminated with the scent. In case of honeybee, the queen substance secreted by the mandibular glands serves also as a sexual attractant during the nuptial flight. In some Lepidoptera are found scent scales or androconia in males, the scent which they produce serves to excite the female to accept the courting male.

Transfer of sperms

The transfer of sperms to the female which takes place at mating is often an extremely elaborate process. The aedeagus or penis of the male after entering the vagina may actually penetrate up the duct of the receptaculum so that the sperms get mixed with the secretion of the accessory glands or ejaculated directly into the spermatheca. Occasionally, the sperms are discharged into the vagina or into the bursa copulatrix which arise from it. In most insects the sperms are not conveyed to the female in a free fluid but are enclosed in a saccular structure or "spermatophore" formed by the secretion of the male accessory glands. In Collembola and Thysanura these sperm bundles are just deposited at random on little stalks. On coming in contact with female vulva, the sperm bundles rupture and the sperms enter the vagina. In Lepidoptera, Orthoptera, Coleoptera, the spermatophore is a complex structure which is formed during copulation and transferred with its contents to the female genital tract. From the genital tract, the sperms or sperm bundles, as the case may be, are transferred to the spermatheca or receptaculum seminis.

Fertilization

Fertilization takes place just before egg laying or oviposition. Few sperms leave the receptaculum and enter the micropyle of the egg just as it passes through the receptacular duct. Several sperms usually enter the insect egg but only one of these is concerned with fertilization. On approaching the egg nucleus the sperm loses its tail and gets converted into male pronucleus. As soon as the male pronucleus is formed the egg nucleus divides twice, the first of which is the reduction division. The two divisions result in the formation of polar bodies and the female pronucleus. The latter then proceeds to fuse with the male pronucleus to form the zygote.

SPECIAL MODE OF REPRODUCTION

1. Viviparity: In those insects in which embryonic development is completed within the body of the female parent and which therefore produce larvae or nymphs instead of laying eggs, are known as viviparous insects. In certain cases in which the eggs are retained after fertilization, the embryo receives nourishment directly from the parent in addition to or instead of that present in the yolk. Such insects are regarded as truly viviparous and some anatomical adaptations are present in the parent or egg which facilitate the exchange of nourishment. In some viviparous insects, the eggs are retained and development occurs in the ovarioles while in others, the vagina is enlarged to form a uterus.

2. Ovoviviparity : Several insect species retain eggs, within the genital tract of the mother until the larvae are ready to hatch. Hatching takes place just before the eggs 'are laid. All the nourishment for the embryo is present in the egg and no special nutritional structures are developed. This type of viviparity is called ovoviviparity and it differs from normal oviparity only in the retention of the eggs.

3. Parthenogenesis: The phenomenon in which the eggs undergo full development without fertilization is known as parthenogenesis. It is very common among insects. Many grasshoppers and some Lepidoptera can occasionally develop from eggs which escape fertilization. Parthenogenesis is very common in Aphids which gives rise to a series of parthenogenetic viviparous generations of females in summer and in autumn, produce sexual forms, males and oviparous females which lay a small number of fertilized winter eggs.

4. Paedogenesis: The term paedogenesis means reproduction by juveniles i.e. immature forms. The ovaries become functional in the larval stage and eggs develop parthenogenetically. Ultimately, the body of the parent larva become filled with small daughter larvae that eventually break to the outside e.g. some flies, aphids, bugs and thrips.

5. Polyembryony : This term denotes the production of two or more embryos from a single egg. Polyembryony is very common in parasitic Hymenoptera. After the maturation of the egg the polar bodies may divide amitotically to form nutritive sheath, meanwhile egg nucleus undergoes complete cleavage and the separated cells each, develop into an embryo. Thus, two or more than two, sometimes hundreds of embryos arise from a single egg. The net effect of polyembryony is to increase the reproduction potential of the insect.

6. Hermaphroditism : In certain insects the male and female reproductive organs are found in the same individual. In the scale insects *Icerya purchasi*, in addition to normal males there are forms which are externally similar to the females of related species but are actually functional hermaphrodites, true females being absent. The hermaphrodite reproductive system is not unlike the usual condition but the gonad which is formed by the anterior fusion of two originally separate organs produce both sperms and eggs. The outer cells of gonad form ovarioles while the inner ones give rise to sperms. The eggs occasionally develop parthenogenetically into haploid males but are usually fertilized in the gonad by sperms from the same individual. The hermaphrodite do not copulate with each other.

7. Castration : Castration means any process which inhibits completely or to a considerable extent the production of mature sex cells by the organism, in this case, insects whose reproductive organs may be greatly atrophied or well developed but not normally functional. Castration may be due to some physiological conditions or due to the development of the parasite within the insect, causing sterility i.e. parasitic castration. The examples of parasitic castration are found in Aculeate Hymenoptera, Auchenorrhynchan Homoptera parasitized by Strepsiptera.

19

THE DEVELOPMENT

The development of the embryo takes place after oviposition but in some cases it has started and partly taken place in the parental body. In the case of viviparous insects, the whole development takes place within the insect body. The period of the egg stage after oviposition varies to a great extent e.g. it is only momentary as a larva emerges immediately in some cases. In others, it varies from a few hours to a few days, but in some extreme cases it is about 9 months and also it may last for two years. The study of the development of insects can be done under two divisions (1) Embryology or Embryonic development and (2) Post embryonic development.

EMBRYOLOGY

A number of workers like Johannsen and Butt (1941) and Hagan (1951) have given a detailed account of the embryonic development. Similarly the physiology of development has been studied by Pflugfelder (1952).

Structure of the egg (Fig. 55) In the eggs of most insects there is a clear distinction between the various regions of it e.g. the anterior and the posterior poles can be differentiated. In the ovarioles, the eggs are placed in such a position that the cephalic end of it is directed towards the head of the parent. Similarly the dorsal and ventral surfaces also correspond with those of the parent. A typical insect egg is contained in a shell or chorion which is variously sculptured or designed. The chorian is secreted by the follicular epithelium and it may consists of several layers. Beneath the chorian is the delicate cell wall of the ovum which becomes the vitelline membrane of the mature egg. At the anterior end of the egg, in most cases there is a minute aperture or sometimes a groups of pores or canals in the chorian, known as micropyle. It is through the micropyle that the spermatozoa enter for fertilization but in some cases the oxygen also diffuses into the egg through the pore or micropyle. The internal mass of egg is made up of two portions (1) Protoplasm and (2) Deutoplasm or yolk.

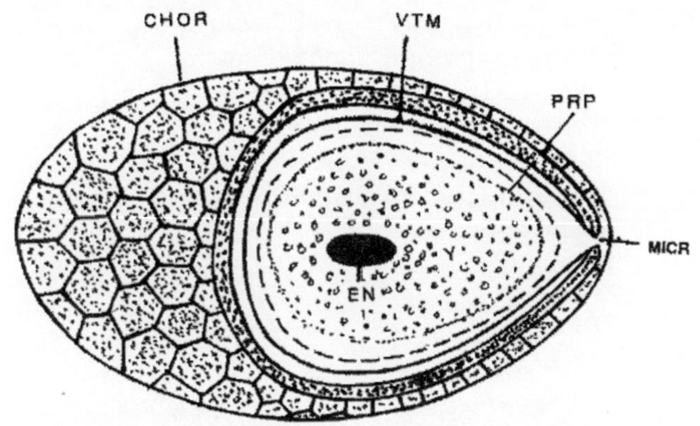

Fig. 55. Diagram of an insect egg.
Abbreviations: CHOR- chorian, EN- egg nucleus, PRP- periplasm, VTM-vitelline membrane, Y- yolk, MICR- micropyle.

The protoplasm forms the boundary layer of the periplasm lying just below the vitelline membrane thus completely surrounding the egg. It forms a reticulum which pervades the substance of the egg.

The deutoplasm or the yolk is present in large amount and is a mixture of fatty and protein materials. It is contained within the meshes of the cytoplasmic reticulum. In many eggs, in addition to these, there are also present some symbiotic micro organisms which are received from the mother. The unfertilized germinal vesicle or nucleus remains in the centre of the egg enclosed in a island of protoplasm known as nuclear cytoplasm. Maturation of the egg usually takes place after the entry of spermatozoa. During this process the egg nucleus moves towards the periphery where it undergoes meiotic division and the polar bodies are formed. After fertilization, the zygote nucleus again moves to the centre and then commences to divide into daughter nuclei. The typical shape of an insect egg, rich in deutoplasm is elongate, oval and that is the usual form of the egg in the ovarian tubule. On being laid, it may take on various shapes ranging from that of an elliptical, globose to that of a flattened disc.

Cleavage and Blastoderm formation

The zygote nucleus divides repeatedly to form a number of cleavage nuclei each one of which becomes surrounded by a mass of protoplasm giving a stellate appearance. These cleavage nuclei are at first scattered throughout the yolk but later on, when a large number of them have been formed,they travel towards the cortical layer or periphery where they arrange themselves in a single cell layer surrounding the yolk. This one cell thick wall, lying just below the vitelline membrane is the blastoderm. In some insects at this stage some of these cleavage nuclei become seggregated at the posterior pole of the yolk to form the future germ cells or which is the first organized form of the embryo.

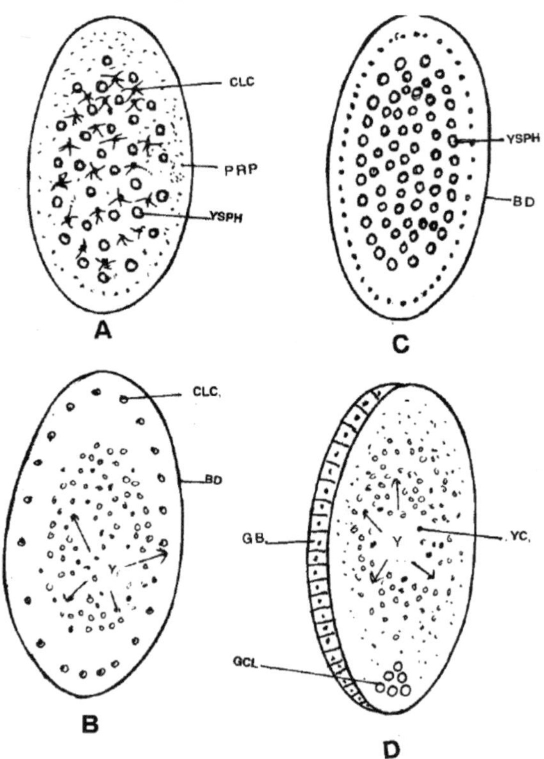

Fig. 56. A.B.C.D. Diagrams showing development of embryo and formation of germ band.

Abbreviations: BD- blastoderm, CLC- cleavage cell, GB- germ band, GCL - germ cell, PRP- periplasm, Y- yolk, YC- yolk cell, YSPH- yolk spheres.

These cleavage cells (Fig. 56 A,B), which remain in the yolk along with other cells, form the primary yolk cells or vitellophages. The vitellophages later on increase by the addition of secondary yolk cells from the blastoderm. These yolk cells are meant to liquify the yolk and thus make it fit for assimilation. In insects, blastoderm has completely surrounded the egg; the embryo is in the blastula stage and its cleavage cavity or the blastocoel is filled with the yolk.

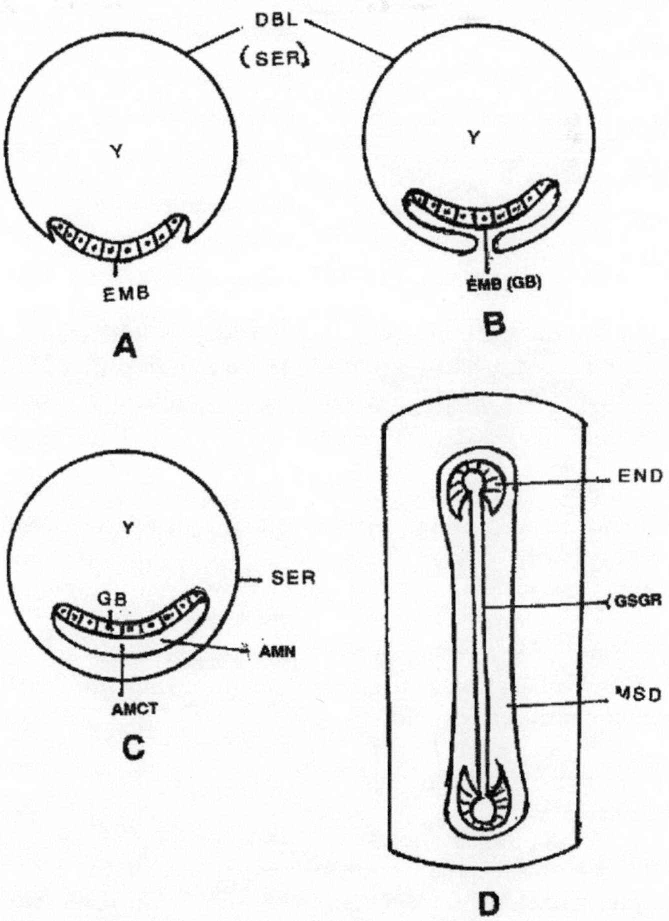

Fig. 57 A,B,C. Diagrams showing formation of embryo and its covering membranes **D**. Endoderm formation in *Leptinotarsa*.
Abbreviations: AMN- amnion, AMCT- amniotic cavity, EMB- embryo, DBL- dorsal blastoderm, GB- germ band, SER- serosa, Y- yolk, GSGR- gastral groove, END- endoderm, MSD- mesoderm.

Germ band formation

The first change that takes place in the formation of the embryo is the thickening of cells in that wall of the blastula which will form the ventral side of the future insect. Thus, in the ventral region, the cells are crowded together to make a thicker area called the "ventral plate" or "germ band" (Fig. 56D) and the rest of the blastoderm forms the extra embryonic area. In some cases, like Diptera, etc. where the egg is poor in yolk content the germ band is quite extensive so much so that the two extremities nearly meet on the dorsal aspects of the egg. In other cases, where the yolk is very abundant, the germ band forms only a small plate like area that may be termed as "germ disc" rather than a germ band. The extra embryonic area is composed of small flat cells and is on the dorsal aspect of the embryo. It may be termed as the dorsal blastoderm or serosa (Fig. 57 A-C).

Embryonic coverings and blastokinesis (Fig. 57 A-C)

The germ band becomes covered by amniotic fold arising from its edges. These folds gradually grow towards each other and ultimately meet and fuse thus enclosing the germ band in a space called amniotic cavity. Thus the germ band becomes covered with the double cellular layers, outer serosa and inner the amnion. It is the formation of the embryonic envelops and the various movements associated with it, which is termed as Blastokinesis. As a result of the formation of these coverings, the embryo gets separated from contacts with the egg shell either for a portion or throughout the whole period of develop-ment. There are 4 methods of separation and they are (1) by invagin-ation of the embryo, (2) by involution of the embryo, (3) by the form-ation of cellular protective membranes and (4) by the production of cuticular coverings.

Formation of the germ layers

In the insect embryology, the formation of the germ layers is more difficult to study. The main difficulty lies in the fact that the apparent gastrulation of insect embryo gives rise directly to a cell layer that forms the tissues. The primitive method of the forming of mesenteron by the typical invagination process of gastrulation does not exist. In the insect embryo blastopore is never an open aperture. There are different processes of formation of (i) endoderm and (ii) mesoderm in insects.

Endoderm: In most cases it appears as two widely separated masses of cells which are applied to the blind end of the stomodaeum and proctodaeum. These masses become `U' shaped, the arms of the U elongate and grow so that the two masses are just together in the median line and also extend laterally so as to eventually extend around the yolk and form a complete tube which constituted the mesenteron or primitive intestine (Fig. 57 D).

Mesoderm (Fig. 58 A-F) : In the Collembola, the first formed inner layer is the mesoderm. It is formed by a general tangential division of the blastoderm cells. The divisions taking place either at the time of cleavage or afterwards. But in the Pterygota, the surface of the germ band becomes differentiated into a median area or middle plate and into two lateral areas known as the lateral plates. From the middle plate is formed the mesoderm in the following ways: (1) in some cases by an infolding or invagination of the middle plate, (2) by an over growth of the median plate by the lateral plates, (3) by a proliferation of the cells from the inner surface of the middle plate.

In any case, the formation of mesoderm evidently represents a modified gastrulation, from which the endoderm appears to have been eliminated. Actually the middle plate more probably represents a common mesoderm endoderm rudiment, since a part of the endoderm is usually generated from the mesoderm or in close association with it.

Segmentation of the embryo : The two layered germ band of the embryo as it is now being called, at an early stage in the development becomes divided by means of transverse furrows into a number of segments. The segmentation may take place simultaneously with the formation of the gastral groove but generally it becomes clear after the separation of the inner germ layer. These segments ultimately number upto 20 in all. It is first of all divisible into two regions only viz., a protocephalic or primary head region and a protocormic or primary trunk region. The protocephalic segments are permanent, because of the larger lateral lobes, it is divisible into three segments viz. (1) preantennary, (2) deutocerebral or antennary and (3) tritocerebral or intercalary.

The first three primary trunk segments unite with the protocephalic segments giving rise to head appendages. Except the preantennary segments, all other five give rise to head appendages. The 2nd segment forms the antenna while the 3rd one becomes a pair of rudimentary appendages, and the first three trunk segments which form the head region bear the rudiments of the developing mandibles, maxillae and the labrum.

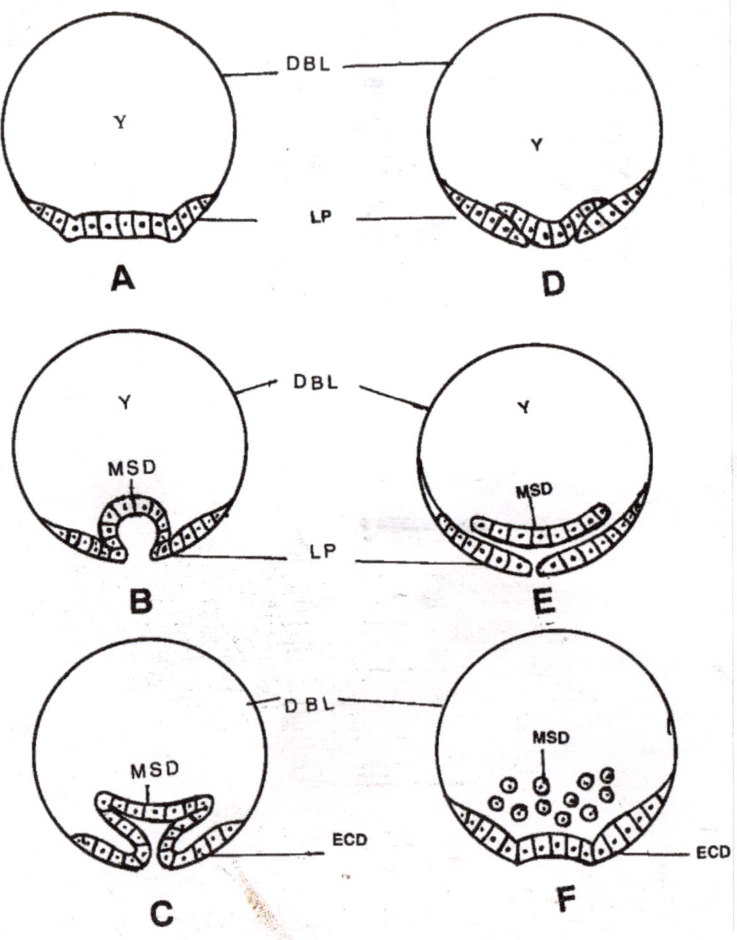

Fig. 58 A-F. Diagrams showing three methods of the formation of mesoderm. *Abbreviations:* DBL- dorsal blastoderm, ECD-ectoderm, LP- lateral plate, MSD- mesoderm, Y- yolk.

The next three protocormic segments i.e. the 4th, 5th, 6th one bear the rudiments of the future three pairs of thoracic legs and ultimately form the thorax. The remaining segments constitute the abdomen of the embryo. In most insect embryos, the abdominal region consists of 10 segments together with the telson. However, there are many reasons to believe that this number is 11 and along with the telson forms 12 segments. The abdominal segments may develop a pair of embryonic appendages which are usually reabsorbed before hatching except the last pair, which forms the cerci. In some orders, often the first pair is much more pronounced and later on takes up different shapes in the different groups and are known as pleuropodia and perform a variety of functions e.g. secretion of enzymes to dissolve the egg shell before hatching. In some it forms a sheath or a pair of elongate filaments (Hagan, 1951).

In the development of some insects, three embryonic stages can be distinguished on the basis of the degree of the development of the segmentation and the appendages of the abdomen. These appendages succeed one another more or less clearly (Berlese, 1930). The earliest stage is "protopod phase" where the segmentation is absent or indistinct and the appendages are present only on the head and thorax in rudimentary condition. The next stage is the "polypod phase" in which the abdomen is distinctly segmented. Each segment bearing a pair of appendages. The 3rd is the "oligopod phase" which is also clearly segmented but the abdominal appendages also appear except the Ist pair called pleuropodia and a few which modify into genital organ.

Organogenesis: With the development of the germinal layers, differentiation of the various organs and systems takes place.

Ectoderm : This gives rise to integument, wings, stomodaeum and proctodaeum, nervous system, cephalic glands, endocrine and exocrine glands, sense organs, oviduct and spermduct.

Mesoderm: This layer gives rise to circulatory system, fat bodies, oviducts, sperm ducts, muscles, connective tissues etc.

Endoderm: This layer forms the epithelium of ventriculus and hepatic caecae.

Eclosion: When the embryonic development is completed the hatching of embryo takes place which is known as "eclosion". At the time of eclosion, to facilitate emergence of the embryo, extra embr-yonic membranes, the serosa and amnion both rupture. The egg shell also ruptures. After breaking up of the egg shell, the larva or nymph, as the case may be, emerges with the help of the movement of the appendages.

20

THE METAMORPHOSIS

Insects like other Arthropods attain their maximum size by undergoing a succession of moults or ecdyses. The number of moults through which an insect passes is fairly constant for a species. The form assumed by the insect between any two ecdyses is termed as instar. Thus the life history of insects is made up of succession of instars culminating in the adult form. In many of the simple and generalized insects, the several instars are very similar to one another and only differ from their appropriate adults in the absence of wings and the incomplete development of their reproductive system. However, in the majority of the winged insects the adult is in sharp contrast to the wingless young stage. It is these forms that are said to undergo metamorphosis. It is thus clear that the term metamorphosis relates to the development of these insects in which the young stage is completely different in form, size and structure from the adult.

During larval development, there is usually no marked change in body form, each successive instar being essentially similar to the one preceding it, but the degree of change from last instar to the adult varies considerably and may be very marked. This change is called metamorphosis (Snodgrass, 1952, 1954, Wigglesworth, 1965). In physiological terms, it is a change which accompanies a moult in the absence of juvenile hormone. According to Highnam and Hill (1969) the overall change in form between the first larval instar and the adult is called metamorphosis.

TYPES OF METAMORPHOSIS

In the most primitive insects such as Collembla and Thysanura body proportions and internal organs remain similar after each moult or cedysis. Species under going such a development are categorized as Ametabela indicating no change in metabolism.

Most insects undergo changes in shape or form during their development, a process of metamorphosis. Some species have only minor modifications, but a gradation may be seen up to very drastic alterations e.g. from legless worm like immature individuals to the adults with legs and wings as seen in flies. The metamorphosis has been categorized in different ways. One classification includes insects that have immature with external wing pads, the exopterygota and species that have internal wing pads as larvae, the endopterygota. Another classification agrees with the previous demarcations but substitutes hemimetabola for exopterygota and holometabola for endopterygota. Still another view separates the exopterygota into paurometabola (for those whose immatures develop terrestrially) or hemimetabola (whose immatures inhabit water) and retains the endopterygota as a single unit, the holometabola. For the sake of convenience, here the terms (1) ametabola (2) hemimetabola and (3) holometabola have been used to illustrate development and metamorphosis.

1. Ametabolous insects

In most primitive insect groups such as Collembola, Thysanura and other Apterygota, no distinct external changes are evident except in size i.e. the youngs are practically identical with adults except in size and reproductive ability. So little change occurs in them during their growth that they are considered as having no metamorphosis. Hence, this entire group is called Ametabola. Examples of ametabolous insects are Spring tail, Silver fish etc. (Fig. 58 G).

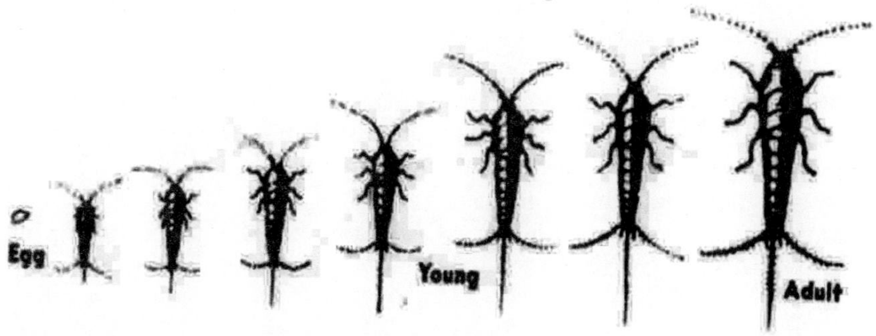

Fig. 58 G. Silver fish

2. Hemimetabolous insects

In hemimetabolous insects the metamorphosis is incomplete or gradual, the immature instars are usually similar to the adults in general appearance of body, mouth parts, antennae, compound eyes and legs and also in feeding habits. The immature forms differ from the adults in lacking wings and well developed reproductive organs. In winged hemimetabolous insects, the wings become visible as external wing pads, which gradually increase in size with each moult. In hemimetabolous type of metamorphosis the immature insects are called nymphs. Thus a nymph is a young insect which differs mainly from the adult in that the wings and reproductive organs are only present in an incompletely developed condition. Examples of Hemimetabolous insects belong to Hemiptera, Dermaptera, Dictyoptera, Orthoptera etc. (Fig. 58 H)

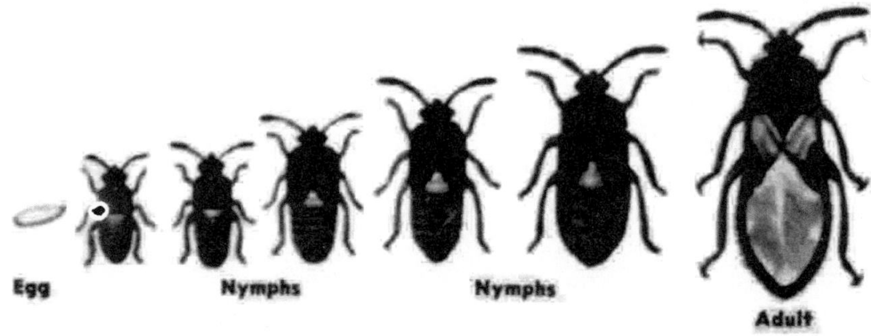

Fig. 58 H. Chinch bug

3. Holometabolous insects

In holometabolous type of metamorphosis, the wings are developed internally until the preadult instar, in which stage the wings are everted as large pads. The holometabolous metamorphosis is represented by three distinct growth froms during the post embryonic period viz. (1) the larva or a feeding stage, (2) the pupa or resting stage and (3) the adult the reproductive stage. The adults have fully developed wings and reproductive organs. Examples of Holometabolous insects belong to Lepidoptera, Diptera, Coleoptera, Hymenoptera, etc. (Fig. 58 I).

TYPES OF LARVAE

The larvae of Hemimetabolous insects essentially resemble the adults and they are sometimes called nymphs to distinguish them from the larvae of holometabolous insects. The most conspicuous difference between Hemimetabolous and Holometabolous larvae is in the development of the wings.

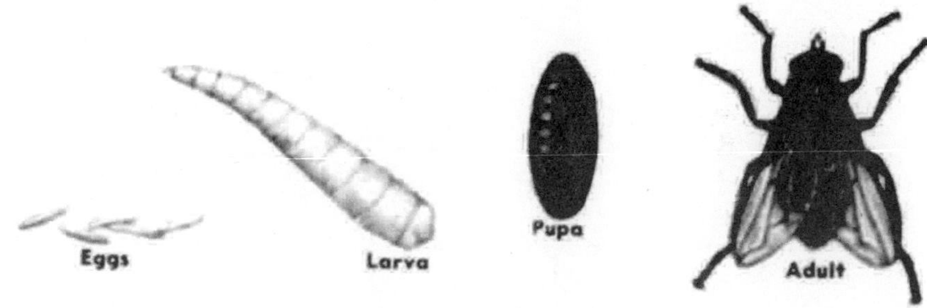

Fig. 58 I. House fly

In Hemimetabola, the wings develop as external buds which become larger at each moult finally enlarging to form the adult wings. In Holometabola, the wings develop as invaginations beneath the larval cuticle and so are not visible externally. The invaginations are finally everted so that the wings become visible externally when larva moults to pupa. There are many different larval forms amongst the Holometabolous insects viz.

(1) Protopod larva (Fig. 59A) : The protopod larva has many different forms and is often quite unlike a normal insect. The protopod larva resembles the protopod phase of the embryonic development and has incompletely segmented abdomen. These larvae hatch from eggs which contain very little yolk hence the young larvae hatch at a very immature stage of development e.g. parasitic Hymenoptera.

(2) Polypod larva (Fig. 59B) : This larva assumes the polypod phase of the embryonic development and in addition to thoracic legs, has abdominal prolegs. It is generally poorly sclerotized and is a relatively inactive form living in close contact with its food. The larvae of Lepidoptera, Mecoptera, are of the polypod type.

(3) Oligopod larva : This corresponds to oligopod phase of the embryonic development. The larva of this type has a well developed head capsule and mouth parts similar to adult but has no compound eyes. They are of two subtypes viz.

(a) Campodeiform larva: The larva is well sclerotized, dorsoventrally flattened and is usually a long-legged predator with a prognathous head. The compodeiform larvae occur in the Neuroptera, Trichoptera, Strepsiptera and some Coleoptera (Fig. 59C).

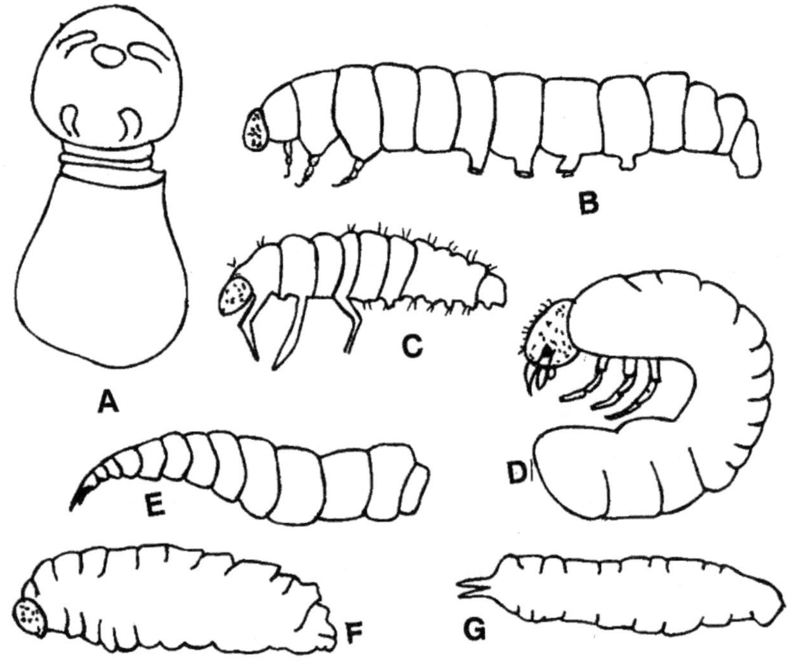

Fig. 59. Different types of larvae.
A. Protopod, **B**. Polypod, **C**. Campodeiform,
D. Scarabaeiform, **E**. Acephalous, **F**. Eucephalous, **G**. Hemicephalous.

(b) Scarabaeiform larva : The larva is fat with a poorly sclerotized thorax and abdomen and which is usually short-legged and inactive burrowing in wood or soil. The scarabaeiform larvae are found in the Scarabaeoidea and some other Coleoptera (Fig. 59D).

(4) Apodous larva : As the name indicates, the larva has no legs and is very poorly sclerotized. They have been classified into 3-subtypes on the basis of sclerotization of the head capsule viz.

(1) Acephalous: Without a head capsule, characteristic of Cyclorrhapha (Diptera) (Fig. 59 E).

(2) Hemicephalous : With a reduced head capsule which can be retracted within the thorax e.g. Tipulidae and Brachycera (Fig. 59 G).

(3) Eucephalous : With a well sclerotized head capsule as found in Nematocera, Buprestidae, Cerambycidae and Aculeata (Fig. 59 F).

TYPES OF PUPAE

The pupa is a resting or quiescent stage, a characteristic feature of all holometabolous insects. Pupae of some of the insects are quite active e.g. mosquitoes.

Three types of pupae have been recognized (Fig. 60 A-C).

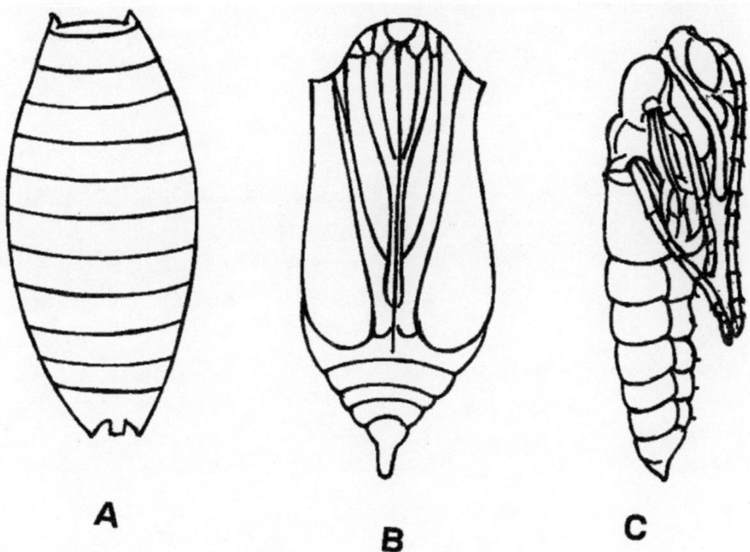

Fig. 60. Different types of pupae.
A. Coarctate, **B**. Obtect, **C**. Exarate.

(1) Coarctate pupa : (Fig. 60A) It is a free or exarate pupa enclosed in a puparium, formed from the preceding larval exoskeleton (cuticle). The puparium is barrel shaped and is segmented but there is no outward sign of appendages e.g. Cyclorrhaphous Diptera (Flesh flies, Blow flies, House flies etc.).

(2) Obtect pupa (Fig. 60 B) : In this type, the appendages are glued down to the body of the pupae by a secretion produced at the larval/pupal moult. The obtect pupae are usually more heavily sclerotized than the free pupae or exarate pupae e.g. Lepidoptera, some Coleoptera, (Chrysomelidae, Staphylinidae) Lower Diptera.

(3) Exarate pupa (Fig. 60C) : In this type, the appendages i.e. legs, antennae and wing-pads lie free and are not glued to the body e.g. Hymenoptera, Coleoptera, Neuroptera and Trichoptera.

Further, a number of insects retain parts or nearly complete portions of the previous moulted skin, this phenomenon is called the *pharate* condition. The pupae in general exhibit only limited movement and most are sessile, found in protected environment such as in the ground, under bark, in cases or retreats, fashioned by the larvae or in silken cocoons produced by last instar larvae.

Part - II

Applied Entomology

21

CHEMICAL CONTROL

Recent discoveries of new synthetic insecticides have sparked exciting advances and major breakthroughs in the control of insect enemies. Chemicals have subdued pests that once caused wide spread crop destruction, death of domestic animals and epidemics of insect borne human diseases. The modern insecticides are both effective and reliable. The whole world is resorting to them more and more for the solution of many insect problems.

The chemicals which kill the insects by their chemical action are termed as insecticides. They are used for the protection of men, domestic animals, crops, agricultural products from the attack of insects when other methods fail to control the pests. Insecticides are seldom used in full strength but are formulated in a variety of ways to make them easier for application.

FORMULATIONS OF INSECTICIDES

Following are the common formulations of insecticides viz., (1) dusts, (2) granular formulations, (3) insecticide-fertilizer mixtures, (4) wettable powders, (5) solutions, (6) emulsifiable concentrates, (7) aerosoles, (8) fumigants and (9) miscellaneous formulations.

Whatever may be the formulations, the poisonous chemical present in an insecticide must penetrate the vital organs and tissues of the insect and ultimately kill it.

(1) Dusts : Insecticidal dusts are those powders which are used dry and mixed with or impregnated with certain organic materials or pulverized minerals (powders) such as talc, pyrophyllite, bentonite etc. These minerals are called carriers or vehicles since they carry the insecticide. Dusts are blown to deposit on plants by dusting machinery or blowers. Ground to a fine size, most dust will pass through a 325 mesh screen and range in size from 1 to 40 μ. The finished dust may be 0.1 to 25% active material. In dust form, in general, the toxicity of an insecticide increases as the particle size decreases.

(2) Granular formulations : These are similar to dusts except for larger particle size. The range of particle size in a granular product is designated by a two figures mesh classification e.g. 30/60 means that virtually all the insecticide granules will pass through a standard 30-mesh sieve while a negligible quantity will pass through a standard 60 mesh sieve. Some of the common granular formulations are 16/30, 20/40, 24/48 and 30/60. Granular insecticides are generally used as dressings on or in the soil and may be applied with fertilizer spreaders or special granule applicators.

(3) Insecticide fertilizer mixtures : Insecticide fertilizer mixtures may be formulated by adding granular insecticides to commercial fertilizers or by spraying insecticides directly on to the fertilizer. Such mixtures are applied at the regular fertilizing time to provide both plant nutrients and control of soil insects.

(4) Wettable powders : Theses are similar to dusts but they are meant to be diluted and suspended in water and used as spray. To make an insecticidal dust act in this manner a dispersing and wetting agent is added to the formulation. They are more concentrated than dust as they may contain as high as 75% toxicant.

(5) Solutions : Many of the modern synthetic insecticides are insoluble in water but soluble in organic solvents. These soluble insecticides in solution form are used directly for insect control. They are however seldom used on plants because of their phytotoxic reaction.

(6) Emulsifiable concentrates : The most common and versatile formulation is the emulsifiable concentrate. This formulation consists of an insecticide, a solvent (for the insecticide) and an emulsifying agent. Mixing the concentrate with water forms an emulsion. The solvent used may evaporate quickly after spraying, leaving a deposit of toxicant after the water has evaporated.

The use of an emulsifying agent serves several purposes; it makes possible the diluting of a water insoluble chemical with water, it reduces the surface tension of the spray thus allowing it to spread and wet the treated surface more effectively, helps the spray and make a better contact with the insect cuticle.

Generally oil soluble emulsifying substance is used. Normally emulsions are unstable and break up into their component parts. This action is termed breaking. For spraying on plants a quick breaking mixture is preferred. Since this results in heavier deposits of toxicant.

(7) Insecticidal aerosols : Aerosols are minute particles suspended in air or fog or smoke. The diameter of these particles range from 0.1 to 50 μ. The dispersion of insecticide into aerosol form may be accomplished by burning, vapourising with heat, automizing mechanically or releasing through a small hole an insecticide that has been dissolved in a liquified gas. The last method of aerosol preparation gives the popular aerosol bomb.

(8) Fumigants : Insecticides used in the gaseous form are known as fumigants. Fumigants are most often formulated as liquids under pressure and are held in tanks. When the liquid is released in open air it changes back to a gas, quite often fumigants are a mixture of two or more gases.

(9) Miscellaneous formulations: There are certain insecticides which are special formulations meant for specific uses. Often insecticides are stuffed in large pills and capsules and introduced into the stomach of animal. Insecticides may be mixed in shampoos, intended for use on house pets, waxes for use on floors may contain an insecticide. Poison baits consist of toxicants combined with a food stuff attractive to the insect pests. All these types of insecticides are special formulations designed for special purposes.

SYNERGISTS

Some chemicals have the property of greatly increasing toxicity of certain insecticides. When the increased toxicity is markedly greater than the sum of the two used separately, it is termed a synergistic action. Most synergists have been used with pyrethrum or allethrin etc. Synergistic action is important because it provides a means for a more effective insecticide and it reduces the cost of control.

CLASSIFICATION OF INSECTICIDES

The foregoing types of insecticide formulations are all lethal, effective in killing the insects sooner or later. Generally the insecticides are broadly classified on the basis of the mode of their entry into the insect body as stomach, contact and fumigant poisons. Modern classification however does not follow this basis since the present day synthetic insecticides enters the insect's body in more than one way. To avoid this overlapping, modern classification is now based on the chemical nature of insecticides and the major divisions are therefore inorganic and organic. The organic insecticides are further divided into oils, botanicals and synthetics. Oils are petroleum products used primarily against fruit tree insects. Botanicals are of plant origin and include such insecticides as Rotenon, Pyrethrum etc. Among the synthetic group of organic insecticides, several chlorinated hydrocarbons, organophosphorus compounds etc. used in the past are now banned.

With regards the older classification, stomach poisons are materials which are ingested by the insects and which kill primarily by the action on or absorption from the digestive system. Usually they are limited to the control of chewing insects. Contact poisons are absorbed through the body wall and must come in direct contact with the insects to kill them. They are usually adopted for destroying sucking insects. Fumigant poisons, enter the tracheal system of the insects in the form of gas. Insects such as stored grain pests and certain soil insects are killed by the fumigants.

1. INORGANIC INSECTICIDES

Though inorganic insecticides have been largely replaced by organics, some of the forms are still used to control certain insects. Most of the inorganic insecticides are stomach poisons. Inorganic materials used in insect control are chiefly compounds of Arsenic, Fluorine and Sulphur.

Note: Most of the arsenical and fluorine compounds have been banned for domestic and agricultural use in many countries. The main compounds are: 1. Lead arsenate, 2. Calcium arsenate, 3. Paris green or aceto arsenate of copper, 4. White arsenic, 5. Zinc arsenate, 6. Sodium fluoride, 7. Barium fluosilicate and 8. Sodium fluoaluminate.

Sulphur Compounds

Amongst the sulphur compounds finely ground sulphur has been widely applied as a dust for control of mites and other insects.

Sulphur is boiled with freshly slaked or hydrated lime to produce a mixture known as liquid-lime sulphur, one of the chief constitutents of this mixture is calcium polysulphide. Liquid lime sulphur is used in the control of scale insects, aphids, etc. on fruit trees.

Another important inorganic insecticide is Bordeaux mixture which is prepared by mixing copper sulphate, lime and water in a definite proportion. This is chiefly used as a insect repellent against cucumber beetles and leaf hoppers.

2. ORGANIC INSECTICIDES

1. Oils

These form one group of organic insecticides which in their natural stage are highly phytotoxic. However, when they are used in an emulsion they may under certain conditions be safely applied to plants. They vary greatly in the degree of refinement and in physical properties according to their intended use. Oils are employed in a number of ways; they may be used as solvents or carriers for insecticides. Oils by themselves are insecticidal and there are different types of oils.

(a) Summer oils are highly refined and less phytotoxic. They are generally used to control mites and scale insects.

(b) Dormant oils are less refined and are also applied to control scale insects.

(c) Superior spray oils are chiefly meant for controlling Aphids.

(d) Emulsible oil containing 95-99% oil and emulsifier.

(e) Tank mix oils are made up of oil, emulsifier and water added separately to a spray tank. The last two are used for the protection of agricultural crops.

The use of oils has several advantages, initially it is cheap, have a good spreading capacity, are easy to mix and quite safe to animals life. Another important consideration is the fact that insects have not developed so far any resistance to oils. However, there are certain disadvantages with oils e.g. they have a low toxicity to insects. They are unstable and extremely phytotoxic.

2. Insecticides of Plant Origin

These are insecticides derived from plants and majority of them are contact poisons. They have several uses in insect control. The most important insecticides of plant origin are nicotine, pyrethrum, rotenone, sabadilla, ryania and neem, all of which serve as insect toxicants.

(1) Nicotine : It comes chiefly from two plants, *Nicotiana tabacum* and *Nicotiana rustica.* They are extracted from the leaves and stems of these plants. Nicotine is highly toxic to a great number of insects, acting on the ganglia of the central nervous system possibly at the synapses. It is toxic when ingested, absorbed through the body wall or taken in through the trachea. The most commonly used form today is nicotine sulphate at 40% content. Free nicotine solution is also used as a fumigant or greenhouse spray. Nicotine insecticides are used as 0.04% sprays, 3% dust, fumigants and dips. Water or summer oil may be used as a diluent in sprays and hydrated lime, ash or sulphur for preparing dust. Nicotine sprays and dusts are safe to use on plants. They are compatible with most other insecticides.

(2) Pyrethrum : This derives its toxicity from 4-esters: pyrethrin 1, pyrethrin 2, sinerin 1 and sinerin 2. It is a plant product extracted from the flowers of *Chrysanthemum cinerariaefolium.* A synthetic compound recently developed known as allethrin has the same toxicity as pyrethrum.

It is prepared by extracting the toxins with kerosene, alcohol or ethylene dichloride from flowers of the *Chrysanthemum.* This concentrated form, may be diluted with petroleum oil and used as sprays. Concentrated extracts as acetone, alcohol or a hydrocarbon solvent are available for spraying plants. These are diluted with water to form aqueous emulsion. Pyrethrum is harmless to warm blooded animals and man but is highly toxic to most insects. It is used for controlling the houseflies and mosquitoes. It acts rapidly as a nerve poison and the affected insects fall quickly on to the ground. This is termed 'knock down' and the insects thus knocked down need not necessarily die, some may recover after a lapse of time. Often the toxicity of pyrethrum is enhanced by the addition of chemicals such as sulphoxide or piperonyl butoxide to the insecticide. Such chemicals which increases the toxicity of an insecticide are known as "Synergists".

(3) Rotenone : This is prepared by grinding the dried roots and bark of several species viz. *Derris, Lonchocarpus* and *Tephrosia.*

Rotenone may be extracted with a solvent like chloroform or carbon tetrachloride and a concentrated extract is then prepared. This may be diluted in sprays or mixed with a dust carrier and used as an impregnated dusts. It is nontoxic to warm blooded animals and man. It is also non-phytotoxic. But it is poisonous to cold blooded animals. Alkaline carriers such as lime should not be used with it. Rotenone sprays and dusts loose their toxicity rapidly on exposure to light and air. It is a selective insecticide and can kill only some insects i.e. rotenone is very specific in its toxic action to insects. It is believed to kill the insects by inhibiting the utilization of oxygen by body cells or depriving the tissues of oxygen through a depressing reaction on the nerve and muscles connected with tracheal breathing. They are used in combating cattle grubs, lice, aphids and beetles.

(4) Sabadilla : It is derived from the seeds of *Schoenocaulon officinale*. It is sold as a finely ground powder diluted with an inert filler. A 50% concentrate is manufactured. This should be diluted and applied as a dust at 10 to 20% sabadilla content. Sabadilla dust is strongly irritating to the nasal memberanes causing sneezing and eye irritation. These effects are however transitory and soon disappear after exposure. Sabadilla is most useful as contact dust to control Hemipterous insects which are resistant to other insecticides. It is very specific in its action and is ineffective against many other common insects.

(5) Ryania : It is obtained from the stems and roots of a plant called *Ryania speciosa*. This is used for the control of sugarcane borers and certain Lepidopterous insects. It acts both as a contact insecticide, as well as a stomach poison. It has residual effect and is relatively stable.

(6) Neem (*Azadirachta indica* A. Juss): A lot of work has been done in the past few decades and several seminars and symposia were held to evaluate the insecticidal properties of various parts of the neem tree viz. leaves, flowers, fruit kernels in the powder or oil form to control insect pests infesting various crops and crop-products. It has been stated that the Neem is an important source of insecticide and all its parts viz., leaves, bark, flower, kernel and kernel oil are variously used as insect-repellents. Since ancient times, its leaves are kept amongst woollen and silken clothes and books to protect them against insects. Farmers throughout the country are aware of the need to protect the grains from spoilage during storage and use of neem leaves in grain heaps.

The oil is also used as a disinfectant and emulsifying agent in insecticides. The maximum insecticidal activity has been found in neem kernel. Several workers have tried the neem leaves and kernel powder as grain protectant. Mixing of seeds with neem powder at 1.0 and 2.0% effectively protected from pest infestation. Neem oil at 0.5 and 1.0% provided the best surface protection when applied to pulses against the pulse beetles *Callosobruchus chinensis* and *C. maculatus*. Repellent effect of extracts of neem leaves, flowers and kernels against rust red beetle *Tribolium castaneum* and Khapra beetle *Trogoderma granarium* has been evaluated and found that neem kernel extract was more repellent than the leaf or the flower. Adults of *T. castaneum* failed to reproduce when fed with flour treated with neem kernel extract. Neem oil was found effective in preventing adult emergence and prolonging the development from egg to adult stage of *Rhizopertha dominica*.

Chopped leaves of neem were found effective in controlling the *Sitotroga cerealella* in stored rice. In addition to stored grain pests, numerous other pests of agricultural importance have been controlled by neem products in the form of powder, seed cake or aqueous and solvent extracts.

Several commercial formulations have been evaluated against pest of tobacco, groundnut, black pepper, pulses, cotton, oil seeds, vegetable and other agricultural crops. Neem plant parts, also act as growth regulator, repellent and antifeedant, thus help to control crop pest population without altering the quality and nutritive value (For details see : Narwal *et al.*, 1997, and Koul and Wahab, 2004). One of the most potent derivatives of Neem is Azadirachtin.

Powdered flowers and fruit kernel powder treated rice, caused mortality of *Sitotroga* adults, 59-64% after 7 days, 2-8% powder was used. As pointed out above neem oil 1-2% was found most effective in reducing adult population of *Sitophilus, Rhizopertha* and *Ephestia*. The neem kernel powder when mixed with cowpea *Vigna sinensis* at 3%, caused 85-95% mortality of *Callosobruchus* (For details see : Singh *et al.*, 1997).

3. Synthetic Insecticides

(A) Chlorinated Hydrocarbons: Some commonly used chlorinated hydrocarbon insecticides are described as follows :

(1) Dichloro-diphenyl trichloroethane (DDT)* : It is a very powerful insecticide, which exercises a prolonged lethal residual effect till covered by the dust or dechlorinated. It is a creamy white, crystalline powder having a faint unpleasant odour. It is waxy to touch, quite stable, resistant to oxidation and light. It is almost insoluble in water but soluble in organic solvents like castor oil, cotton seed oil, xylene, benzene, kerosene, dioxane, ether, acetone, chloroform, carbon tetrachloride, liquid paraffin etc. It is not a repellent hence, insects do not avoid it but it acts as stomach as well as contact poison. The effect is gradual but sure and persists for a long time. It is non-phytotoxic except to cucurbits. It is very effective against most insects such as beetles, weevils, caterpillars, thrips, whiteflies, and leaf hoppers. It is compatible with most insecticides and fungicides, except lime sulphur. Considering the effective insecticidal dose, it is comparatively less toxic to mammals (a human fatal dose taken orally is 30 gms). Insolubility in water renders it harmless to use on human skin in dust or aqueous emulsion but when dissolved in oils, its large doses may cause a number of diseases like convolution, cerebral depression, and necrosis of liver. Its inhalation is also harmful but due to its high toxicity for insects, large areas can be treated with small quantities of insecticides by ordinary hands or power sprayers. It is never used in the pure form but compounds are made with certain bases to form dust or spray. The dose of DDT ranges between 0.5 to 20% in mineral oils, 0.1 to 5% of emulsified dispersible or wettable powder in water and 2 to 10% in dust mixed with talc, clay, chalk, sulphur, and other carriers. Most plants, except cucurbits do not suffer by treatment with DDT. The DDT analogue, DDD (dichlorodiphenyl dichloroethane) is less toxic to mammals and is used for flies and vegetable pests. The acute oral LD50 for male rats is 113 mg/kg. DDT is stored in the body fat of birds and mammals and excreted in the milk of mammals.

(2) Benzene hexachloride (BHC) Gammexane* : It is a crystalline tasteless, perfectly stable powder of sand colour, having a camphor like odour, practically insoluble in water but soluble in oils and paraffins. It is available as a pure product or mixed with a carrier like lime, gypsum or chalk. It is perfectly compatible with sulphur and DDT but it decomposes, when mixed with alkalies like lime, Bordeaux mixture and nicotine etc. It is five times toxic than DDT hence acts more quickly but being volatile, the residual effect is less

---

* Now banned.

lasting. It can serve both as stomach and contact poisons for most of the insects such as thrips, leafhoppers, aphids, woodlice, and earwigs but less effective against bollworms, citrus fly and is more or less ineffective against flour moths. It can be used as a slow acting fumigant for stored products alone or in combination with carbon disulphide, carbon tetrachloride etc. It is less toxic to mammals when taken orally or absorbed to skin. A fatal oral dose for man is 30 gms (its gamma isomer is more toxic and 15 gms is fatal dose for man).

(3) Lindane : It is the coined name for the gamma isomer of BHC of a purity of not less than 99%. Using lindane in formulation, eliminates the disagreeable odour and tastes on some crops which result when they are treated with BHC. It exhibits strong stomach poison action, persistent contact toxicity and fumigant action, against a wide range of insects. It is non-phytotoxic at insecticidal concentrations. The technical BHC causes "tainting" of many crops but there is less risk of this with lindane. It is stable to air, light, heat and carbon dioxide and is not attacked by strong acids but can be dehydrochlorinated by alkalies. It occurs as colourless crystals and is practically insoluble in water, slightly soluble in petroleum oils, soluble in acetone, aromatic and chlorinated solvents. The acute oral LD50 for rats is 88 mg/kg. It is effective against many soil insects e.g. beetle larvae and adults, fly-larvae, ants and many biting and sucking insects e.g. aphids, psyllides, whiteflies, capsids, midges, thrips etc. This insecticide is now restricted in use. Trade names: Lindane, Lintox etc.

(4) Aldrin*: It is a white crystalline solid, insoluble in water but is quite soluble in mineral oils and highly soluble in most organic solvents like acetone, benzene, xylene etc. It is stable to heat, alkaline and mild acids and can be mixed with alkaline soils, fertilizers and other insecticides like sulphur, Bordeaux mixture, calcium arsenate, nicotine, paris green, rotenone etc. It is a broad spectrum, persistent, nonphytotoxic insecticide. It acts as a stomach as well as a contact poison. Toxic effect on insects is cumulative and therefore it is not advised to use on edible crops. It is used against all soil insects like termites, beetles (both larvae and adults) fly larvae, cutworms, wireworms, crickets etc. Aldrin is dusted on soil at the rate of about 2.5 kg per acre. Aldrin can control a variety of soil insects at the depth of about 10-11 cm. It is used as 30% emulsifiable concentrate, 2.5-5% dust, 5 and 20% granules, 20-25% wettable powder, or liquid

* Now banned

seed dressing. The treated seeds should not be used for human or animal consumption.

Insecticidal preparations containing aldrin, should be handled with care. Respirators should be worn by those applying this material and excessive skin contact should be avoided. Acute oral LD50 for rats is 67 mg/kg. it is absorbed through the skin.

(5) Endosulfan : It does not accumulate in milk or fat but has a high mammalian toxicity. In plants and animals both, it is metabolized to a sulphate. It is extremely toxic to fish but is safe for bees. It is formulated as emulsifiable concentrate, wettable powder, dust and granules. Trade names: Thiodan, Endosel etc.

Note: Other products viz. - dieldrin, endrin, chlordane, toxaphene and methoxychlor etc. were used in the past but currently banned.

(B) Organo Phosphorus Compounds : These are extremely poisonous to man and involves special care, precautions, apparatus and protective clothings and rubber gloves for handling them. All systemic insecticides belong to this category. Those still in use are given below:

(1) Malathion: It is a non-poisonous, non-phytotoxic insecticide. The chemical is yellowish brown liquid slightly soluble in water and miscible with most of the organic solvents. It has a strong penetrating, unpleasant smell and is broken up into non-toxic ingradients when mixed with alkaline media. It can be used as stomach and contact poison, besides as a strong fumigant for mites and other insects which are not killed by other poisons. It is non-toxic to man, 100 times less than parathion either orally or by skin absorption or by vapours. It can be used as contact poison for aphids, scale insects, mealy bugs, jassids and caterpillars in 5% dust or 25-50% emulsion for spraying. Two or three applications fortnightly will control white flies, mealy bugs and scale insects. Acute oral LD50 for rats is 2800 mg/kg. Trade names - Malatox, Cythion etc.

(2) Methyl Parathion : It is a white crystalline substance in pure state, sparingly soluble in water and well soluble in most organic solvents. It has a low thermal stability. It hydrolyzes rapidly in alkaline medium. Due to its low stability, toxicity is lost soon from the treated surface. In soil, micro-organisms decompose it within a week. It is a contact and stomach poison with mild fumigant action. It has penetrating action also. Within 3-4 days 80-90% of it decomposes. It can be used against aphids, mites, leafhoppers, whiteflies, scale insects etc. It is available as emulsifiable concentrate and dust. Trade names: Folidol, Dalf etc.

(3) Trichlorphon* : It is initially used for controlling house flies, acts as a contact as well as a stomach poison. It is a white crystalline powder, quite soluble in water, insoluble in petroleum oils, poorly soluble in carbon tetrachloride, diethyl ether, benzene and ethanol. It is stable at room temperature but is decomposed by water at higher temperatures and acid media to form dichlorvos. Manufactured in various forms like wettable powder (50%) soluble powders (50 to 80%), granules (2 to 5%), emulsifiable concentrate (50%) and mixed with sugar crystals. Used for controlling ticks, mites, house flies and fly maggots and bugs. Trade names: Dipterex, Tugon etc.

(4) Diazinon: It is a non-systemic insecticide with some acaricidal action. It may be phytotoxic at higher dosages. It is a colourless oil almost insoluble in water but is miscible with ethanol, acetone, xylene and is soluble in petroleum oils. It decomposes above 120°C and is susceptible to oxidation.It is stable in alkaline media but is slowly hydrolyzed by water and dilute acids. It is compatible with most insecticides but should not be compounded with copper fungicides. The acute oral LD50 for male rats is 108 mg/kg. Effective against flies and fly-maggots, mites, thrips, aphids etc. Formulated as aerosol-solutions; 25% emulsifiable concentrate, 40 and 25% wettable powder, 4% dust and 5% granules. Trade name: Basudin.

(5) Dichlorvos : This insecticide has combined actions of contact, stomach and fumigant. On account of its high vapour pressure, it has greater penetration and it gets readily decomposed. When applied as emulsifiable concentrate, it is effective against sucking and leaf-mining insects. It is non-phytotoxic and non-persistent. It is also used as a fumigant against mosquitoes and other dipterous insects. It has low dermal toxicity but it is toxic to bees. It is marketed in the form of an emulsifiable concentrate. Trade names: DDVP, Nuvan etc.

(6) Monocrotophos : It is a crystalline solid with mild ester smell. It is miscible with water, soluble in acetone and ethyl alcohol, but insoluble in kerosine and diesel. It is corrosive to iron, steel and brass.

---

* Now banned

It is a fast acting stomach, contact and systemic insecticide. It can be used against a wide range of pests. It is absorbed by roots and then translocated to stem and leaves. Scorching of edges occur in some varieties of sorghum, apple and cherry. Monocrotophos is recommended for the control of aphids, white flies, leaf miners, thrips, shoot borers, boll worms, caterpillars, folige feeder-beetles and a number of other pests. It is formulated as emulsifiable concentrate. Trade names: Nuvacron, Corophos etc. Its use is restricted.

(7) Fenitrothion : It is a brownish yellow liquid with faint odour. It is insoluble in water but soluble in most organic solvents. Fenitrothion is slowly hydrolyzed in the presence of alkali. Its thermal stability is also low and may explode at 100°C. Iron container encourages decomposition. It is a contact and stomach insecticide with penetrating property. It is a low toxic poison. It decomposes in plants quite actively. It is highly effective in controlling sucking pests like aphids, jassids, bugs as well as caterpillars and flies. Used in the form of 5% dust or 50% emulsifiable concentrate. Trade names: Folithion, Fenitox etc.

(C) Carbamate insecticides

Carbaryl : It is a contact insecticide with slight systemic action. It is widely used against the insect pests of fruits, vegetables, cotton and other crops. It does not deteriorate in storage and has a long residual effect on the crops. It is comparatively safe for insect pollinators, parasites and predators. It is formulated as dust and wettable powders. Trade names: Sevin, Carbaryl etc.

3. SYSTEMIC INSECTICIDES

These are the insecticides which are absorbed into the sap stream by the plant tissues on application to leaf or root system and are then transmitted to various parts in such quantities as to render its juices sufficiently toxic to certain insects specially sap suckers for several weeks. These are therefore termed as systemic insecticides. Their use in food and fodder crops cannot be recommended but are extremely useful for cotton and ornamental plants. The systemic insecticides can be safely used against phytophagous insects but not against predators and parasites. They may be applied as sprays or granules to either soil or foliage. The concentration is determined with respect to specific pests.

It is important to note that a large number of organophosphorus compounds act as systemic insecticides. These, as a rule, are highly poisonous substances hence full protective clothing must be always worn. Although, majority of them are non-phytotoxic but many are harmful to bees, livestock, fish, game and wild animals. They are mostly used against sap sucking insects and phytophagous mites. Following are some of the common systemic insecticides

(1) Dimethoate

This insecticide is used mainly against fruit flies and aphids. The pure compound is a white solid, only slightly soluble in water but soluble in most organic solvents except saturated hydrocarbons. It is stable in aqueous solution and to sunlight but it is readily hydrolyzed by aqueous alkali. It is incompatible with alkaline pesticides. The acute oral LD50 for rats is 250-265 mg/kg. It is effective against aphids, psyllids, some flies, sawflies, mites etc. It is marketed in the form of 20-40% emulsifiable concentrate, 20% wettable powder and 5% granules. Trade names: Rogor, Dimetox etc.

(2) Demeton-o-methyl (Methyl demeton)

It acts as a systemic as well as a contact insecticide and acaricide, with a fast killing ability and moderate persistence. It is a clear yellowish oily liquid insoluble in water and soluble in most organic solvents. The acute oral LD50 for male rats is 65 mg/kg. It is effectively used against aphids, mites on most crops, some flies and has only a limited effect on brassica aphids. It is used as emulsifiable concentrate. Trade name: Metasystox. Its use is restricted.

(3) Phorate

It is a persistent insecticide used in granular and emulsifiable concentrate formulations for protection of seedlings from sap feeding and soil insects. It is a clear liquid only slightly soluble in water but miscible with carbon tetrachloride, dioxane, xylene, and vegetable oils. It is hydrolyzed by alkalies in the presence of moisture. The acute oral LD50 for male rat is 3.5 mg/kg, for females 1.6 mg/kg. Used in the form of granule. Trade names: Thimet, Foratox etc.

(4) Phosphamidon

This insecticide is readily absorbed by the plants but with only a little contact action. It is non-tainting and non-phytotoxic except to some cherry varieties and sorghum varieties. It is a pale yellow oil miscible with water and is readily soluble in most organic solvents except saturated hydrocarbons. It is stable in neutral and acid media but is hydrolyzed by alkali. Compatible with all but highly alkaline pesticides. It corrodes iron, tinplate and aluminium and is packed in poly ethylene containers. Acute oral LD50 for rats is 28.3 mg/kg. It is effectively used against sap feeding insects and leaf eating ones, particularly aphids, bugs many caterpillars (but not Noctuidae) rice stem borers, thrips, beetles, grasshoppers, saw flies, fly larvae and mites. Used as 100% emulsifiable concentrate. Trade names: Dimecron, Daron etc. Its use is restricted in agriculture.

(5) Quinalphos

It is a white crystalline solid without any odour. It is sparingly soluble in water but readily soluble in most of the organic solvents. It gets readily hydrolized in alkaline medium. It is a strong contact and stomach insecticide and acaricide. It is effective against a wide range of insects, particularly sucking and caterpillar pests. Used in the form of 4% dust, 25% emulsifiable concentrate and 5% granule. Trade names: Quinatox, Ekalux etc.

4. FUMIGANTS

These are respiratory poisons taken into the tracheae in gaseous form and exert their influence when absorbed into the tissues from tracheoles. The other fumigants are suffocating agents which simply clog the spiracles and slowly kill the insects by cutting oxygen supply. Such chemicals can be applied to plants in air tight chambers, green houses and nurseries and to plant products and grain houses, mills, granaries etc. Time for fumigating granaries, mills etc. may be prolonged for 6 to 32 hours. It is extremely essential that no one must be allowed to enter the chamber in which operations are being carried out and provision must be made to open several doors, windows and shutters from outside without entering it.

(1) Carbon disulphide

This gas is highly inflammable and explosive when mixed with air, hence its use involves a great risk. Even a lightened cigarette or a spark from an electric switch is enough to ignite it. It is a yellowish or water white, heavy liquid with a disagreeable odour. Because of the fire and explosion hazards, the carbon disulphide should not be used in fumigation, if other fumigants can be obtained. It is very toxic to insects and plants. For seed and grain fumigation carbon disulphide is applied to the top layer of grain, most effectively in the form of a coarse spray from a ground spray pump. As the vapour is heavier than air, it sinks down through the grains killing all form of insect life. The time of exposure is 36 to 60 hours.

(2) Nicotine

It is used for fumigating green houses in humid conditions at dark hours. It can be obtained as such by heating nicotine to liberate the poisonous gas or paper soaked in liquid nicotine and dried may be burnt. Tobacco plants on roasting also emit the same gas. The acute oral LD_{50} to rats is 50-60 mg/kg. It can be used as contact insecticide with some ovicidal properties. It is effective against aphids, capsids, leaf miners, thrips, sawflies. Also used for fumigation of glass houses. Marketed as 95% alkaloid or as nictotine sulphate (40% alkaloid) also as 3-5% dust.

(3) Ethylene dibromide

It is very heavy liquid which is used as a soil fumigant to control wireworms. It is sold in solution with a petroleum oil and contains either 20 to 40% by weight. It is poisonous to man when inhaled as vapour or in contact with skin causing burning and irritation. It is inflammable so should be stored in light container in a cool place. Its use is restricted.

(4) Methyl bromide

It is a colourless gas which is heavier than air. Under pressure, it becomes a heavy colourless liquid. The gas is non- inflammable, almost odourless, very penetrating and poisonous to man. It can be used on a wide range of fruits and vegetables to destroy insects. It has been successfully used in fumigating cotton seeds to control the pink bollworms. Its use is restricted for general use.

(5) Naphthalene

It has been used as an insecticide for controlling the clothes moth. It is also used as a repellant to combat the carrot rust fly and as a greenhouse fumigant. It is phytotoxic so it should be used only by experienced persons. An unrefined or crude naphthalene has been found effective as a soil fumigant for controlling wire worms. It is white crystalline compound with a penetrating but not unpleasant odour. The fumes are very irritating to the eyes.

(6) Aluminium phosphide

It is a yellow crystalline solid, stable when dry but in the presence of moist air phosphine gas is librated which is inflammable in air. The phosphine gas is highly toxic and can be used for grain fumigation to control storage pests. Commercial product is a mixture of Aluminium phosphide and ammonium carbonate which is available as 3 gm tablet.

Note: Like chlorinated hydrocarbon insecticides, a number of organophosphorus compounds, systemic insecticides and fumigants which were used in the past, are not in use now.

5. ANTIFEEDANTS

The chemicals which possess the property of inhibiting the feeding of insects are termed as antifeedants or antifeeding agents. These chemicals neither kill nor repel the insect but only inhibit feeding by making the food distasteful. A number of carbamates have antifeedant properties, in addition to the insecticidal action.

Many compounds e.g. nicotine will temporarily inhibit the feeding activities of insects, but in the case of antifeeding agents this inhibition is more or less permanent and is the main cause of death. The fungicides : fentin acetate and fentin hydroxide have been shown to inhibit feeding in a wide range of plant feeders including *Spodotera (Prodenia) litura* larvae and the adults and larvae of *Leptinotarsa decemlineata.* It is relevant to note that these compounds also exhibit chemosterilant activity and insecticidal toxicity against a number of pests. Some of the common antifeedants used for controlling a wide variety of insect pests like beetles, weevils, Lepidopterous larvae etc. are Triazene, Brestan, ZIP (a complex compound of zinc salt), Baygon, Pyrethrum.

6. ATTRACTANTS

The attractants are the materials used to lure insects into traps and poisoned baits. The use of geraniol in sugar syrup as ant baits and fermenting syrup for trapping various moths are good examples. Other substances like eugenol, sugar, molasses, yeast extracts, milk, sweet fruits, amyl acetate and ethyl alcohol are the common insect attractants. The natural or synthetic pheromones mixed with poison baits have also been used as attractants for the control of insects.

The insects recognize the food and oviposition sites by olfaction and to a certain extent communicate with each other by the use of specific odours, the pheromones produced by a variety of glands and comprise a range of chemical types. The only pheromones which have been exploited on a large scale for insect control are the sex attractants which bring about mature sexes together for mating purposes. The gypsy moth, *Porthetria dispar* is the only such species in which sex attractant was employed as a pest control method. The synthetic sex attractants i.e. pheromone analogue, named Gyplure was almost as active as Gyptol - the natural benzene extract of the abdomen of virgin females, but the Gyplure was much easier to synthesize. In field trials, quantities as small as 10^{-7} µg were able to attract males.

A great deal of success has been achieved, particularly with synthesized chemicals which are attractive to various species of fruitfly. One of the first compound to be found and used successfully to attract the Mediterranean fruitfly, *Ceratitis capitata* was Siglure. This fruitfly was also attracted and eliminated by another compound, Trimedlure mixed with insecticide. In a similar manner the Oriental fruitfly, *Dacus dorsalis* has been eliminated by methyl eugenol while a related compound eugenol (I part) mixed with geraniol (9 parts) has proved an effective lure for the Japanese beetle, *Popillia japonica.* Trimedlure and methyl eugenol have been combined with a third compound Cue-lure, which is an attractant for the melonfly, *Dacus cucurbitae.* In this way three of the most important fruit fly pests are being brought under control by the use of attractants. Likewise, anethol attracts the adults of the codling moth, *Cydia pomonella.* Butyl sorbate has been used to monitor a number of Chafer beetles.

7. REPELLENTS

These are the chemicals which repel or drive away the insects although they do not kill them, thus they have the effect of keeping insects away from the food sources. Insects exhibit a phenomenon of negative chemotropism hence they keep away from articles smeared with repelling agents. Different types of repellents are needed for different types of insects because all insects cannot be repelled by the same chemical.

Repellents must be safe, economic and not too unpleasant to use, their efficiency depending on their ability to deter the greatest number of species from feeding for the longest period. The duration of protection afforded by a chemical depends to a great extent on where it is applied e.g. if placed on the skin, it is liable to be diluted by perspiration, to be oxidized, hydrolyzed or volatilized or simply removed by abrasion. Thus, skin is usually protected for only a few hours but clothing may retain repellents for a few days or a week or more. As soon as repellents ceases to be potent enough to offset the attractive qualities of host, insect pests will be able to return to their normal activities with consequent lack of protection. Some of the important chemicals which drive away the insects are :

Bordeaux mixture : It is useful in driving away fleas, beetles, and leaf hoppers. It also destroys fungus and insect eggs. It is prepared with 2.5 kg of copper sulphate, 2.5 kg of hydrated lime and 250 litres of water. The copper sulphate may be dissolved in water in a glazed vessel and lime is diluted in another container. The inner water is then poured into copper sulphate and stirred vigorously. Finally the whole mixture may be diluted with the remaining quantity of water. The Bordeaux mixture renders sprayed plants distasteful and acts as a stomach poison for chewing insects, when mixed with 1/2 to I kg. of lead arsenate.

Creosote or coal tar : The creosote, coal tars and other derivatives of coal tar like naphthalene and carbolic acid etc. are successfully used as repellents to bugs, clothes moths, flies, beetles, and termites. These derivatives of coal tar can be mixed with soap and used as sprays.

A number of oils are used as repellents e.g. kerosene, cedarwood oil, fish-oil, castor oil, tar oil, pine oil, resin oil, citronella oil etc. are useful to protect the bodies of animals from insect attack. Similarly powdered alum, naphthalene, camphor, turpentine, calcium carbide, sulphur, tobacco dust, dry leaves of neem and powder of dry neem seeds can be used for dusting purposes. Benzyl and benzyl benzoate were found repelling to chigger and skin mites, dimethyl phthalate repells mosquitoes and mites and trichlorobenzene repells termites and ants.

The most widely used repellents these days are synergisted pyrethrins applied as sprays. In this case they act as repellents and not toxicant and their efficiency increased by the addition of compounds such as furfuryl alcohol (MGK 11). The compound furfuryl alcohol is also repellent to cockroaches and other household pests.

PRECAUTIONARY MEASURES NEEDED WHILE USING AND HANDLING THE INSECTICIDES

Before using any insecticide the operator must read carefully all the instructions given on the label of the packing which provides complete information about the nature of the poison, directions of their uses and warnings etc.

It is well known that most of the insecticides used now a days are poisonous to man and all other animal life including domestic animals. The insecticide enters the body by numerous ways and vitally influence the various systems of the man and other animals e.g. poisoning is caused by absorption through skin, inhaling fumes, or dusts, or by ingestion through several sources. (1) Hence care should be taken that no insecticide gets in the eyes, nose, mouth or skin. (2) They must be stored in safe places where no children can reach. (3) After the use of insecticides, the operator should wash hands etc. (4) Hands should not be dipped in any preparation of the insecticide and all the equipments used in operation should be washed. (5) Some of the insecticides may be absorbed through the skin. The preventive measures consists of using goggles, rubber gloves, washing of the contaminated clothes etc. (6) The hazards of inhaling dusts, mists and gases may be overcome by using respirators. (7) The clothes of the operator should be changed and the empty containers of the insecticides should be burried deep in the soil and away from the water sources. (8) When treating with fumigants, avoid exposure to heavy concentration or inhalation of gas for longer duration and use gas masks. (9) After the required exposure for fumigation, the room should be opened and entrance be allowed only after thorough airing.

22

INSECTICIDE APPLICATION EQUIPMENTS

Broadly speaking the insecticide application equipments can be classified into (a) sprayers, (b) dusters and (c) granule applicators.

(1) Sprayers : There are several types of sprayers (Figs. 61 A-E, 62 A,B,E) and they are mainly used for applying insecticides in liquid form. They may be either hand operated (hand sprayers) or electrically operated (electric sprayers) or operated by power (power sprayers). The most common hand sprayer is the hand atomiser, fine mist or spray can be obtained by this machine which carries a small tank at the base of a syringe and sends out a spray through a nozzle. Compression of air results in throwing out of the liquid from the end of suction tube extending into the tank with the blast of air breaking up the liquid into a mist form. This is the most widely used apparatus for domestic purposes, gardens and nurseries. Special nozzles are available for emitting a jet at right angles for spraying under leaf. Another common sprayer used for domestic purpose is the bucket pump sprayer. This equipment essentially consists of a long stout pump, a long delivery tube and an air chamber to keep reserve pressure during the upward stroke of the piston. This apparatus generally involves the presence of two operators one for handling the pump and the other for directing the spray.

A more efficient and convenient hand sprayer is the Knapsack sprayer. This apparatus is a combination of a tank pump, delivery tube, nozzle and straps by which it can be hung over the shoulders. One of the hands of the operator works on the pump and the other directs the spray. This equipment serves very well for spraying insecticides in fruit gardens and small agricultural plots as well as hilly tracks for valuable crops such as tea and coffee.

Fig. 61A. Tractamount sprayer.
(By Courtesy of American Spring and Pressing Works Ltd., Bombay)

For large scale operations the most convenient machinery is the compressed air sprayer. This equipment consists of a pump fitted in a tank which is operated by hand to generate considerable pressure inside the tank which in turn results in producing a continuous spray of the insecticidal liquid through the nozzle. The entire equipment can be handled by one operator. A pressure gauge is also fitted to this machinery to read the amount of pressure generated in the tank. Normally such equipment is operated for applying residual deposits of insecticides in houses and in other buildings to control mosquitoes and insects of medicinal importance.

Household electric sprayers are operated by electricity and are often used by professional pest control operators to control insect pests in public buildings.

With regards the power sprayers, these are used for very large tracks of land with great advantage. In such equipments the insecticidal liquid stored in a large tank is pumped out in the form of spray by a delivery tube through several nozzles by means of a motor which may be worked by petrol, oil or electricity. Some of the common power sprayers are (1) multipurpose hydrolic type, (2) low pressure low volume sprayer, (3) high pressure high volume sprayer, (4) blower sprayer and (5) aerosole generator or fog machine. Many of these power sprayers are installed in aeroplanes for aerial application of insecticides over large tracks of land.

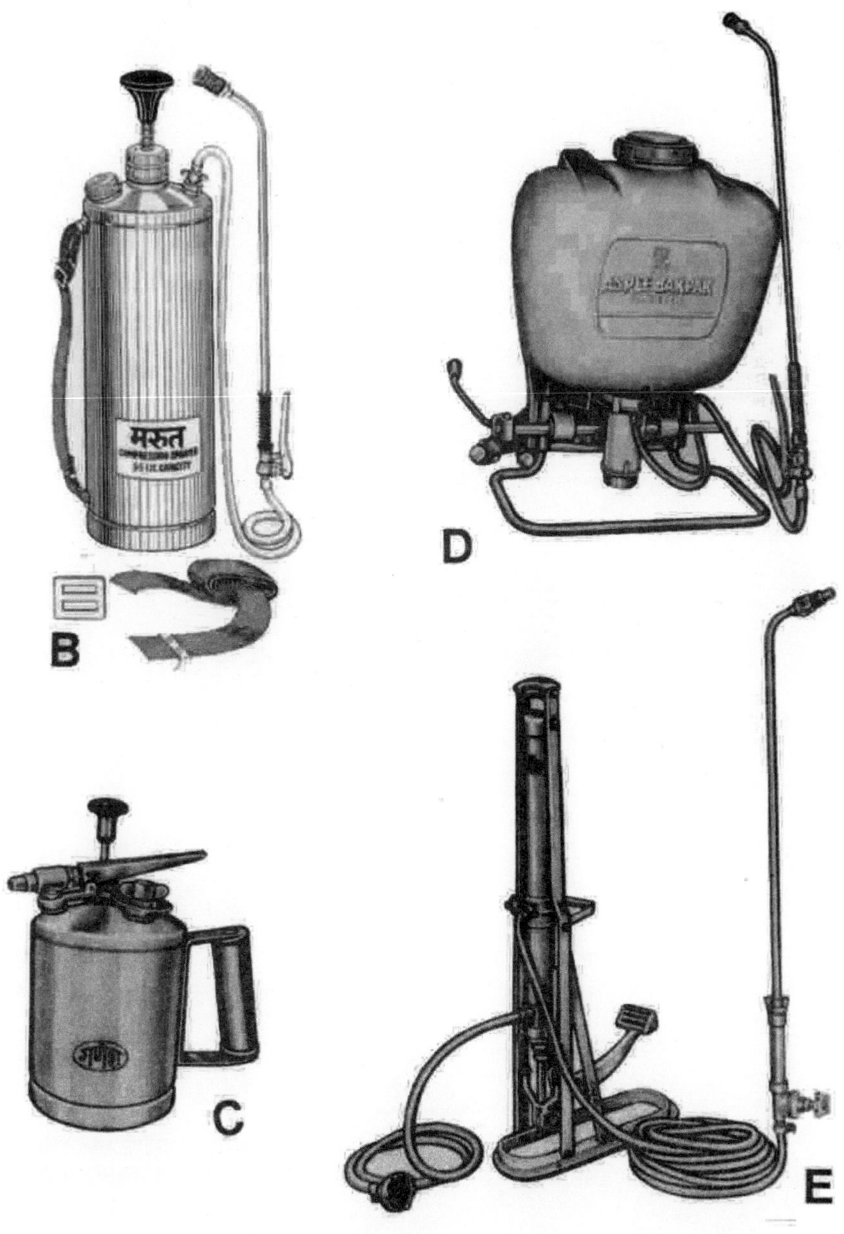

Fig. 61. B. Hand compression sprayer, **C**. Hand sprayer, **D**. Pesticide sprayer, **E**. Foot sprayer
(By Courtesy of American Spring and Pressing Works Ltd. Bombay)

Care should be taken to ascertain that the material of the tank and other parts of the power sprayer are all safe which must be insecticide proof. Besides the parts of the power sprayer must be detachable and should afford easy cleaning and replacement.

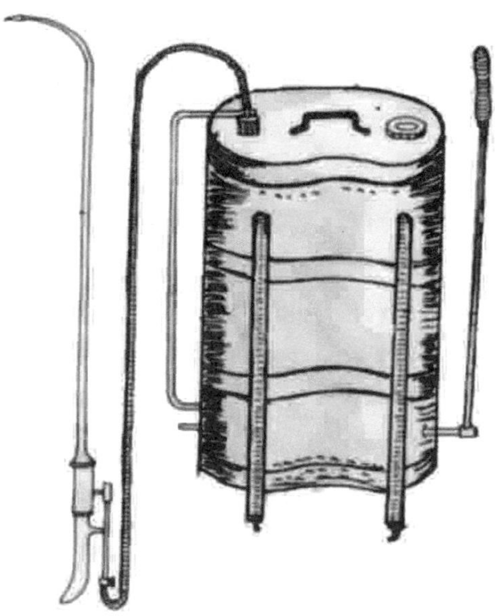

Fig. 62 A. Knapsack sprayer

(2) Dusters (Figs. 62 C,D, 63 A) : There are various types of machinery available for applying insecticides in the form of dusts. Some of the common hand dusters are the syringes or hand blowers. Knapsack dusters are also available for operation on a large scale. The most commonly used duster for large scale operation is the Rotary blower (Fig. 63A). A fan is installed inside the machinery which is operated by one hand and this expels a cloud of dust through one or two exits. Dusters are also driven by motors for dusting large tracks. Many of them are installed in aeroplanes for aerial dusting.

(3) **Granule applicators:** (Fig. 63 B,C) These are newly developed insecticidal equipments designed to provide a means for accurately and uniformly distributing granular formulation of insecticides (Fig. 63B). The most common applicator is the rowcrop granular applicator. This serves as a convenient tool for applying granular insecticides for control of insects on plants arranged in a row.

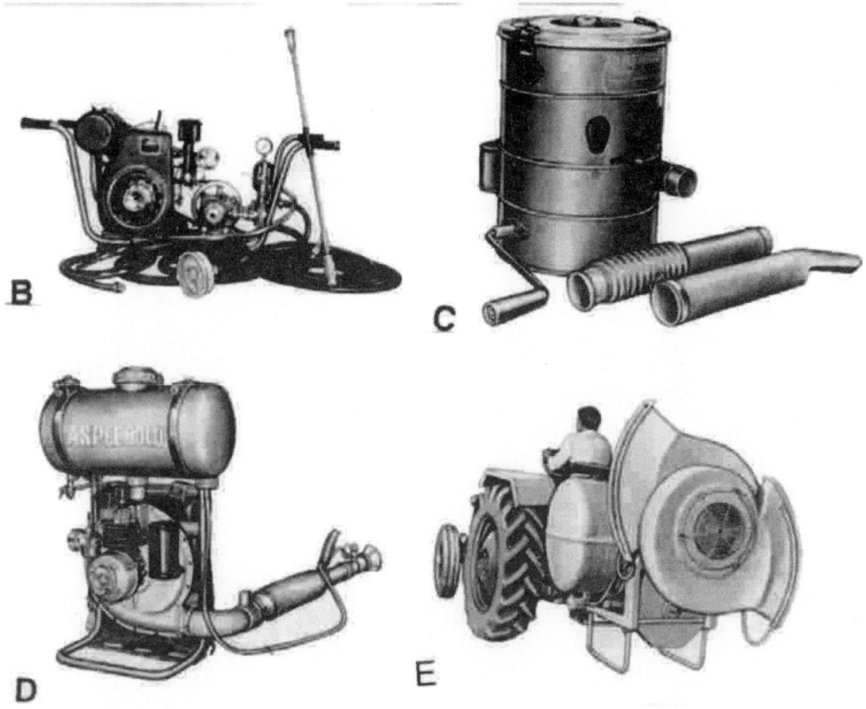

Fig. 62 B. Power sprayer, **C**. Hand rotary duster, **D**. Mist blower cum duster, **E**. Tracta mount air assisted sprayer.
(By Courtesy of American Spring and Pressing Works Ltd. Bombay)

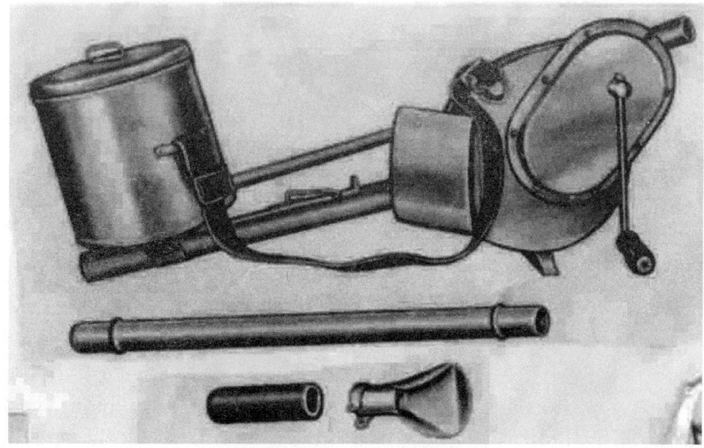

Fig. 63. **A**. Rotary duster
(By Courtesy of American Spring and Pressing Works Ltd. Bombay)

Another type of granular applicator is the trailor type applicator widely used in large farms. They are equipped with metering devices for regulating the dosage.

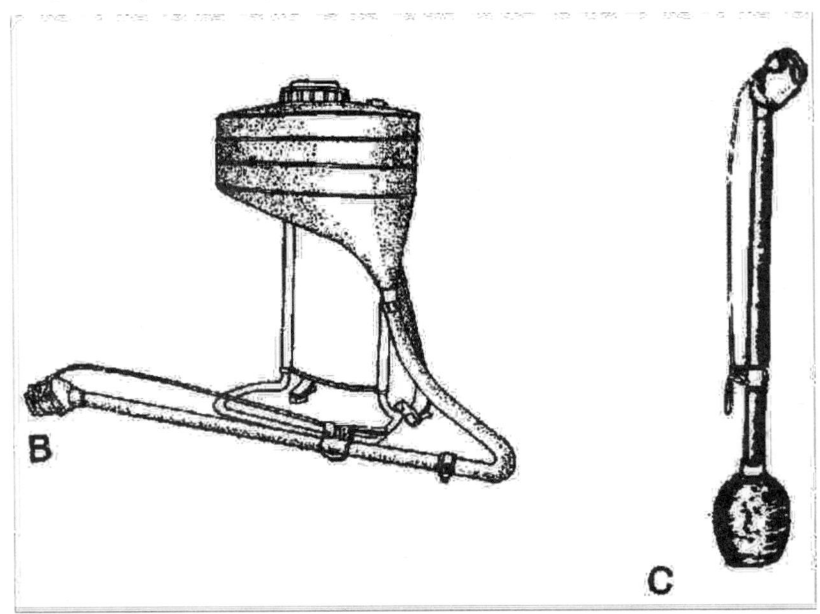

Fig. 63. B. Knapsack granule applicator, **C**. Hand granule applicator

Thus, it is clear from the information given in the chapter 21 and 22nd that the insecticides and their applications for the control of harmful insects, forms one of the most interesting branches of study in applied Entomology. While the botanical or inorganic insecticides are applied for the control of insects, it is the synthetic group of insecticides which dominates the field of toxicology. The mode of action of insecticides may be in more than one way on an insect and therefore no distinct classification of insecticides based on their mode of action is possible. Constant research is going on actively all over the world to develop newer and newer types and more effective kinds of insecticides so as to counter act the resistance developed by some insects for certain insecticides. Concurrent with the attempts to develop new insecticides the efforts and investigations are carried out by man to device novel and more modern type of insecticidal application equipments which may be in the form of sprayers, dusters or granular applicators (as shown in Figs. 61, A-E; 62 A-E and 63 A-C).

23

CULTURAL CONTROL

By cultural control is meant those methods of planting, growing and harvesting a crop which will prevent or lessen insect damage. This method consists of a slight departure from the usual time of planting, sowing or harvesting the crops or changing the plan of crop rotation, weed control, disposal of the crop remnants after harvest. Planting good seeds and resistant varieties are very important in controlling some insect pests.

Since cultural methods are usually economical, they are specially useful against pests of low unit-value crops. Practices which reduce the chances of buildup of pest populations may hold them below the level which will cause economic damage. Such methods are particularly applicable to field crops and forests.

Knowledge of the life history or bionomics of a pest species is essential to the effective use of cultural control methods. The principle of the "weakest link" or most vulnerable part of the life cycle usually applies. The environment is changed by altering farming practices at the correct time so as to kill the pests or to slow down their multiplication. In this way, the method is aimed more at prevention than at cure. If the environment is unfavourable, the pest may not reach a population level which will cause serious damage.

Cultural controls are often used when chemical or biological methods have not yet been devised for an injurious species. Cleanup of the sources of infestation and changes in the planting or harvesting time are particularly important when no effective method of killing the pest is known. However, these methods are also used in combination with other controls.

1. Use of pest resistant varieties: Seeds of healthy plants preferably of a resistant variety only should be sown as far as possible. Certain varieties are naturally resistant or less susceptible to insect attack e.g. certain varieties of wild apples are immune to aphid attacks while most of the cultivated varieties are very susceptible. Hence, by interbreeding wild with cultivated variety a strain can be evolved which will yield good fruits and shall also be immune to pests. Similarly the Indian cotton *Gossypium arboreum* is much more resistant to the attack of jassids, whiteflies and the pink bollworms than the American cotton, *G. hirsutum*. Also the indigenous cotton is resistant to leaf roller but susceptible to pink boll worm and the foreign variety behaves in an opposite manners. Hence, by crossing the Indian species with the American species a new strain, immune to aphids, jassids and pink bollworms etc. can be developed.

2. Tolerance to insect infestation - A variety may be infected by insects yet survive and show less injury than others because of its ability to replace injured parts such as leaves and rootlets.

3. Ploughing in relation to insect control : The plough if used at right time is a good tool for combating many insect pests; it disturbs or kills them, eradicates weeds upon which they might feed and breed, exposes them to natural enemies or to weather control and buries them so deeply that few adults can emerge. A thorough stirring of the soil before planting is an indirect method for controlling the corn root aphids, because, it breaks up ant colonies, kills many aphids and prevents the growth of weeds on which they live until corn roots are available.

4. Clean cultivation : Clean cultivation means the removal of weeds, plant residues, and other materials from the fields and growing only healthy crops. The destruction of crop residues is a very good preventive measure for controlling certain species of insects e.g. caterpillars or beetles. The elimination of remaining effectively stops further insect breeding by cutting of their food supply and shelter.

Many insects infesting crops will develop on weeds which may or may not be related to these crops botanically. Hence, weed control is a good practice for reducing insect infestation. Damage due to *Leptocorisa* is very serious when paddy fields are surrounded by weeds which provide a good shelter to the pests till the ears have been formed in them.

Similarly various species of red hairy caterpillar, *Amsacta moorei, Amsacta collaris* and other species of this insect are polyphagous and lay eggs on a number of weeds and on emergence the caterpillars feed on the weeds and then migrate to the nearby fields of any crop.

5. Crop rotation : Crop rotation is effective as a preventive measures against insects that feed on relatively few plant species or where the insects are incapable of long distance migration but crop rotation is useless against a general feeder. If one and the same crop is grown every year the insect population is bound to increase due to abundance of food material. On the contrary if crops are grown in rotation or alternate years e.g. a crop of one plant family followed by that of different family e.g. barley grown in spring may be followed by legumes or pulses and then wheat in winter and so on. The crops of the same family e.g. cotton, and lady's finger belonging to the same family Malvaceae must not follow same year. Mixed crops are sometimes useful as pests do not thrive in large number.

6. Flooding or irrigation : Where water is available, it is sometime possible to destroy insects by flood or irrigation of the field. Flooding rice fields destroys many sugarcane borers. Similarly other burrowing insects e.g. crickets, grasshoppers, beetles and bugs come out of the ground after flooding the fields. They either die their natural death or readily picked up by birds. Ploughing followed by heavy irrigation within 30 days destroys many pink bollworms. The sugarcane and wheat crops can be saved from the attack of termites by irrigation.

7. Fertilizing : Nitrogenous fertilizers in some soils tend to increase the susceptibility of sorghum to damage by attack of some insects. As a result of proper manuring the plant growth is stimulated. The healthy and vigorous plants can easily resist the attack to which the weaklings would have surrendered.

8. Pruning and thinning : Some pests are normally carried from old crop to the new one. Pruning and thinning decrease the intensity of attack. Infested plant shoots and dead hearts must be immediately removed. Ratooning should be avoided and there should be plenty of interval between harvesting a crop and sowing a new one.

9. Time of planting and harvesting : Time of planting and harvesting has great influence on insect infestation e.g. early maturing cotton is not attacked by pink boll worm in areas where moth donot emerge early. Late planted wheat escapes oviposition by hessian flies in America because the plants do not emerge until most of the flies have disappeared as they are very short lived. Similarly early sowing of rice in the Punjab between the 3rd week of May and mid June is helpful in protecting it from the attack of rice borer, *Tryporyza incertulas* so the time should be so regulated that the crop may not be worth the attack at the active period of the pests.

10. Closed season : An insect species limited in its feeding to one crop may be eradicated by not growing this crop for a year or two. This has been done with success in isolated areas for eradicating the pink bollworm. Such a closed season consists in not growing cotton for a period of several months.

11. Destruction of volunteer crops : Volunteer crops are those which grow from self sown seeds, and stubble crops are those which sprout from roots are stubbles (stumps), may be a source of insect infestation and they should be destroyed by ploughing or other means.

12. Trap crop : A trap crop is a small planting often only a few rows, made somewhat earlier than the main planting for the purpose of diverting insects away from the main crop. The trap crop can either be harvested early or cut and used as fodder before a generation of the insect can be completed. Otherwise it will serve as a breeding ground for the pest which will then attack the main crop in large numbers. The plant species used for trap crop should be one that is very attractive to the insect. The preferred host plants can also be grown around the main crop and when the pest has appeared it can be cut and destroyed.

The other methods of cultural control are early cutting of crop, rate and depth of planting, sowing, separation of complimentary host plants, pasturing the heavy infected crops and drying out the soil etc.

24

PHYSICAL AND MECHANICAL CONTROL

The protection of crops and foods from insect attack by physical and mechanical means is the simplest and most effective method. Such measures consist of destruction of insects by mechanical means, burning, trapping, protective screens or barriers, use of high or low temperature, soaking in water and drying.

1. Hand picking : When only a few plants are infected, certain large conspicuous species may be removed from the plants by hand and destroyed just by pressing the abdomen or dipping them in kerosenized water.

2. Mechanical means : A rotary blow has been found to destroy over 96% of white grubs as well as cut worms, army worms and other caterpillars. The cutting wheel is thirty two inches in diameter and fourteen inches wide and carries sixteen steel blades. It operates by power take off from a tractor at 150 rpm.

3. Trapping : The insects may be turned into a trap by light and baits or the trap may be mobile and drawn across a field.

4. Light traps: The attraction of nocturnal insects to light is a common observation and the light traps have been invented and tested for killing crop pests. The attractiveness of an artificial light depends upon its candle power and colour. A yellow light will attract 60% fewer insects than a white light of the same candle power and brightness. The insects attracted by light may be destroyed as they hit electric grid or they may be drowned in water and oil after they have been turned into the trap. These are effective against noctuid moths.

5. Bait traps: Baits are a form of control whereby insects are attracted to a selected spot on which they can be easily removed. Bait traps have been found to be very effective in controlling the nocturnal insects like cockroaches, crickets, caterpillars and nocturnal moths.

6. Temperature treatments : (a) *Low temperature control* : Low temperature extremes are fatal to insects of stored grains. Usually a temperature of –28°C or lower for at least several hours will kill most of the insects. *Tribolium confusum* can be killed in 24 h when exposed to subzero temperatures. Low temperatures are utilized for the control of insects in mills, warehouses and similar establishments.

(b) *High temperature control* : The *Tribolium confusum* cannot complete its development at 41°C and temperatures of 44°C or above kills the adults. High temperatures are also fatal to other species and can be used to kill them. Heating cotton seeds to a temperature of 65.5°C for 30 seconds will kill all pink bollworms in the seeds. Temperatures of 48.8-51.6°C in all plants of a flour mill for 10-12 h will kill all insects exposed to such temperature.

Exposing the infested grains to the sun in summer also kill all the adults of the stored grain insects.

7. Physical barriers : The protection of crops from insects by barrier is possible in case of crawling and migratory insects. Treated paper barriers and dust barriers are also used. Trenches are also used to stop army worms. Similarly water channels can be used for checking the migration of crawling insects.

8. Burning: Although destroying insects by burning their shelters is sometime practicable, burning pastures and woodland to destroy insects should not be recommended. It has been found that such practices reduce soil fertility and increases soil erosion. Furthermore, it fails to destroy injurious species but does eradicate many beneficial forms.

Both cutting away dead hearts and burning them in time save valuable crops from destruction. Burning the field refuse and stubbles, are the other useful ways of pest control. e.g. Sorghum, stem borer.

9. Sieving and winnowing : It is possible only for domestic purposes. Effective against stored grain insect pests.

10. Flooding: It is well known that a large number of insects breed and dwell in the soil, which may be destroyed by flooding the fields with water. There are several other simple methods through which a number of pest infestation can be controlled but the mechanical method of insect pest control involves a lot of labour like shaking the trees and killing the insects either in the larval or adult stages.

25

BIOLOGICAL CONTROL

The term biological control means control of insects by several biological agencies such as parasites, predators and pathogens. The introduction of sterile males in a normal population and use of sex attractants have been reported to bring about effective biological control. De Bach (1964, 1974), Reay (1969), Huffaker (1974), etc. have given exhaustive account on biological control; a brief account from these books is given here :

The theoretical basis of the biological control depends on the existence of a natural enemy. The effective natural enemy can be deduced to have the following characteristics : (i) high searching ability, (ii) high degree of host specificity or preference, (iii) good reproductive capacity relative to the host and (iv) good adaptation to a wide range of environmental conditions. The most essential characteristic is high searching ability. It should be borne in mind that a really effective enemy may be scarce in its native home because it regulates the host population at low level. There usually is one best enemy for each species in a given habitat and one frequently is sufficient for complete biological control, often, however, a second or third enemy species may add to host population regulation and may in fact be necessary to achieve satisfactory biological control. The best enemy species may differ for different host habitats. Hence there is generally no single best natural enemy extending throughout the range of a pest species. With these qualities the introduced natural enemy by itself or in conjunction with other mortality factors is expected to prevent outbreak of the pest species or atleast to drop down major population fluctuations.

The essential principle of biological control is to maintain a state of biological equilibrium of a living organism. This biological equilibrium is normally maintained by a set of forces called environmental resistance.

Often the biological equilibrium of a living organism is disturbed either due to environmental or biotic factors and consequently a particular organism multiplies and spreads in a unlimited manner so as to assume the status of a pest. The inherent ability of an organism to survive and to reproduce within a given time and under optimal environmental conditions is known as biotic potential. Biotic potential is a force opposed to environmental resistance. It takes into account the number of young produced per female in each reproduction, the number of reproduction in a given time and the sex ratio of the species.

Advantages of biological control

Only a limited number of introduced pests have been successfully exterminated with chemicals. It would seem wiser in many instances to attempt a biological control programme which might reduce pest populations below the level which causes economic damage. Biological control has the tremendous advantage that, if successful, it becomes self-sustaining and integrated into the normal environment of the control area.

If pest populations are reduced to the point where economic damage is negligible, the control programme is a success. Biological controls tend to be particularly useful on low unit-value crops where complete control may not be required or where chemicals are not recommended. Pests of field and forage crops, forests, and range may be economically controlled by biological methods; chemical controls against such pests may be impractical.

A biological control destined to be fully effective will be easily and quickly established. If an imported parasite or predator is not established within three years of careful releases under good conditions for its development, the programme may be justifiably discontinued.

TYPES OF BIOLOGICAL CONTROL

There are 4 main types of biological control viz. (i) introductionof parasites, (ii) predators, (iii) pathogenic organisms such as bacteria, viruses and fungi etc. and (iv) removal of fertile males and creating sterility in males by gamma radiation and introducing these sterile males in a normal population. In addition to these 4 main types, there is a fifth type of biological control where the indigenous parasite of the locality is being utilized for this purpose.

1. PARASITES (Figs. 64 A-C)

Parasitic insects develop as larvae on or in a single host individual from eggs generally laid on, in or near the host and usually consume all or most of the host body, killing the host and then pupate, either within or outside the host. The free living adult parasite emerges from the pupa and starts the next generation a new by actively searching for host in which to oviposit. They tend to attack only one host stage i.e. eggs, larvae, or pupae, although there is also some overlapping in certain cases, adult insects do not serve as hosts. Very often, life cycles are commonly short, ranging from 10 days to 2 weeks or so in mid summer but correspondingly longer in cold weather. In general they all have great potential rates of increase.

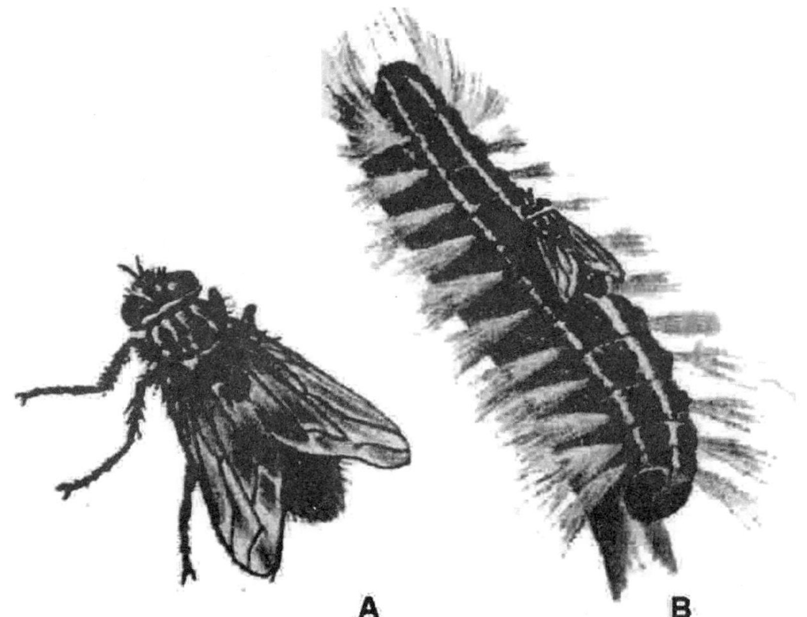

Fig. 64A. Tachinid fly, **B.** Tachinid fly laying eggs on Gypsy moth larva

According to the species one or more parasitic larvae may characteristically develop per host individual, thus we have either solitary or gregarious parasitism. Sometimes two different species of larvae develop from eggs laid in the same host. This is multiple parasitism. There is also ectoparasitism and endoparasitism, depending upon whether the larva develops externally or internally. The major groups occur in two orders of insects viz. Hymenoptera and Diptera.

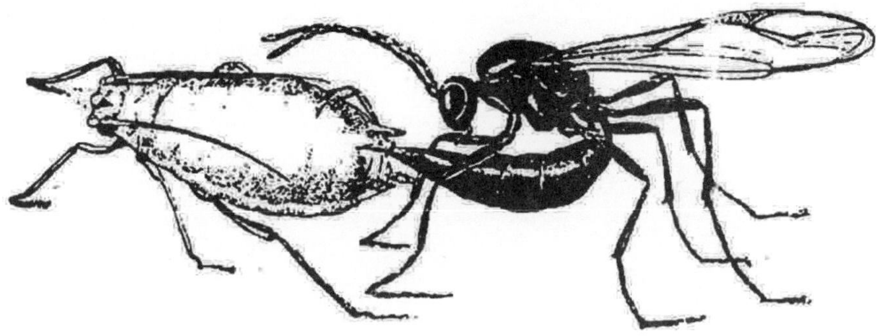

Fig. 64C. A minute parasitic wasp, *Aphidius testaceipes,* laying egg in an aphid (From Pfadt, 1962)

Hymenoptera

This is the dominant order among all entomophagous insects. Over two thirds of the cases of successful biological control of pests species have been achieved by hymenopterous parasites. The most important and numerous hymenopterous parasites occur in the major groups commonly known as chalcids, braconids, ichneumonids and proctotrupeds.

(1) **The chalcids :** The chalcids attack species in nearly all orders of insects, the preferred ones being Coleoptera, Diptera, Homoptera and Lepidoptera which also happen to include the bulk of our chief crop pests. The immature stages of chalcids predominantly develops in host's eggs or larvae or pupae and some develop in adults of Homoptera. The encyrtids and aphelinids include many that specialize in attacking scale insects, mealy bugs, white flies, and related forms and have been so outstandingly successful in biological control.

(2) The braconids: This is a major large family of parasitic wasps. They have been used in biological control. Only an occasional rare species is hyperparasitic hence the family is nearly entirely beneficial. Parasitism may be external or internal, solitary or gregarious. Their host preferences are mainly lepidopterous and coleopterous larvae but also commonly Diptera, including the damaging fruit flies and Homoptera, particularly aphids.

(3) The ichneumonids: The Ichneumonidae comprise about 20 per cent of all parasitic insects. Most are primary parasites, hence the family is predominantly beneficial. Some of the largest and most conspicuous parasite species occur in this family. Ichneumonidae probably parasitize wood and stem boring larvae in the Lepidoptera, Coleoptera or Hymenoptera but many attack lepidopterous pupae too. Some species have a very wide host range. Most of the Tryphoninae are solitary parasites of sawflies. The ophioninae are predominantly internal parasites of lepidopterous larvae although a few develops in the grubs of root feeding scarabs in the soil.

(4) The serphoids : This contains several families, the most important of which are the Platygasteridae and the Scelionidae. All are parasitic on immature stages of other insects and they are predominantly primary internal parasites.

Diptera

In addition to tachinids which comprise the most important dipterous family in biological control, this order contains several parasitic families including - Cyrtidae, Nemestrinidae, Pipunculidae, Conopidae and Pyrgotidae. However, tachinids constitute the major parasitic family both in numbers of species and in economic importance and have been the only group utilized extensively in biological control.

The Florida red scale, *Chrysomphalus ficus* invaded Israel and spread throughout the Citrus growing areas and soon become a major citrus pest. An indigenous ectoparasite, *Aphytis chrysomphali* was the only parasite commonly recorded from the scale in Israel. The attacks of this species were limited mainly to male scale pupae and its effect on the scale population was negligible.

Another species of *Aphytis* viz. *Aphytis holoxanthus* and the endoparasitic *Pteroptrix smithi* were introduced into Israel from Hong Kong during 1956-57 against the Florida red scale; and complete biological control was achieved throughout the Coastal plain, which is the main citrus growing area of Israel, during the first 2 or 3 years following the introduction of the parasite, and the Florida red scale was virtually eliminated from that area as a pest.

Similarly biological control of Olive scale was achieved by introducing two additional species of parasites viz. *Coccophagoides utilis* and *Anthemus inconspicuus* from Pakistan to California. The biological control of California red scale was achieved by introduction of *Aphytis lingnanensis* from South China to California.

A great deal of success has been achieved in Newzealand and other parts of the Australia by using *Pteromalus puparum* imported from United States to control populations of the small white butterfly, *Pieris rapae.*

Although various *Aphidius* species have been used for aphid control, the best known Braconid, is *Macrocentrus ancylivorus.* It has been reared on a large scale in the United States using the larvae of the Strawberry leaf roller as the host and liberated for the control of Oriental fruit moth, *Grapholitha molesta* in peach orchards.

Finally, mention should be made here of the spectacular eradication of the Citrus black fly, *Aleurocanthus woglumi* from Cuba and various islands in West Indies by *Eretmocerus serius* imported from Malaya. Control was achieved so rapidly that when *Aleurocanthus woglumi* become an important pest in Mexico some years later, *Eretmocerus serius* was first choice as control agent. Unfortunately, it was a dismal failure and further investigations indicated that climatic conditions were unfavourable. The pest was eventually brought under control by *Amitus hesperidum* and *Prospaltella* species imported from India.

Factors governing parasite

One of the important factors governing parasite introduction for biological control is the geographical location of the area under experimentation and the conditions met within that area and presence of an indigenous fauna of a peculiar and restricted character. Introduced parasites meet with relatively little competition from indigenous forms. Limitation of the area to be covered by parasitic colonization, resulting in the presence of only few main crops and therefore facilitating adequate organization and centralization of control experiments. Other factors which hamper an efficient programme of biological control are the phenomena of multi-parasitism and hyperparasitism, there is ample evidence to show that the efficacy of introduced parasites has been enormously reduced by these two factors.

Certain other considerations which have a direct bearing on parasitic introduction are suitable adaptation of artificial measures of control so as to put the destruction of more parasites than hosts, proper recognition of the time factor in biological control experiments and proper evaluation of the capacity of a parasite to discover its host

Utilization of indigenous parasites

Biological control is also brought about through the agencies of indigenous parasites of a locality, such operations consist either of conserving or increasing the number of parasites in a given area or of attempting to colonize such parasites by artificial culture methods under laboratory experimental conditions in an area of a country where they did not previously exist.

One of the important features of this type of biological control is the alteration of the host parasite ratio in a given locality. Example of such a biological control is *Trichogramma minutum*, a chalcid egg parasite on notable pests such as *Pyrausta nubilalis* and the sugarcane borer, *Diatraea saccharalis*.

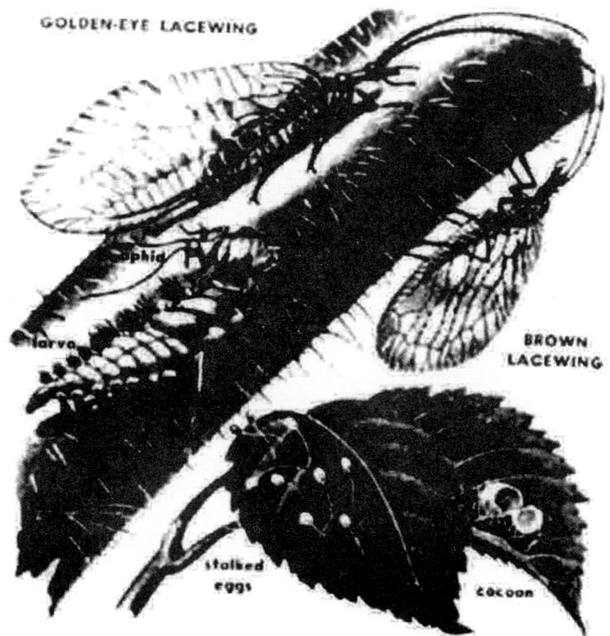

Fig. 64D. Lacewing larva feeding on aphids

2. PREDATORS (Figs. 64 D-H)

Predator insects differ from parasitic ones in that the larvae or nymphs, as the case may be, require several to many prey individuals to attain maturity. The adults generally deposit their eggs near the prey population and after hatching the active mobile immatures search out and consume prey individuals. This need of larvae or nymphs to search is an important distinction from parasitic forms whose larvae develop on a single host individual and thus have the advantage of not having to discover additional hosts. Adults of many species are also predatory. There may be one or several generations to one of the prey. Both larvae and nymphs as well as corresponding adults may be predaceous or only one stage may exhibit the habit.

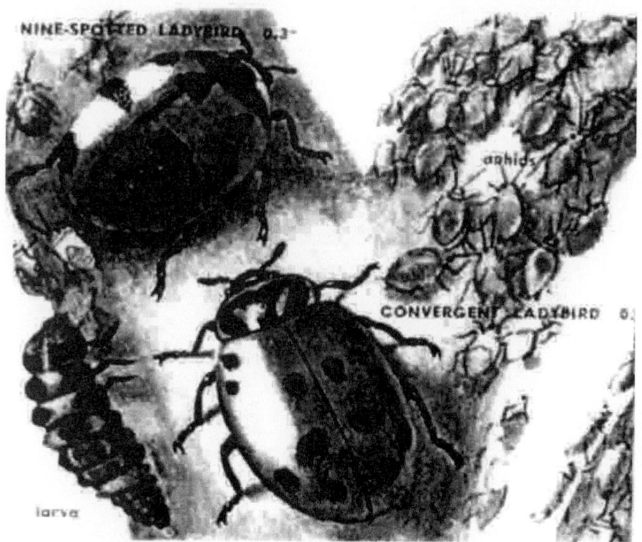

Fig. 64E. Lady bird beetles. Adults and larvae feeding on aphids

1. **Coleoptera:** The main predaceous families of the Coleoptera are - Coccinellidae, Staphylinidae, Histeridae, Lampyridae, Cleridae, Cantharidae, Meloidae, Cicindelidae, Carabidae, Dytiscidae and Gyrinidae. The Coccinellidae and the Carabidae are of most important in biological control of agricultural pests.

2. **Neuroptera:** Most of the species of the order Neuroptera are predatory. The most important families in biological control have been the Chrysopidae and Hemerobiidae which attack many agricultural pests including scale insects, mealy bugs, aphids, white flies, mites etc.

3. Hymenoptera: About one quarter of the families of this order are strictly predaceous. Most of the predatory forms tend to be social and live in colonies. The colonial vespid, wasps provision their paper nests with the bodies of caterpillars and other soft bodied insects. Species of *Polistes* have been used in several biological control projects.

4. Diptera : A few families are entirely predaceous and many others contain predatory species. The more common economic important families include: Syrphidae, Asilidae, Cecidomyiidae, Anthomyiidae, Calliphoridae and Sarcophagidae; of the latter four, many are predaceous in the egg masses of grasshoppers.

5. Odonata : The dragon flies and damselflies are predaceous both as nymphs and as adults. The nymphs are all aquatic and feed on a wide range of aquatic insects.

Fig. 64F. Syrphid fly maggot feeding on aphids

Biological control has been very successful in certain cases where injurious insects have become established in land, they did not previously inhabit and have developed into persistent pests. A classical examples of such a biological control is that of the cottony cushion scale, *Icerya purchasi* in California, by a predatory coccinellid beetle, *Rodolia cardinalis* introduced from Australia. Introduction of this beetle in California resulted in almost complete and permanent destruction of *Icerya*, within less than 5 years. This species was so completely destroyed that the survival become numerically insignificant. Subsequently the predator achieved complete biological control of the cottony cushion scale in 25 additional countries and a substantial degree of control in four other countries (De Bach, 1964, 1974).

Fig. 64G. Syrphid fly, **H.** Syrphid fly maggot feeding on mealy bugs.

Many outstanding successes in the biological control of serious coccid pests have since been recorded. Some notable cases include the complete biological control of the Citrophilus mealy bug, *Pseudococcus fragilis* in California, the coconut scale - *Aspidiotus destructor* in the Island of Fiji, Maritius and Principe, Green's mealy bug, *Pseudo-*

coccus citriculus in Israel; the red wax scale, *Ceroplastes rubens* in Japan; and Coffee mealy bug, *Planococcus kenyae* in Kenya.

A mention may be made of the vertebrate predators like fishes, frogs, reptiles, insectivorous birds and mammals etc. which are the regular feeders and deserve to be provided with suitable conditions to grow and breed profusely.

3. PATHOGENS

Pathogenic microorganisms attack insects and have life cycles more or less characteristic or similar microorganisms developing in other groups of animals. Insects are probably subject to as wide as variety of disease as are the vertebrates. Except for the fungi, disease organisms gain entry in the host via mouth or the digestive tract, that is, the insect host must eat plant or other food contaminated with pathogen. In case of fungi, entrance is gained through the insect integument and free water or very high humidity is generally required. Thus, fungi tend to be restricted to moist environments. However, fungi do have advantage of attacking sucking insects which because of the nature of their feeding on sap, tend to be fairly free of disease caused by microorganisms because they rarely ingest them. Virtually no insect disease organisms occur in mammals and none have been recorded from man. Thus they are safe to use in biological control even in large scale microbial spraying operations. The most common disease of insects are caused by bacteria, fungi, viruses, and protozoa.

1. Bacteria : Because it can be produced fairly easily on a large scale and has a wide spectrum of activity, *Bacillus thuringiensis* was of the first microorganisms to be incorporated into a commercial "microbial insecticide". The most sensitive insects appear to be lepidopterous caterpillars, particularly those in which alkaline conditions prevail in the midgut. This is related to the fact that the spores (it is a crystal bearing spore former and can be conveniently stored and applied in the resistant state without loss of potency) can only enter the insect on ingestion and that the proteinaceous crystals of toxin is only liberated at around pH 9-10.

Those pathogenic bacteria which do not form spores have proved to be of little practical value in the field and indeed only two other groups of spore formers have been employed with any success i.e. *Bacillus popilliae* and *Bacillus cereus*. The former species produces a condition known as milky disease in the Japanese beetle, *Popillia japonica* and has been used extensively in North America for control of this pest and Chaffer beetles. Although *Bacillus cereus* does not produce a crystalline toxin, it also requires closely defined mid gut conditions with an optimum around pH 7. This *Bacillus* is active against the codling moth and certain Hymenopterous pests.

2. Viruses: The classic example of the employment of viruses in pest control is the highly successful use of a polyhedrosis viruses against the pine sawfly, *Neodiprion sertifer* in Canada. A related pest *Neodiprion swainei* and the spruce sawfly, *Diprion hercyniae* are similarly controlled by a polyhedrosis virus.

Another example is the control of alfalfa caterpillar, *Colias eurytheme* by introduction of polyhedrosis virus. The cabbage looper, *Trichoplusia ni* and other brassica caterpillars have been controlled in a similar way. Only one other group of viruses, the granulosis type has been used against pests in field, in this case the Lepidopterans, *Pieris brassicae* and *Pieris rapae.*

3. Fungi and Protozoa

Pathogenic fungi have several properties which should make them ideal organisms for use in biological control. They produce highly resistant stages and are capable of prolific spore production once they became established. However, their great dependence on microclimatic conditions, especially humidity has seriously curtailed their use and only a few have been employed successfully of these *Entomophthora exitialis* has produced promising results against the spotted alfalfa aphid, *Therioaphis maculata.*

Many protozoa appear to cause chronic rather than acute symptoms in their victims and the microsporidian, *Thelohania hyphantriae* takes upto four weeks to exert its full effect which has been shown to cause mortality in the fall webworm, *Hyphantria cunea.*

4. GENETIC METHODS

Genetic pest control implies the manipulation and use of genetic material in a manner injurious to pest insects. The control of insect pest populations by the release of sterile males has been demonstrated with at least five insect species (Lachance *et al.*, 1967). This spectacular technique has been termed autocidal control and involves using an insect species to bring about its own self-destruction. It is accomplished by irradiating laboratory reared males of the species to an extent sufficient to disrupt the genetic function of the sperm nucleus but not appreciably interfere with the normal ability of the male to mate or of the sperm to penetrate the egg of the female. However, such fertilized eggs fail to develop so a wild female mated with a sterilized male produces no progeny. If sterile males are released in large enough number in relation to the wild population they will fertilize more females than will the wild males. The advantage is cumulative in each generation, hence eradication may be achieved within a few generations under ideal conditions. The major success thus far has been with the screw worm which was eradicated from the island of Curacao, from Florida, and from Texas to California by massive annual release of sterile males.

A low infestation of the Oriental fruit fly was eradicated from Guam in 1963; the melon fly was eradicated from Rota island in 1963 using a combination of a bait spray and sterile insect release, and the field cockchafer eradicated from a seventy four acre plot in Switzerland in 1962. The boll weevil is reported to have been eradicated from a small isolated field by the release of chemically sterilized males. Chemosterilants have been under intensive investigation but are not as yet practical. They offer the advantage of treating the wild population directly and thus avoiding the mass rearing necessary with gamma radiation (De Bach, 1964, 1974).

USE OF PHEROMONES

The ability of an insect to locate a mate and to copulate once it has been found, is often directed by natural specific chemicals produced by the insect called pheromones. Natural chemical attractants and mating stimulants have been shown to occur in many pest insects. They are often effective in incredibly small concentrations; one caged virgin female of the introduced pine sawfly attracted over 11000 males. Some have been chemically identified and even synthesized. Its high specificity for particular target insects however makes this approach highly desirable. Such phenomena gave rise to the intriguing idea of using these natural or artificial pheromones to attract and trap insects, to lure them to contact poisons, chemosterilants or pathogens and to mask the location of females by saturating the environment with synthetic sex pheromones. Such techniques could offer highly specific control methods with little or no ecological side effects. An example of possibility is furnished by the eradication of the Oriental fruit fly on Rota island where the artificial pheromone or attractant viz. methyl eugenol lured the male flies to a poisoned coated surface. A similar but more important achievement was the eradication of the mediterranean fruit fly from about one million acres in Florida in 1956-57. It appears that adequate control from pheromones or attractants probably can only result if they are so highly effective as to be able to reduce the pest population to low level in the treated area and are used over a large area so that immigration of the pest from the outside would not negate the effort. Pheromone traps are commonly used for controlling various moths e.g. *Helicoverpa (Heliothis) Spodoptera, Leucinodes, Earias* etc.

26

INSECT HORMONES AS PESTICIDES

It is well known that the growth and moulting in insects is controlled by three main hormones. The brain hormone secreted by the neurosecretory cells of the brain activates the prothoracic glands. These glands in turn secrete one or several closely related steroids, the ecdysones. The ecdysone(s) cause the insect to moult. The kind of cuticle secreted by the epidermal cells at each moult is affected by a third group of hormones, the juvenile hormone, which are secreted by the corpora allata. In the presence of juvenile hormone larvae moults into larvae. However, in their absence in hemimetabolous insects larvae metamorphose into adults and in holometabolous insects, larvae moult into pupae and then into adults.

Moulting is controlled by regulating the release of brain hormone. Maturation and metamorphosis are controlled by regulating the release of juvenile hormones. In simplest terms, the ecdysones or their analogues stimulate the synthetic activities necessary for moulting, whereas juvenile hormone analogues influence the kinds of synthetic activities that occur in response to the ecdysones. Out of the three the last two hormones and their analogues or mimicking substances are of much value as for as the control of insect pest population is concerned.

A. Use of Juvenile Hormones as insecticides

As is known, the juvenile hormone inhibits the metamorphosis in all insects and as long as juvenile hormone is present in high concentration larvae will moult into larvae and they may die in this way without attaining adulthood.

It was the discovery of the Slama's "paper factor" (Slama and Williams, 1965, 1966) that opened a new chapter on the methods of the control of insect pests, by inhibiting metamorphosis and causing several other abnormalities. The greatest attention has been paid to juvenile hormone analogues with reference to their possible utilization as entirely new types of insecticides. The juvenilizing effect of juvenile hormone analogue is very great. The application of 0.001 mg of the juvenile hormone analogues to the body surface of the last instar nymph of *Pyrrhocoris apterus* or very brief contact with the ``paper" leads to a complete juvenilizing effect. It has been also pointed out that the "paper factor" named `juvabion" extracted from the North American trees such as the balsam fir (*Abies balsamea)* normally used for manufacturing the paper has been extremely useful in controlling the *Pyrrhocoris apterus* and other members of the family Pyrrhocoridae. It has been clearly demonstrated that the application of 0.01 mg of the crude extract of juvabione-, a juvenile hormone analogue extracted from balsam fir, is quite effective and inhibited metamorphosis of the last instar nymphs of *Pyrrhocoris apterus.*

The delayed action of juvenile hormone on metamorphosis after application to the late embryo was also seen in *Pyrrhocoris.* Normally during the moult to the fith instar the corpus allatum stops secreting juvenile hormone and the larva moults to an adult 7 days later. If however, juvenile hormone analogue is applied at the outset of fifth instar, then the individual moults after 5 days to a supernumerary sixth instar larva. Similarly appliction of juvenile hormone analogue(s) to the *Pyrrhocoris* embryo after blastokinesis also produced sixth instar larvae. Some of these giant larvae moulted again to seventh instar larvae but they never metamorphosed (Riddiford, 1972).

It has also been calculated that about 1 gram of the crude extract of the balsam fir tree would be sufficient to kill about 100 million bugs. Likewise, it has been found that topical application of 0.1 mg juvabione to *Pyrrhocoris apterus* sterilizes the adult females. Slama and Williams (1966) reported complete inability of certain insect species to reproduce in contact with a juvenile hormone analogue or "paper factor" present in the American paper products. After the discovery of the "paper factor" there occurred a rapid progress in the research of juvenile hormone analogues. As remarked by Slama *et al.* (1974) the general principles concerned with practical use of juvenile hormone analogues, in the insect control occur to be quite simple. The basic idea is that the development disturbances caused by juvenile hormone analogues at different ontogenetic stages may lead to mortality of the affected species/or at least may prevent insects from reproduction. Since a real potency of insect pest gradiations is primarily associated with their tremendous capacity to reproduce, it is easy to realize as pointed by Slama *et al.* (1974), that now serious consequences for development of insect pest population such an arrest of reproudction might have with juvenile hormone analgues. The applied Entomologists get an access to chemical growth regulators which have no direct toxicity to insect's and other form of life i.e. human beings, domestic animals, birds etc. Thus juvenile hormone analogues form a new class of chemicals which can be characterized as specific regulators of insect growth and development. Williams (1967) suggested the term "third generation pesticides" to the methods of the hormonal control of insect pest population particularly the use of juvenile hormone(s) as insecticides and a realistic account of the practical use of juvenile hormone analogues in insect control has been expressed by Stall (1971).

Further, juvenile hormone analogues not only inhibit metamorphosis in adults but also block the maturation of imaginal discs which are the primordia of many adult integumentary structures in holometabolous insects. The imaginal discs occur in larvae as tiny sacs of epithelial cells which come directly from embryo e.g. in Lepidoptera, each imaginal wing disc is set aside in the embryo as a small nest of cells destined to form the pupal and adult wing. These imaginal disc cells divide during larval life but fail to secrete a cuticle until the time of pupation.

Pupation is brought about by a falling concentration of juvenile hormone and results in the rapid morphogenesis of wing disc and the secretion of a cuticle. If juvenile hormone is applied to a Lepidopteran larva prior to pupation, it prevents both the morphogenesis and cuticle synthesis in the wing disc. When juvenile hormone disappears, the imaginal wing disc continues its development (Slama and Williams, 1966). Riddiford and Williams (1967) and Riddiford (1970) have demonstrated that the application of juvenile hormone analogues to young embryos interferes with their development. Thus juvenile hormone blocks not only the metamorphosis of the larva to adult but it also blocks the equally profound metamorphosis of the embryo to larva. Thus it is clear that juvenile hormone analogues play a significant role in the embryonic development.

An interesting feature of juvenile hormones action is that they act directly on the target cells and do not require mediation of any specialized organ. An other feature is that the juvenile hormone analogues rapidly penetrate the cuticle of insects.

It has been found that the immature insects which have been affected by an exogenous juvenile hormone usually fail to moult spontaneously, die soon after moulting or fail to reproduce (Schneiderman 1972). Juvenile hormone analogue treated insects are intermediates between larva and pupa, pupa and adult, or larva and adult.

Bhatnagar-Thomas (1972) has studied the potentiality of juvenile hormone analogue as pesticide in controlling *Trogoderma granarium*. She has reported the inhibition of metamorphosis and has recommended the use of juvenile hormone analogue for control of Khapra beetle (*T. granarium).*

Mariamma-Jacob and Prabhu (1979) have reported various types of abnormalities in the embryo of *Dysdercus singulatus.* Awasthi (1985) has reported that juvenile hormone analogue treated last instar larvae of *Philosamia ricini* indicated numerous abnormalities e.g. 5-7 days duration of larval life was prolonged to 24 days and they died without metamorphosis. The pupae retained larval characters; the head, wings, antennae etc. were abnormal in the moths which emerged from the pupae of treated larvae and died soon after emergence

Schneiderman (1972) has observed that young or immature insect larvae are not killed by juvenile hormone analogues. In order for these agents to kill insects they must act during last larval instar or during metamorphosis where they cause the production of intermediates. Alternatively, juvenile hormone analogues must act upon adults where they function as chemosterilants. Furthermore, juvenile hormone analogues must be applied at the proper time and must persist for several days to be effective in producing non-viable intermediates.

As pointed out by Schneiderman (1972) this matter has been analyzed by many workers. The main argument is that : to produce maximum kill with minimum dose, juvenile hormone analogues must be applied when some parts of the insect's body are sensitive to these substances whereas other parts are not e.g. it has been found that bugs, locusts and other exopterygote insects usually become insensitive to juvenile hormone analogues a few days after the last larval moult. Consequently one must apply juvenile hormone analogues shortly after the last larval moult, generally during the first third of the last larval instar or much later, after adult emergence, where juvenile hormone analogues will have an ovicidal effect (Schneiderman, 1972). Riddiford (1972) has also studied the ovicidal action of juvenile hormone analogues and has remarked that the juvenile hormone analogues can act as an ovicide if given to adult female or to her freshly laid eggs and this action of juvenile hormone would seem to be the most practical means of using juvenile hormone for insect control. One can eliminate the potentially harmful larval stages as well as next generation and even if the application is not early enough to prevent hatching the delayed effects will prevent metamorphosis into a normal reproductive adult.

B. Use of moulting hormones as insecticides

A brief mention may be made here also of the use of moulting hormone in pest control. Studies of several workers have shown that most insects die soon after they have been induced to moult by an injection or topical application of an ecdysone. Alpha ecdysone appears to be less toxic than the other ecdysones, perhaps because the insect regulates its rate of conversion to the active form, beta ecdysone. The most striking abnormalities produced by ecdysones are juvenile hormone-like effects e.g. if a pupa of *Tenebrio* is injected with 25 micrograms of beta-ecdysone it moults into a second pupa instead of into an adult. It has been also found that exogenous ecdysones usually kill insects raises the question of whether ecdysones may be suitable as insecticides. They affect insects at all stages from newly hatched larvae to adults. They can be applied topically in various solvents and penetrate the cuticle and several volatile derivatives have been synthesized so that ecdysones can be used as fumigants. Some analogues are effective when fed to insects. Moreover, many ecdysones are quite stable. Also various conjugated ecdysones are being synthesized in an effort to produce ecdysones which the insect cannot break down. Robbins *et al.* (1968) have demonstrated that various conventional insecticide synergists like piperonyl butoxide synergize the action of ecdysone analogues.

Another attractive feature of the potential use of ecdysones and related substances for insect control is that they are already widespread in the environment. It was discovered that substances with ecdysone activity are found in many vascular plants in some cases accounting for as much as 1% of the dry weight of certain plant tissues. Thus for more than 40 different steroids with ecdysone activity in insects have been isolated from higher plants by Nakanishi *et al.* (1966), Takemoto *et al.* 1967), Siddall (1970) etc. These phytoecdysones include authentic alpha and beta ecdysone.

Thus, it is hoped that in future synthetic hormone analogues both juvenile hormone analogues and moulting hormone analogues would be utilized for controlling the pest populations after solving some of the practical difficulties.

27

PLANT QUARANTINE AND LEGAL CONTROL

No attention was paid to the insect pests being carried from one place to another with the transportation of plants or seeds till it was noticed that the insects introduced into the foreign countries thrive best and increased at a very high rate in their new homes e.g. when American grapes were brought to France, *Phylloxera* caused such a heavy damage to French vine yards that the necessity of checking this pest from being carried to various parts of the world was seriously felt. Hence legislation had to form laws for this purpose. India and western countries like USA have passed a number of Acts to regulate the pest and to transport plant and their products. Under these Acts all imported goods should be carefully inspected to be insect free or should be fumigated and sterilized before allowing them to disembark. The export and import laws of the country now enforce the compulsory fumigation of food stuffs before and after embarkation.

The purpose of plant quarantine is to prevent by legal restrictions the entrance and establishment of a plant disease or insect pest in an area where the disease or pest does not exist. The area to be protected by this means may be a country, a state or it may be a smaller unit such as a part of the state. There are four methods of plant quarantine viz. (1) inspection at the point of destination, (2) inspection and (3) certification at the point of origin, (4) complete embargoes and controlled introduction of plants.

1. **Inspection at the point of destination:** This inspection is necessary because many insect pests can be discovered and their accidental introduction thereby eliminated. Often it is physically not possible to examine all materials likely to carry insect pests with sufficient care to detect them. Also because some insects and plant

diseases have inconspicuous stages they are extremely difficult to locate.

2. **Inspection at the point of origin** : Most countries and many states follow this plan i.e. inspection and certification at the point of origin. In this case materials are allowed to enter the state or the country as the case may be, provided they bear a certificate issued by the plant quarantine officer which mentions that the materials have been inspected and found to be free from insect infestation or plant disease infection. Again here also it is not possible to seperate physically all the materials likely to carry pests.

3. **Embargoes** : By this we mean the exclusion of all plant materials or commodities which are classed as hosts of the insects. This is also not possible in practice. Because the area where the pest exists cannot always be definitely determined and the material listed in embargo may not include all of those which might carry or harbour the pests.

4. **Controlled introduction** : This consists of allowing the importation of only a very limited amount of material for propagating purposes, obtaining it from pest free or disease free areas and from plants believed to be insect or disease free, inspecting the shipment very carefully then planting in an isolated place for a preliminary period. After obtaining healthy stock in this way, it can be released for multiplication and distribution.

Enforcement of plant quarantine

Foreign plant quarantine: Plant quarantine inspectors are stationed by the Government at all of the principal ports of entry. Materials which must be watched are those which can carry the pest or diseases. They not only include the food plants of the insects but any article on which any stage of it might be found. Such as cotton used for packing, gift might contain seed infested with the pink boll worm and straw which could be infested. The central and state Governments have enforced several laws for this purpose, from time to time.

The Government of India have enforced several Acts like sea customs Act of 1876 to stop the entry of Mexican cotton boll weevil, Destruction Insect Pests Act of 1914, Poison Act of 1919, Drug Act of 1940 etc. for preventing the entry of infested seeds and not allowing any material without proper inspection and certification that the material is free from all sorts of insect infestation as well as fungal, viral and bacterial diseases. As a rule all the food stuffs are fumigated.

28

INTEGRATED PEST MANAGEMENT

Definition and Concept

A mention may be made here that when a single method of insect pest control fails, then several other methods are adopted at one time. According to Hill (1975) integrated control or the Integrated Pest Management (IPM), is now defined as a pest population management- system that uses all suitable techniques (physical, cultural, chemical and biological) either to reduce the pest populations and maintain them at levels below those causing economic injury or to so manipulate the populations that they are prevented from causing such injury.

Integrated control achieves this ideal by hormonizing techniques in an organized way by making the techniques compatible and by blending them into a multifaceted flexible system and at the same time ensuring the least possible ecological disruption. This definition of integrated control now generally accepted is much broader than earlier difinitions, which stressed the complimentary use of chemicals together with biological control.

It was pointed out by Smith and van den Bosch (in Kilgore and Doutt, 1967) that the previous attitude was logical because these two approaches are our primary resources in the fight to control insect pests and as used in many instances they are in direct conflict with each other. However, practical experience and logic have made it clear that we must integrate not only just chemical and biological control but all available procedures and techniques into a single pattern aimed at profitable crop production and minimal environmental disturbance. Conway (1972) found that in Sabah, excessive use of pesticides was severely aggravating the local pest situation and the most effective immediate measure was to curtail spray applications to allow survival of the predators and parasites which normally keep most of the pest populations in check. Rabb (1970) arrived at another difinition where he stated that originally integrated control tended to refer to the modification of insecticidal control in order to protect and enhance the activities of benificial insects (predators and parasites). Subsequently, integrated control interpretations have become more comprehensive until now some definitions of integrated control embody most of the essentials of pest management. Rabb preferred the term pest management because it connotes a broader ecological basis and a wider variety of opinions in devising the solutions to the pest problems.

According to Reay (1969) the need for an integrated system has arisen form the short comings of each method. Chemical insecticides are essentially drastic and short term in their effects and result in the deposition of toxic materials on food stuffs and livestock. Moreover, they often have a broad spectrum of activity and can so alter the biotic components of the ecosystem as to cause the emergence of new problems. He further remarked that the biological agents on the other hand are safe to use but long term in action and may permit a pest to destroy a substantial part of a crop before they exert their effect. The correct use of integrated systems presupposes that a great deal is known about the behaviour of the organisms which are found in the larger area and involves the very careful use of insecticides. Only these compounds or formulations which will affect the pest more than the rest of the fauna can be used and their timing and placement must be related to the biology of any beneficial species present. Thus, Reay (1969) adds that e.g. the use of DDT formulations in which the toxic particles are covered with a protective coating ensures that the compound will act only as a stomach poison to phytophogous insect pests and will not affect beneficial species as a contact toxicants. Similarly, the timing of insecticidal spray can ensure that parasites or predators will not be harmed because it is known that they are elsewhere or in a resistant state at that time. Even broad spectrum compounds may be used as long as they do not have prolonged residual activity and are employed as "spot" treatments or if sufficient areas are left untreated to maintain as reservoir of beneficial organisms. Reay (1969) has further mentioned that because of the lack of truly selective chemicals and the existence of what frequently amounts to scanty knowledge of the ecology of the pests and their enemies, integrated system have found only limited application. However, there are certain exceptions too.

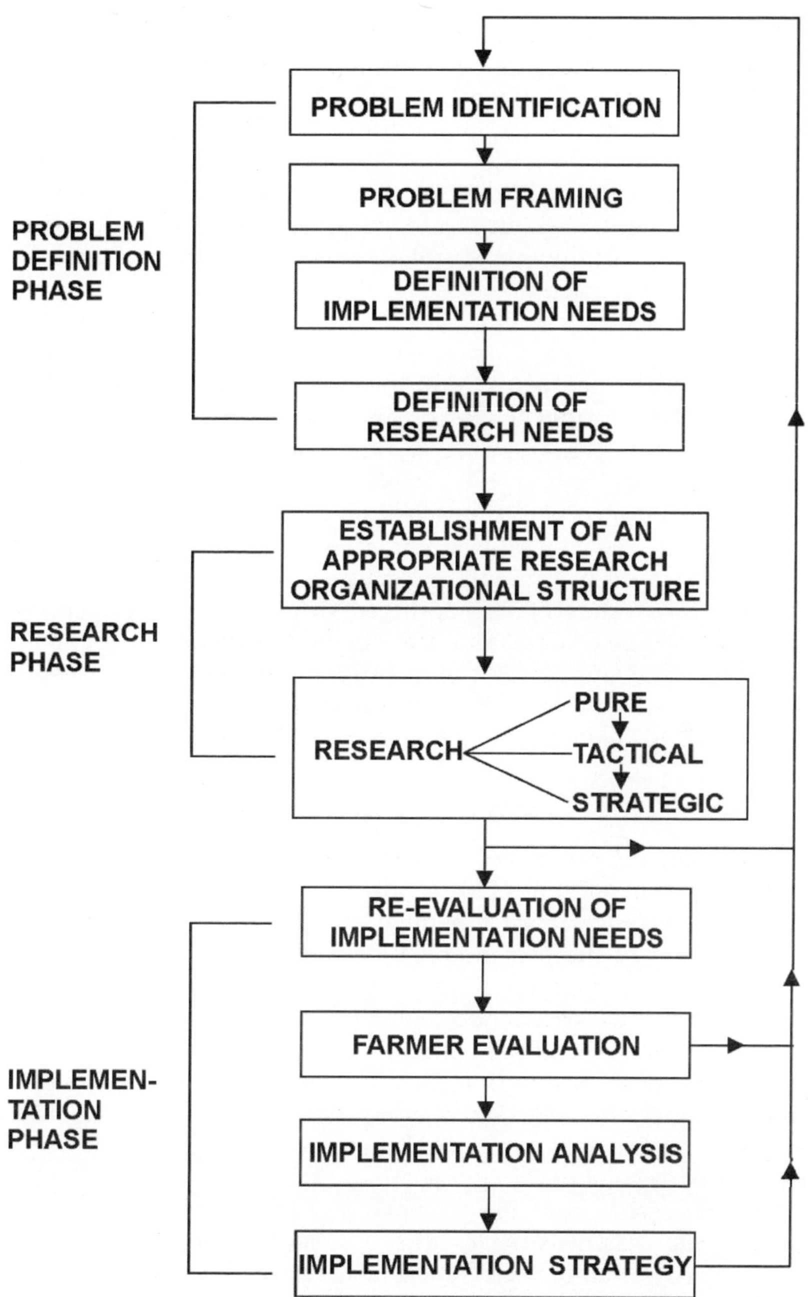

Chart showing the three phases in the development of an IPM programme (From Dent, 1991).

Dent (1991) recognized three distinct phases in the development of an IPM programme: problem definition, research, and implementation **(see chart)** and suggested that the success of a programme depends largely on the time and effort given to the first of these. Clearly, the more information that can be collated for the definition phase, the easier it will be to identify the research needs and implement the management strategy that will provide the maximum economic and socially acceptable returns. In addition to basic ``scientific" information on a pest's ecology and current methods of control, there are important socioeconomic aspects to be considered; i.e., losses in yield must be placed in the context of the farming system in which they occur, and any changes in strategy must be formulated in consultation with, and meet with the approval of, farmers. In many early projects, relatively little information was available on which to base IPM strategies, and these were developed largely by intuition, despite which they were remarkably successful. Now a days the considerably greater body of information generally available is stored in computers for use in constructing and testing mathematical models of the agroecosystem. Once developed, such models can be used to examine the effects of varying one or more of the factors that influence the population density of a pest and therefore, to determine the optimum method of reducing and maintaining this density below the economic injury threshold.

Agroecosystem

The following are some of the questions that might be asked in the development of an IPM strategy for the insect pests in a particular agroecosystem (remember that insects typically are only one component of the pest problem, and that any strategy for their management must be compatible with those developed for other pests and with overall socioeconomic goals). (1) What is the economic injury threshold? It has been observed frequently that insecticide users were maintaining pest population density at levels such below the economic injury threshold, increasing their operational costs in both the short and the long term. (2) What are the best parasitoids and predators to use and how might their abundance be increased? (3) Are some of the pests amenable to control by microbial insecticides? (4) What are the best insecticides to use, when should they be applied, and what are suitable doses? These aspects should be considered not only for the pest complex but also with respect to the beneficial insects in the system and thus require knowledge of the specificity of action, stability, and possible side effects of the insecticides. (5) Are suitable cultural procedures available? (6) Are resistant strains of the crop available? (7) Do the farmers already have the resources, human and technological, to adopt the new strategy? (8) Will the new strategy be acceptable to farmers, both economically and socially? Culture, tradition, and religion are important factors in determining farming practice. Thus, changes in planting and harvesting dates may be resisted if they interfere with other customs. Likewise, a suggested change in the cultivars used may not be acceptable if the new cultivars' taste, texture, or yield is different. It will be apparent from this sample of questions that IPM is interdisciplinary in nature and may require the collaboration of experts in widely different areas, e.g., entomologists, plant breeders, chemists, toxicologists, meteoro-logists, computer modelers, economists, anthropologists, and sociolo-gists. (Dent, 1991; Gillott, 1995).

Many examples of pest management might be cited that differ in complexity, geographical location, and crop pest being controlled. In its simplest (and rarest) form, pest management involves only one method of control and, strictly speaking, cannot be described as IPM. Thus, biological control is entirely satisfactory for control of sugarcane pests in Hawaii and coconut pests in Fiji (DeBach and Rosen, 1991). More often, two or more control methods are employed. For example, at the height of the pesticide syndrome, apple growers in Nova Scotia used a battery of fungicides and insecticides (especially broad spectrum synthetics) against apple pests. Far from achieving the desired effects, this led to even greater problems. By about 1950 earlier pests such as codling moth (*Cydia pomonella*), eye spotted bud moth (*Spilonota ocellana*), oystershell scale (*Lepido-saphes ulmi*) and fruit tree leafroller (*Archips argyrospilus*) had become pests because of destruction of their natural enemies. The need to reduce the escalating costs of this programme stimulated intensive study of apple orchard ecology, including the effects of pesticides on the natural enemy complex. The outcome of this work was the development of an integrated control programme that utilized fewer but more specific insecticides (as well as different fungicides, some of which had killed some natural enemies) and allowed for recovery of many of the natural enemies. By 1955 the scale insect had practically disappeared, and the density of the eye-spotted bud moth had fallen significantly.

As a final example, the integrated control programme for cotton pests in the San Joaquin Valley, California, will be outlined. This region produces about 10% of the cotton grown in the United States. As with many other crops, cotton growing was initially a small scale agricultural activity, not requiring an organized programme of crop protection. With the development of synthetic insecticides, a massive increase in acreage devoted to cotton growing occurred. Insect pest outbreaks were easily dealt with through the regular but infrequent application of an organochlorine insecticide. With time, however, it became necessary to increase the frequency and dose of insecticide, or to change over to using organophosphates or carbamates to combat population resurgences and increased resistance in the two major pests, the corn earworm (*Heliothis zea*) and a mirid bug (*Lygus hesperus*), as well as outbreaks of secondary pests such as cabbage looper (*Trichoplusia ni*), beet armyworm (*Spodoptera exigua*), salt-marsh caterpillar (*Estigmene acraea*), pink bollworm (*Pectinophora gossypiella*), tobacco budworm (*Heliothis virescens*), and spider mites (Tetranychidae) following destruction of their natural enemies. At the peak of this ``crisis phase'', almost 50% of all insecticides applied to field crops across the United States were being used on cotton, and control often required 10 to 20 applications of insecticide per growing season (Gillott, 1995).

Early in the development of the integrated programme, it was realized that no reliably established economic injury thresholds existed for either of the major pests and that, for *Lygus*, time of insecticide application was critical. For *Lygus* it was determined that the generally accepted injury threshold of 10 bugs/50 net sweeps was invalid; rather, the major period of importance was June 1 to July 20 and only if the density exceeded 10 bugs/50 net sweeps in this period should insecticides be applied. After July 20, *Lygus* densities of 20 or more bugs/50 net sweeps would not affect the quantity or quality of cotton produced, provided that the plants had flowered normally in June and early July.

Economic threshold

It was noted that corn earworms seldom became pests except in fields treated for *Lygus* outbreaks (using the old value for economic injury threshold) where the corn earworms natural enemies (the bugs *Geocoris pallens* and *Nabis americoferus*. and the lacewing, *Chrysopa carnea*) were destroyed by insecticide. With the new threshold for economic injury due to *Lygus*, tied in with the plant's seasonal development, the use of insecticides was reduced, especially after late July, and the earworms natural enemies survived. Further reduction in insecticide usage came about with the raising of the economic injury threshold for corn earworms from 4 earworms/plant to 15 *treatable* earworms (first-and second instar larvae)/plant. This arose, in part, from realization that larvae often feed on surplus buds, flowers, and small bolls that are not picked.

Cultural control is also used to reduce the effects of *Lygus* on the cotton crop. *Lygus* prefers alfalfa (*Medicago sativa*) to cotton and will remain on the former plant when given a choice. Traditionally, however, alfalfa fields were solid cut (all of the field is cut at once) provoking mass migration of *Lygus* into adjacent cotton fields. Alfalfa growers were persuaded to practice strip cutting to prevent such a migration, particularly following the discovery that strip cutting also reduces pest problems in the alfalfa itself by favoring *Aphidius smithi*, a parasitoid of the pea aphid (*Acyrthosiphon pisum*). An earlier proposal that cotton growers plant strips of alfalfa through their fields has not become practical because of difficulty in managing both crops. It was also noted that safflower could be a major source of *Lygus*, the bugs moving into cotton as the former plant began to flower. Careful monitoring of the bug populations in the safflower, and a single insecticide treatment when 70% of the insects are third to fifth instar nymphs, largely eliminated the threat to adjacent cotton fields. Monitoring by means of light traps and pheromone traps (using ``gossyplure", the latter have been particularly valuable for assaying pink bollworm movements into the San Joaquin Valley) is now also routinely employed to determine the relative abundance of adult pest Lepidoptera (Dent, 1991).

Another cultural method that may yet become commercially feasible is the use of food sprays in attempts to prolong the seasonal effectiveness of natural enemies whose populations frequently decline after midseason. For example, in experimental plots adult *Chrysopa* were attracted to the food sprays and produced more eggs, resulting in significantly less injury to plants by corn earworms.

Microbial control has become an important method for use against some lepidopterous cotton pests. For example, *Heliothis* nuclear polyhedrosis virus against bollworm and tobacco budworm in California and other states in lieu of broad spectrum insecticides.

Integrated control results in substantial savings, both financial and in terms of pollution, through the greatly decreased use of insecticides. For example, in the San Joaquin Valley, the traditional insecticide based control programme for cotton pests cost about twice as much per acre as the integrated programme, and even greater savings may be incurred as the programme is further improved. From 1976 to 1982, the average quantity of insecticide applied per acre to cotton decreased by 72% for the United States as a whole. This, in combination with a decrease from 60 to 36% in the proportion of acres treated, resulted in a drop in cotton's share of the field crop insecticide market from 49 to 24%.

Despite these remarkable statistics, many crop growers and government officials remain convinced that the unilateral use of insecticides is best. Persuading these individuals that to continue with this approach is not only economically unsound but will lead to long-term, perhaps irreversible deterioration of environmental quality is perhaps the greatest challenge that scientists have ever faced (Gillott, 1995).

29

PESTS OF STORED GRAINS

1. ***Sitophilus oryzae*** Linn. (Fig. 65) : This is a true weevil, belonging to the family Curculionidae of the order Coleoptera. It is known as Rice weevil. It is one of the most destructive stored grain pests. These insects frequently cause complete destruction of grains and the infested grains will usually be found to be heating at the surface, and it may be damp, sometimes to such an extent, that sprouting occurs. Normally the kernels of the grain are eaten. Adults feed upon whole seeds or flour but the larvae develop only in seeds or in pieces of seeds large enough for them to live within.

Fig. 65. Rice weevil, *Sitophilus oryzae* **A.** Grub, **B.** Adult

Life history

The beetle in appearance is small, reddish brown in colour and is about 3.5 mm long. The female deposits small white eggs in a hole in the grain bored by the powerful, short jaws. One female can lay as many as 250-400 eggs during its life time. The eggs usually hatch in about 3-4 days into soft white legless larvae. The larvae bore their way into the grains until full grown. In about 3 weeks to a month, the larvae pupate inside the grains and live in this stage from three to six days. At the end of this period the adults emerge, immediately mate and a new family is thus started. The entire life cycle of this insect, under favourable conditions, ranges from 4-7 weeks. There are 5-6 generation in a year.

2. *__*Rhizopertha dominica*__ Fabr. (Fig. 66A) : This pest belongs to the family Bostrychidae of the order Coleoptera and is known as Lesser grain borer. It does almost as much damage as the rice weevil. It feeds in both larval and adult stages in the interior of nearly all types of grains. It can also ingest flour. It is most common in wheat and is one of the most destructive wheat insects.

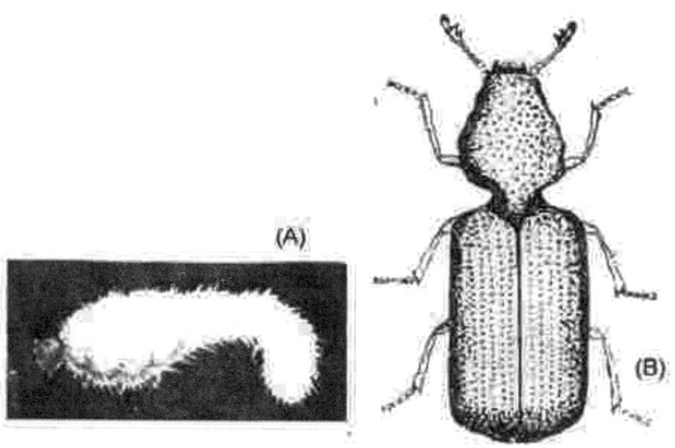

Fig. 66A. Lesser grain borer, *Rhizopertha dominica.* (A) Grub, (B) Adult

Life history

The insect is dark brown-black in colour and is about 3.5 mm long. The female lays eggs on the grains, but some times it will lay on bags, walls and among other food grains in the store. One female can lay as much as 500 eggs. The eggs hatch in 4-10 days into white larvae. The larvae are weak and are unable to attack the grain as the adult does. Consequently they search for soft corners or small holes in the grain. The larval period ranges from 40-50 days before they pupate. Duration of pupal stage varies from 6-9 days. The whole life history of this insect lasts about two months. There are 5-6 generations in a year.

3. ***Trogoderma granarium*** Everts (Fig. 66B) : This insect belongs to the family Dermestidae of the order Coleoptera. It is known as Khapra beetle. This is perhaps the hardiest among the insects that infest stored grains in India. This beetle feeds on most grains and

* Also written as *Rhyzopertha*/Bostrichidae

sometimes on pulses. Wheat flour is its best food and only the larvae of this insect are destructive while the adults are harmless.

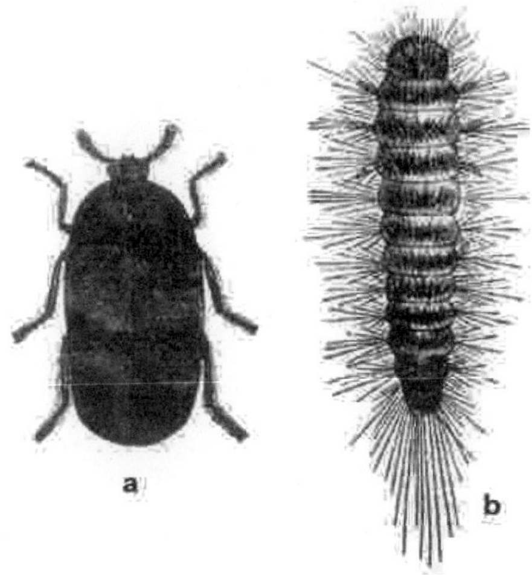

Fig. 66B. Khapra beetle, *Tragoderma granarium* a. Adult, b. Larva (Grub)

Life history

The adult is about 2-3 mm long, oval in shape and has a sooty black colour. The female lays as many as 125 eggs during its life time among the grains. The eggs hatch in about a week into new brownish white larvae whose body is covered with reddish brown hairs that form a tail like structure at the end. These larvae normally feed on flour and cracked grains. The male larva takes 20-30 days and the female 20-40 days to fully fed when they are about 4 mm long. The male larva lives generally for about 50 days under favourable condition after which they pupate. The pupal period varies from 3-6 or 7-21 days depending on climatic conditions. Under unfavourable conditions, the length of larval life can extend to 4 years. There are 4-5 generations in a year.

4. ***Tribolium castaneum*** Herb. (Fig. 67) : This pest belongs to the family Tenebrionidae of the order Coleoptera and is known as Rust red flour beetle. It feeds mostly on broken grains or grains damaged by other pests. It is also a pest of prepared cereal products like Atta, Maida and Suji. A heavy infestation of products by these insects gives a yellow colour to the food which becomes mouldy with the excreta of larvae and adults. Both larvae and adults attack the grains.

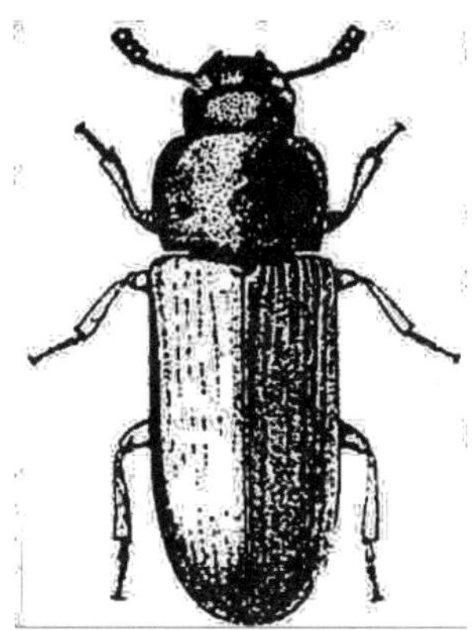

Fig. 67. Rust red flour beetle, *Tribolium castaneum*

Life history

The beetle is reddish-brown in colour and is about 3.5 mm long. The female lays 400-500 clear white sticky eggs scattering them in the flour or food stuffs. The eggs hatch in about 4-12 days into small brownish-white larvae which become full grown in 27-90 days under favourable condition and then pupate. The pupae are white in colour and the pupal period ranges from 1-2 weeks. The pupae hatch into adults which normally live for about six weeks.

5. ***Callosobruchus chinensis*** Linn. (Fig. 68) : This beetle belongs to the family Bruchidae of the order Coleoptera and is known as Pulse beetle. It is a serious pest of stored pulses. All kinds of pulses are eaten by this pest and stored pulses may be completely destroyed.

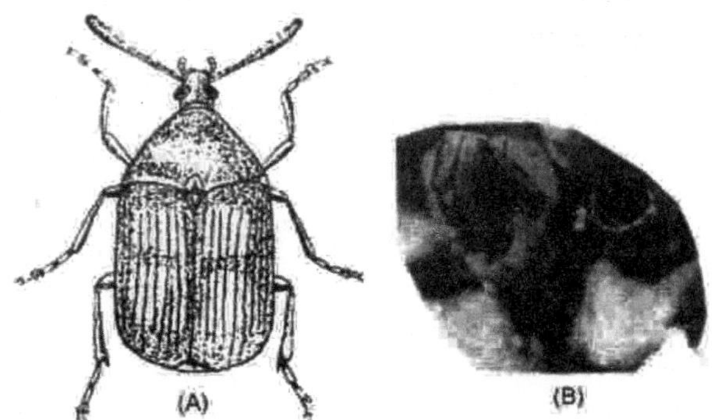

Fig. 68. A. Pulse beetle, *Callosobruchus chinensis* **B.** Damaged seed and the pulse beetle

Life history

The adult is small roundish about 3.5 mm long, brownish in colour. The female lays about 80-100 eggs which are elongate and yellowish white in colour. These eggs are glued to outside of the grain and there may be as many as 1-12 eggs on a single grain. The eggs hatch 5-18 days later into tiny- yellowish larvae which bore into the pulses. The hole made by the larva is so small that it is usually not seen. Inside the grain the larva grows slowly consuming most of the material inside. The larval duration is about 3 weeks then the larva pupates inside the grain and in about 4-7 days it comes out from pupa as an adult. Immediately after emergence mating follows and another family is started.

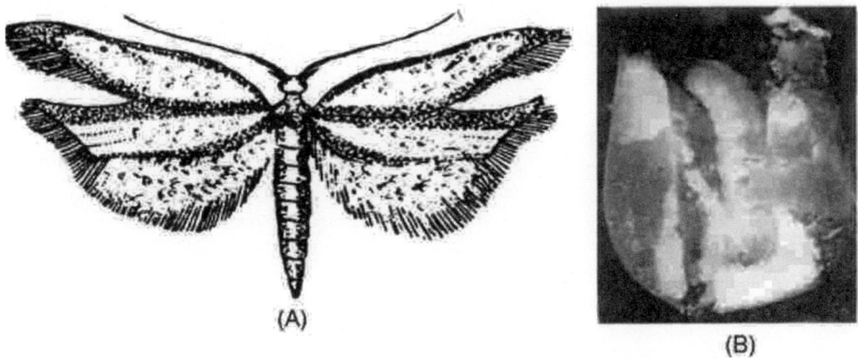

Fig. 69. A. Angoumois grain moth, *Sitotroga cerealella* **B.** Larva of the Angomois grain moth feeding in a kernel of corn.

6. ***Sitotroga cerealella*** Olive. (Fig. 69 A,B) : This insect is known as "Angoumois grain moth" and belongs to the family Gelechiidae of the order Lepidoptera. This is the most destructive grain moth which is very difficult to detect in the early stages of its attack as the damage occur inside the kernel.

Life history

The adult is a small, buff coloured moth, very active in flight and is about 10-15 mm from wing tip to wing tip. The single female lays several hundreds of small ovoid eggs, pinkish to reddish in colour, singly or in clusters of as many as 20 eggs on the grain. The eggs hatch in about a week and the tiny larvae about 8 mm long immediately bore into the kernel. The larval period ranges from 2-3 weeks, after which pupation starts inside the grain in a silken cocoon. After a pupal period of about a week within the grain they emerge as adult moth which do not feed on grains but live for a few days only. Breeding continues throughout the year except during the cold weather. The whole life cycle takes about 5 weeks at 30°C. There are several generations in a year.

7. ***Corcyra cephalonica*** Staint. (Fig. 69C) : This pest belongs to the family Galleriidae of the order Lepidoptera and is known as Rice moth. It is a serious pest of stored paddy, rice etc. It is slightly larger than the grain moth. It thrives well on broken food grains.

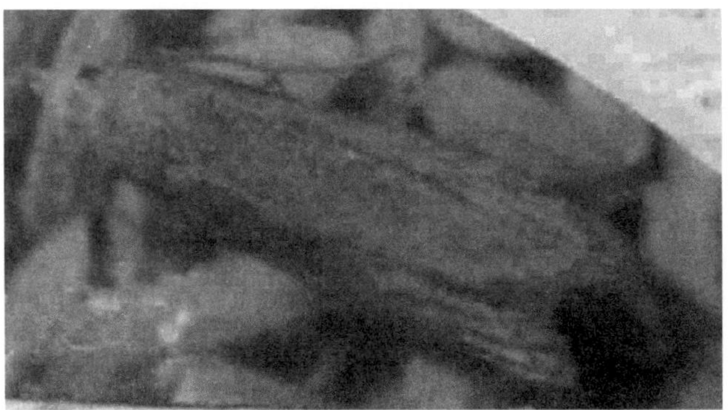

Fig. 69C. Rice moth, *Corcyra, cephalonica.*

Life history

The female lays as many as about 200 eggs loosely on the bags, walls or grains. The eggs hatch in 4-5 days into tiny creamy white larvae which become fully grown in about a month. At the end of this period the larvae spins a tuff silken cocoon and pupates inside this. The pupal period varies from 6-10 days after which the adult emerges and breeds. There are about 5-6 generations in a year.

CONTROL OF STORED GRAIN PESTS

Control of stored grain pests has been a subject of extensive study for a very long time. The methods adopted in recent times to control these insects have been very simple and very effective. Generally all these insects may be destroyed by heat and cold treatment as well as by treatment with fumigants, powerful contact insecticides, in smoke, fog or spray forms or by desiccating or by use of abrasive dusts.

1. Destruction by heat and cold treatment

Heat treatments generally give a good result in mills, stores, store rooms, houses etc. with the result that practically all the insects are killed. The expense involved in such a method of control is lower than that of any other method. Heat treatment of any place infested by stored grain pests should normally be carried out during the period when the outside temperature is high. This method essentially consists of heating up the entire room by a system of tubes through which hot water is continuously circulated. The hot water from these tubes should be able to maintain a temperature ranging 49°-66°C for several hours. Such a range of temperature is fatal to all insects exposed to this heat. Similarly the cold treatment for destruction of insects involves lowering the temperature of infested room to as low as 0°C. This low temperature is equally fatal to insect life.

It may be mentioned here that the above methods of destruction of stored grain pests are normally practiced in countries where facilities exist for central heating and cooling systems e.g. Europe and America. In our country chemical control is the method largely employed for destroying the insect pests of stored grains.

2. Chemical Control

One of the chief ways of chemical control is by fumigation. There are a number of fumigants useful for treating stored grains. Most of them are heavier than air and are applied on the top of the grains.

(a) Carbon disulphide : It is the most economical fumigant. It is preferable to use a 20% mixture in carbon tetrachloride which is effective at a doses of 10 kg./300 m.

(b) Ethylene dichloride : It is also effective when diluted with 25% carbon tetrachloride. It is used as 20 kg./300 m.

(c) Methyl bromide : It is another effective fumigant. It is effective at a level of 500g./300 m. It is outstanding for bagged materials such as flour on account of its great penetrating power.

(d) Chloropicrin : Another important fumigant is chloropicrin which is usually applied at 2.5 kg./300 m.

(e) Aluminium phosphide: It is widely used as fumigant for stored grain protection being marketed in the form of tablets.

There are some less volatile fumigants such as ethylene dibromide and acrylonitrile which give excellent control of insects in granaries.

Stored grain pests are also controlled by vacuomax fumigants. Acrylonitrile is best used for this purpose at a dose of 1/2 kg/1000 ft. This fumigant is almost as toxic as Hydrogen cyanide and is much more penetrating. Some other fumigants viz. ethyl dibromide are general fumigants much less toxic than methyl bromide. It is readily absorbed as a grain fumigant. The breads made from the grains have no bad odour or taste, due to its low vapour pressure and high toxicity, the fumigation of entire godown kills insects both in the grains as well as godown.

Often residual wall treatments with pyrethrum sprays have been found to have surprisingly lasting effect especially when they are deposited in cool, dark warehouses, the toxic effect remaining as long as 100 days.

A novel method of proofing grains against insects is the addition of abrasive powders such as mineral oxides or silicates which scar the cuticle and cause desiccation of the insects e.g. certain abrasive materials are magnacite, silica, hydrated lime, carborandum etc.

Many insecticides have been used as impregnants for food packages to prevent infestation by stored product insects. They include rotenone, fixed nicotine, dinitrobenzidine etc. Sometimes these insecticides contaminate the food stuff itself and render them useless for consumption. Consequently, in recent times, preference has been given to certain inorganic substances such as NH_4Cl which functions as gustatory repellent.

Another method of protecting grain from insects attack is by the use of grain treatments which are applied either as sprays or as powders to uninfested grains. Some of the modern grain protectants are pyrethrine synergids piperonyl butoxide, malathion (1%). Grain protectants have certain advantages over fumigants in that : (a) the grain can be treated while it is uninfested, (b) protectants are effective when the grains are stored in containers too loose to be fumigated, (c) protectants are less dangerous than fumigants and (d) protectants do not effect germination adversely. Various parts of the neem plant have been found very effective in controlling almost all stored grain pests (See: Insecticides of plant origin).

30

PESTS OF COTTON AND FIBRE CROPS

1. ***Dysdercus koenigii**** Fabr. (Fig. 70): This belongs to the family Pyrrhocoridae of the order Heteroptera. The common name is "red cotton bug". It is a serious pest of cotton and sucks juice with its long beak from the seed of either green or ripe cotton bolls. Both the nymphs and adults are voracious feeders and cause extensive damage to the plant. The main symptoms of attack by this pest is the shrivelling of bolls in the plant. A severe infestation by this pest causes serious impairment to the vitality of the plant and destroys the germinating power of the seed by the loss of its oil contents. Besides these features, the formation of lint is badly affected.

Fig. 70. Red cotton bug, *Dysdercus koenigii*.

Although *D. koenigii* is a major pest of cotton, this insect has been reported to thrive well on other malvaceous plants like Hollyhock, lady's finger or in other nonmalvaceous plants like sweet potato, tobacco etc.

---

* Same as *D. cingulatus* Fabr.

Life history

The eggs of *D. koenigii* when freshly laid are soft to touch, they are almost round yellowish-white in colour and a little over 1 mm long. They are laid in loose clusters about 80-100 per cluster normally in cracks on the surface of the moist soil or under fallen leaves round about the plants.

The eggs hatch out under favourable conditions in about 7 days. Small, active yellowish-red first instar nymphs hatch out which resemble the adults except in size and absence of wings. They are about 1.5 mm long dorsally flattened and convex possessing 4- jointed antennae and two jointed tarsi. The nymph generally cluster round the cotton bolls. In course of time, moulting occurs and there are normally 4 moults or 5 instars before a fully developed adult individual is formed. The average time taken for all the 4 moults is 30-35 days under favourable conditions. The duration between any two successive moults is dependent on many factors one of thembeing food. Under unfavourable conditions, there may be 5 moults and 6 instars and the moulting period may extend to 82 days. Generally the entire life cycle of this pest is completed in about 42-45 days.

This insect normally breeds on cotton from August to November, on Hollyhock form November to March or April, and on lady's finger from April to July each year. Many of them hibernate in the adult stage from December to middle of March.

Control

There are several ways of controlling this pest. Often the mechanical methods such as hand picking, trapping etc. are adopted. Chemical control chiefly advocates the employment of certain synthetic insecticides like dust 5% malathion or carbaryl as spray.

2. *Earias fabia* * Stoll. (Fig. 71): This pest belongs to the family Noctuidae of the order Lepidoptera, and is known as "spotted bollworm" It is equally a serious pest of cotton and the young larvae devour the tender shoots, leaves, flower buds and ripening bolls of the plant. The lint and seeds are completely destroyed and the yield is reduced by 10-50%. Like *D. koenigii* this pest also infests other food plants like lady's finger, Hollyhock etc.

---

* Same as *E. insulana* (Boisd.), *E. vitella* (Fabr.)

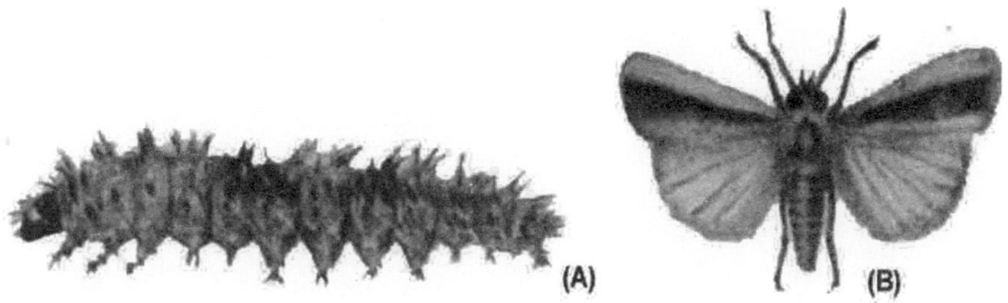

Fig. 71. The Spotted bollworm., **(A)** Larva **(B)** Adult (From Pruthi, 1969).

Life history

The female lays round eggs about 2-4 hundred in number singly on bolls, terminal leaves or buds or flowers. These eggs hatch in 4- 7 days into dark pale greenish larvae with black dots and short spine like hairs on either side of the back. As soon as they hatch, they bore into the tender shoots and later on reach the flowers and bolls. The full grown larvae having a brownish white body, spotted with different colours like brown, green, black and orange. Hence it is called "spotted bollworm". The larval period lasts for 9-16 days depending upon atmospheric conditions. At the end of the larval period they move out of the boll and pupate in a tough grey silken cocoon on plant or soil. The pupal period ranges from 10-12 days after which fresh moth emerges. The entire life cycle spreads over a period of 26-35 days.

Control

While the mechanical control is occasionally employed by hand picking the caterpillars and destroying them, the most common method of control is spraying of the insecticides like endosulfan 35EC @1.25-1.50 litres/ha in 1000 litres of water, 0.1% carbaryl or 0.03% monocrotophos at 15 days intervals 2-3 time controls the pests. Removal of damaged parts of the plant is also useful.

3. ***Pectinophora gossypiella*** Saund. (Fig. 72) : This belongs to the family Gelechiidae of the order Lepidoptera. The common name is "Pink bollworm". The adult moth is dark brown in colour with irregular black margins on the forewings. The hind wings are silvery grey and both wings are fringed with long hairs posteriorly. The tips of hind wings are sharply pointed.

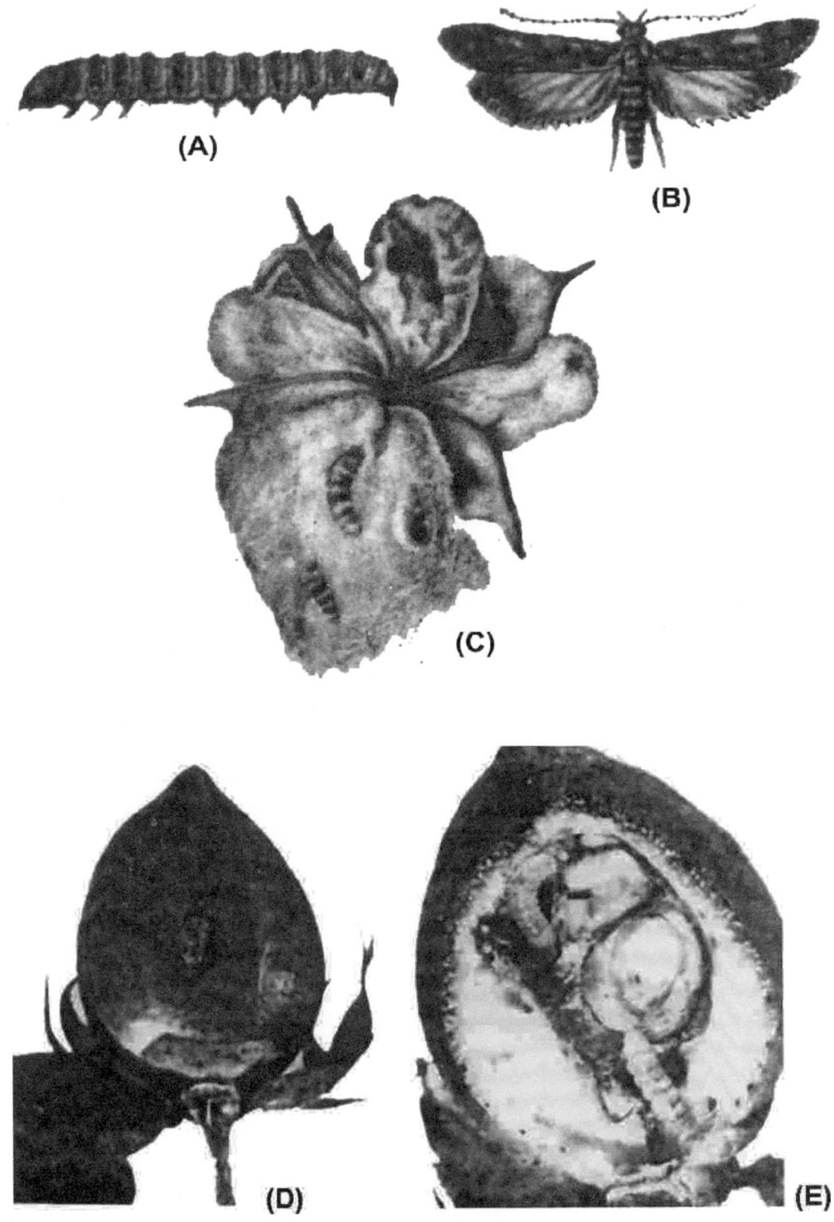

Fig. 72. Pink boll worm, **A.** Larva, **B.** Adult, **C.** Infested cotton boll (From Pradhan, 1969), **D.** Attacked boll by the pink bollworm of cotton, **E.** The pink bollworm feeding inside a boll (From Atwal, 1976).

This is another serious pest of cotton, and the caterpillar damages the bolls vehemently, consequently the seeds inside the bolls are completely devoured and the formation of lint is affected. In fact the infestation of a boll commences with the tiny caterpillar entering the boll at its base through a minute hole. Inside the boll the caterpillar excavate a long sinuous tunnel eating away tissues on its ways and damaging the lint badly. It finally reaches to seed which it tunnels and feeds on the nutrient available in the seeds. It should be pointed out here that the caterpillars of this pest unlike those of *Earias fabia* never attack the shoots. In addition to cotton, lady's finger, Hollyhock etc. also serve as suitable hosts for this pest.

Life history

The female lays elongate, oval eggs about 0.5 mm long and 0.25 mm broad. They have minute longitudinal zigzag ridges and they are laid either singly or in groups on or near a bud or boll. The eggs which are green at first turn yellowish brown gradually and hatch in about 4-10 days into small active yellowish caterpillars. Soon after emergence the young caterpillars starts attacking the crops. They gradually become pink in colour and they bore into bolls and webs together two seeds to make a resting place especially for hibernation. This feature of two seeds attached together is known as double seeds and in this condition the entire larval life is passed lasting for about 4 weeks. This is followed by pupation inside a flimsy silken cocoon in the soil or within the boll but outside the seed. The pupal period lasts about 10-14 days when the adult moth emerges. There are several generations in a year.

Control

The most effective method of physical control is the heat treatment in which the infested cotton seeds are heated to a temperature of 65°C for 30 seconds. This will kill all the caterpillars inside the seeds.

Chemical control involves the use of endosulfan 35 EC @1.25-1.50 litres/ha in 1000 litres of water. Rest of the insecticides used are same as for spotted boll worm.

4. *Spodoptera exigua* Hubner (Fig. 73): It is a polyphagous insect pest of jute plant. It belongs to family Noctuidae of the order Lepidoptera. The damage is caused by caterpillars which are gregarious, moving in swarms and destroy the young leaves of jute. It is also a pest of rice, cotton as well as tobacco.

Fig. 73. Indigo caterpillar - *Spodoptera exigua* **A.** Laarva, **B.** Adult.

Life history

The eggs are laid in clusters (up to 200) on the leaves which hatch in about 2-4 days. The larvae are blackish and there are six instars. Larval development takes 15-20 days. Pupation takes place in the soil and lasts for about 6 days. There are several generations in a year.

Control

Cultural methods of control such as ploughing and burning the crop stubbles, flooding the infested field and the removal of weeds, all help to lower the pest population. The adult caterpillars can be controlled by spraying endosulfan 35EC @450-1500 ml/ha mixed with 500-1000 litres of water depending up on the severity, growth and type of caterpillars.

5. ***Utetheisa pulchella*** Linn. (Fig. 74A) : This belongs to family Arctiidae of the order Lepidoptera. The caterpillar is known as "sunnhemp hairy caterpillar". The caterpillar feeds on the leaves and capsules. The adult moth is creamy white with black and red spots on the fore wings and dark black spots on the margin of the hind wings. The full grown larva measures about 3.8 mm in length and is provided with red, white markings on the body, the head is slightly brown in colour.

Fig. 74A. Sunnhemp moth, *Utetheisa pulchella.*

Life history

The female moths lay small whitish eggs on the tender leaves and shoots. Newly hatched larvae feed on the leaves. In the mature crop caterpillars feed on pods. Pupation takes place in the leaves as well as in the soil. The entire life cycle is completed in about 5 weeks. There are several generations in a year.

Control

The caterpillars can be controlled by spraying the crop with endosulfan 35 EC @1.5 litres/ha in 1000 litres of water. The caterpillars and moths can also be destroyed by hand picking.

6. *Oxycarenus laetus* Kirby (Fig. 74B): This belongs to family Lygaeidae of the order Heteroptera and is commonly known as "dusky cotton bug". It is a minor pest of cotton. The adult bugs are very small about 4 to 5 mm in length, brown in colour. Both the young and adults sucks the sap of immature seeds of the cotton. Lint is stained when they are crushed in factories. There are several other host plants belonging to Family Malvaceae.

Life history

Each female lays white 15 to 40 eggs singly or in groups of 4 to 18 on the half open bolls which become pink later on. After about one week or more nymph emerges. There are seven instars. The total time taken for development ranges from 30 to 40 days. There are several generations in a year.

Fig. 74B. Dusky cotton bug, *Oxycarenus laetus*

Control

The dusky cotton bugs can be controlled by applying methyl demeton 25EC @1 litre/ha in 1000 litres of water in the form of spray on the crop.

7. *Anomis sabulifera* Guen. (Fig. 75): The jute semilooper belongs to family Noctuidae of the order Lepidoptera. It is a specific pest of jute and is the most destructive in the jute tracts of India and neighbouring countries. It is considered to be the most serious pest of jute foliage. The infestation is said to come about in three successive waves of which the middle one is the most injurious. The apical buds of the plant are most vulnerable and the intensity of infection tends to vary inversely with the age of the foliage.

Life history

The pest hibernates during winter in pupal stage in the soil and the moths emerge with the first moist warm weather. The moths lay eggs on the jute plant singly particularly on the underside of the young leaves. The eggs look very much like a tiny drop of water on the leaf. There may be several such eggs on each leaf. A single moth can lay more than 150 eggs. The egg period lasts for about 2 days after which a small, green caterpillar hatches out. The tiny caterpillars on emerging start feeding on the apical leaves and the buds. The full grown larva is about 4 cm. and it attains the size after 5 moults during a period of about 2-3 weeks. The pupation may take place on the plant or in the soil. The pupal period lasts for only a week or so after which the moths emerge and begin to produce the next generation. The whole life cycle takes abut one month, thereafter several generations are completed in a year.

Fig. 75. Jute semilooper, *Anomis sabulifera* **A.** Larva, **B.** Adult

Control

The pest can be successfully controlled by spraying the crop with endosulfan 35EC @ 450-1500 ml/ ha mixed with 500-1000 litres of water is also very effective.

31

PESTS OF SUGARCANE

1. ***Pyrilla perpusilla*** Walk. (Fig. 76) : This belongs to family Fulgoridae of the order Homoptera. This is a serious pest of sugarcane and it prefers the broad leaved succulent variety as the food plant. Both the adults and nymphs are voracious feeders which ingest the sap from the underside of leaves. This results in the shrivelling and drying up of the leaves. A serious attack affects the quality and yield of the crop and the sucrose content is reduced to 50% and Gur becomes tasteless.

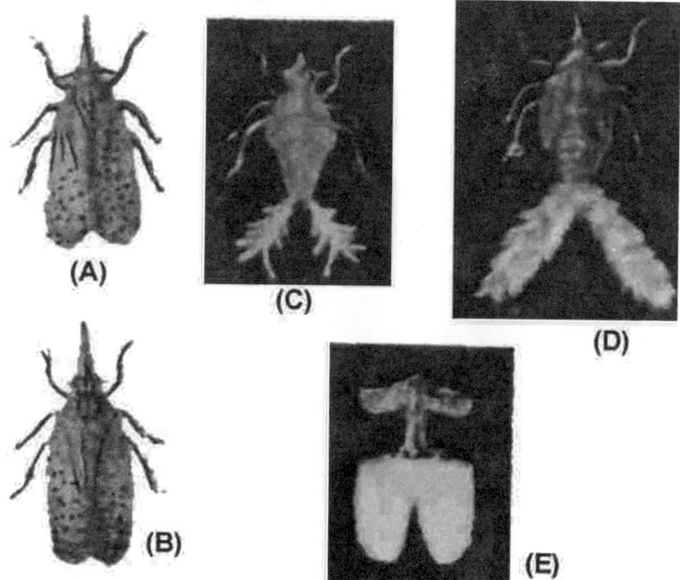

Fig. 76. Sugarcane leaf hopper, *Pyrilla* **A.** Male moth, **B.** Female moth, **C-D.** Nymphal stages, **E.** Anal tuft of female adult (From Pradhan, 1969, 1983)

The adult hopper is normally dull yellow in colour with a head produced anteriorly into an upwardly turned snout. The two pairs of wings are folded and rest over the back, a characteristic feature of Homoptera.

Life history

The female lays greenish yellow eggs about 300-530 in clusters on the lower surface of leaves or in leaf-sheaths or on stem. The number of eggs in each cluster vary from 20-60. All these eggs are covered over with white cottony waxy filamentous threads as a safeguard from parasite attack. Hatching occurs in about 7-29 days depending on season and pale brown nymphs are released. These are covered on their anal processes with waxy filaments. The nymphs undergo five moults after which fully developed adult individuals are formed. The entire life cycle takes about 46-59 days. In a year the first brood crops up in April to June, the 2nd in July to September and the 3rd in September to November.

Control

Occasionally mechanical control by collection and destruction of eggs between fingers on the leaves or by burning is applied. Adoptation of biological control through the medium of certain Hymenopterious parasites and the larvae of certain coccinellides is another successful measures of lowering the pests incidents. Chemical control involves the use of endosulfan or dimethoate or fenitrothion 0.025% spray or malathion 50 EC or endosulfan 35 EC @ 1.25 litres/ha in 1000 litres of water.

*2. **Aleurolobus barodensis*** Mask. : This pest belongs to the family Aleurodidae of order Homoptera and is commonly known as "sugarcane white fly". It is quite a serious pest damaging atleast 50% of the crop from July to November. The insect which are generally found on the lower surface or on the apices of the unopen leaves, desap the plant, bringing down their vigour and turn them sticky and pale. As in *Pyrilla*, a heavy infestation reduces the sucrose content of the plant. The adults are active creatures with prominent black eyes and a pair of mealy white wing.

Life history

The female deposits dull brown or orange eggs in a line near the mid ribs or in any part of the lower surface of leaves during winter.

The number of eggs laid vary from 60-65; the eggs have a broad pedunculated base and a pointed tip. They are found in groups of 15-20. They hatch in about a week or so into dull yellow nymphs. In course of time the colour of nymphs changes to shiny black. The locomotory legs of the nymphs which are functional for a short duration atrophy immediately after the first moulting, and the insect fixes itself to some point on a leaf and remains there till it completes its life history. There are usually 4 moults after which the 5th instar nymph enters into a pupal stage an unusual feature in the metamorphosis of Hemiptera-Homoptera lasting from 10-15 days. Finally the pupa transforms into the adult white fly. The whole life cycle normally occupies 14-25 days under the optimal condition but may prolong to 25-55 dyas depending on climatic conditions. This pest is very active from August onwards.

Control

Both mechanical and biological controls are commonly employed. The pest can be controlled by spraying the crop with malathion 0.1% or endosulfan 35 EC or monocrotophos 40 EC @ 1.5 litres/ha mixed with 800 litres of water.

3. ***Scirpophaga nivella**** Fabr. (Fig. 77): This insect belongs to family Pyralididae of the order Lepidoptera. It is a major pest of sugarcane attacking the crop at all stages of growth and causing a heavy damage in July and August. The larva usually attacks the top-shoots of the cane and hence is known as `top shoot borer'. This results in the curling and drying up of leaves and formation of spiral structures known as dead hearts.

The adult insect is a silvery white moth with dark black eyes and tufts of yellow, orange, brown or red hairs over the anal segment in females.

Life history

The female lays 60-80 eggs, oval or elongate in shape, in five to eight clusters on the underside of leaves. Hatching occurs in about 6-12 days and very small shiny caterpillars are released. Immediately after emergence, these caterpillars start attacking the rolled up leaves. Gradually they bore into the mid rib of an upper unfold leaf and enter the terminal shoot through the rolled base of the top leaves.

* Same as *Tryporyza nivella* (Fabr.)

The entire larval period in this manner is passed either on leaves or on top shoots and in about 22 days the larva becomes full grown possessing a creamy white colour with dark streaks and a few scattered hairs all over the body. It is in this larval stage characteristic tunnels are made inside the terminal shoots.

Fig. 77. Sugarcane top shoot borer - **A.** Larva, **B.** Adult, **C.** Attacked shoot of sugarcane showing larva inside; note dead-heart and egg-mass on leaf (From Pradhan, 1969, 1983)

A full grown larva cuts off portions of the plant at right angle to the original tunnel and makes necessary arrangements for the ensuing pupa to rest. The pupal shelter is chiefly constructed by the caterpillars by chipping up a small circular portion from the bark of the cane. This is gummed in its place by the secretion of the mouth, thus forming a close exist hole. The cavity thus formed is lined with silken threads secreted by larva. It is in this cavity the larva changes into a pupa and remains in this condition for 7-11 days. At the end of this period the adult moth emerges. The whole life cycle takes about 40-50 days but varies when conditions become unfavourable. The pest completes 4-5 generations in a year.

Control

As in the case of other pests, both mechanical and biological controls are employed. Biological control is brought about by the introduction of Hymenopteran parasites like a Braconid wasp or a

Chalcid wasp e.g. *Trichogramma minutum*. Spraying the crop with 1.5 litres of endosulfan 35 EC or monocrotophos 40 EC mixed in 1000 litres of water on leaf whorl from the top is very effective.

4. *Emmalocera depressella* Swinh. (Fig. 78): This belongs to family Pyralididae of the order Lepidoptera. It is commonly known as "sugarcane root borer". The greatest amount of damage is caused by the caterpillars. The attack is confined to the base of the newly formed stem. Presence of the dead hearts or dry central shoot among the young cane is the first symptom of this pest. The damage done by root borer is very characteristic because the caterpillar enters at the base of the stem, close to the surface soil. These borers are confined to base of the shoot which is bored across side ways or round about. On pulling the whole of the affected shoot or even the entire cane breaks exposing the caterpillars. In grown up canes, they may bore the thick roots but seldom attack the base. Damage is serious from the end of April to the end of September or middle of October. The young canes die and the older ones dry up and fall down. Sugar contents is greatly reduced. The moth is cryptically coloured hence cannot be easily detected even in worst infested fields. The female moth measures about 30-35 mm along wing expanse. The hinder wings are shorter and wider having a dark longitudinal stripe.

Life history

Creamy white, scale like flattened eggs are laid singly or in groups on leaves and stems in the first week of May. A female lays as many as 277-355 eggs. Young larvae hatch out in about 4-8 days which are 2 mm long with yellow body and yellowish brown head. They bore the base of the young seedlings, getting entrance through sides near or below the surface of the soil. They are confined to the new stems where they grow, and fully mature larva measures about 30 mm in length with a creamy white wrinkled body with brown head. The larval duration is about 30-50 days. The borer when left underground, attacks other side shoots and plants. The caterpillar is active in cane from the time, the cane is three weeks to a month old. Before pupation the caterpillars come up above the ground level in the stem, makes an exit hole in the form of silken tube which reaches the surface of the soil. The pupae are reddish brown in colour. Pupal stage lasts about 9-14 days. They hibernate in winter months in the larval stage in roots. The complete life cycle takes about 56-69 days. There are about 4 generations in a year.

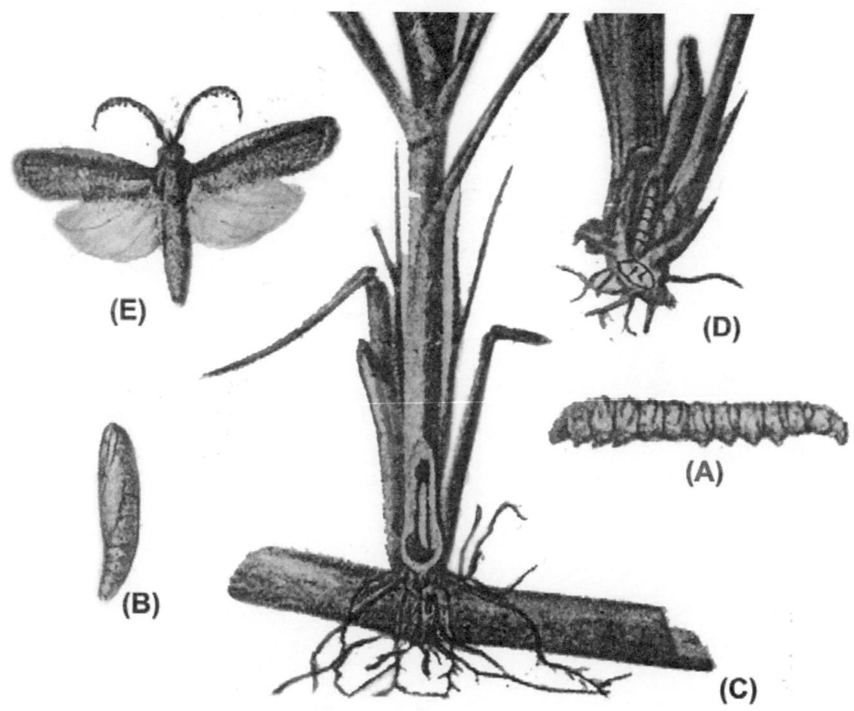

Fig. 78. Sugarcane root borer.
A. Larva, **B.** Pupa, **C.** Cane shoot cut open to show attack by larva,
D. Cane plant cut open to show pupa. **E.** Adult (From Pradhan, 1969)

Control

Prompt removal of dead hearts is quite effective and the moth can be destroyed by light traps. Ratooning should be avoided by burning the stubble in the fields. Chemical control is very difficult. However, spraying 1.5 litres of endosulfan 35 EC or monocrotophos 40 EC in 600 litres of water when eggs or moths are visible in the field, control the pest.

32

PESTS OF OILSEED CROPS

1. *Achaea janata* Linn. (Fig. 79): This belongs to family Noctuidae of the order Lepidoptera. The caterpillar is commonly known as "castor semilooper" which has a hump behind the head. This is one of the most destructive pests of castor crop. Damage is caused by the caterpillars which feed voraciously on the delicate parts of the leaves and at the end only the veins remain. Due to considerable loss of leaves, the castor seed yield is also greatly reduced. The peculiarity of *Achaea janata* is that in the larval stage, it is a serious pest of castor and in adult stage it is a fruit sucking moth and is a serious pest of *Citrus* fruits, a rare feature among moths, which are normally harmless.

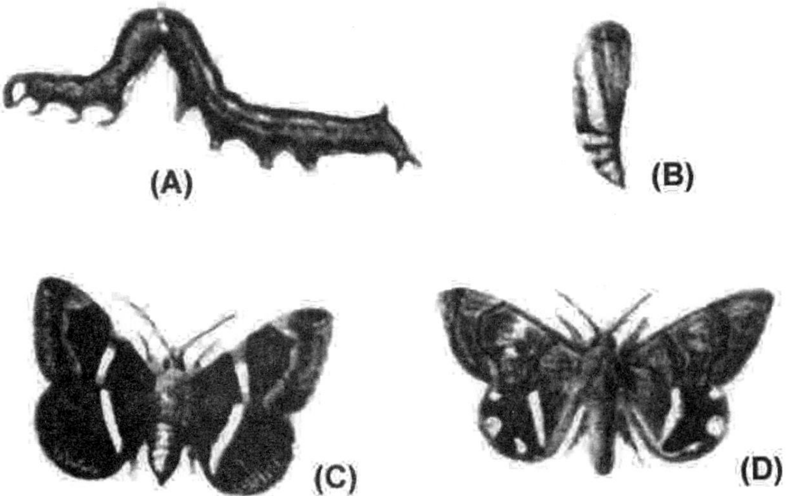

Fig. 79. Castor semilooper, *Achaea janata*.
A. Larva, **B.** Pupa, **C-D.** Adult - forms

The adult of *Achaea janata* is a pale reddish-brown moth with a wing expanse 60-70 mm. The wings, both the fore and the hind, are decorated with broad zig zag markings and a large yellowish area and dark brown patches. The mature larva is dark and is marked with blue black, yellow and reddish stripes and a dorsal hump near the head end of the body. There is a characteristic white mark on the head. The colour pattern is very different in this insect and some workers recognize four different varieties of caterpillars in the 3rd and 4th instars. They are very distinct against the background of the plants.

Life history

Each female moth can lay several hundred eggs. The total number of eggs laid by a female varies from 450-500 which are deposited singly on the leaves. The egg is about 1 mm in length and has numerous ridges and furrows on the surface, radiating from a circular depression at the apex. On account of its cryptic blue-green colour, it is extremely difficult to spot out the eggs on the leaves. Hatching of the larva takes place within 2 to 5 days. The first instar larva is extremely small and measures about 3.5 mm in length. The larva emerges by biting a hole into the egg shell; the latter also forms the first meal of the newly hatched larva before it begins to feed on the leaves. The larva feeds and moults 4 or 5 times and become full grown in about 15-25 days. The mature larva measures about 60 to 65 mm in length. The larva after searching a suitable hiding place prepares a loosely knitted cocoon of silk. Occasionally soil particles also get attached with the cocoons. The pupal duration is about 10-15 days and the moth, on emergence feeds on soft fruits of citrus, mango etc. There may be several generations in a year.

Control

Semiloopers can be controlled by spraying thiodan (endosulfan) 35 EC @ 1.25 litres in 800 litres of water/ha. The use of insecticides is not recommended in the case of castor crop meant for the rearing of the erisilkworms. The semiloopers should be regularly collected and destroyed.

2. Aphids (Fig. 80): These belong to the family Aphididae of the order Homoptera. They are the chief pests of mustard although they have been reported to severely attack other plants such as cabbage, raddish, cauliflower etc. The attack commences on plants even when they are in seed-beds as well as at all subsequent stages of their growth. They suck the sap of the plant by thrusting their needle like stylets into the tissues. Such a damage renders the attacked crop pale, weak, exhausted and incapable of bearing any seed. Leaves are most susceptible, those which are affected curl, shrivel and form cups, and finally drop. The insect pours a good deal of sugary excreta known as honey dew which makes the infected plant sticky to touch and facilitate the growth of a black fungus on them. This gives the appearance of dirty black leaves with shrivelled up lamina on the plant.

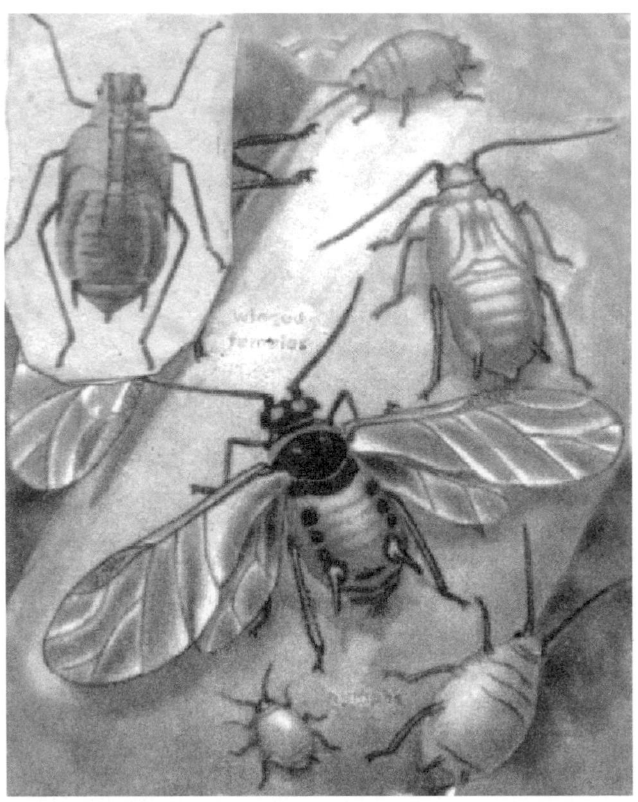

Fig. 80A. Aphids (inset mustard aphid)

The adult insects are very small having light green or slightly yellowish colour. They have a soft body and are of oblong in shape. One of the special morphological features in these insects is the presence of a pair of short tubes known as honey tubes which project from the upper surface of the hind end of the body. It is through these tubes that honey dew is excreted out. Generally apterous forms are present but during migration of the insect species, winged forms are produced.

Life history

Normally they hibernate in winter as small black fertilized eggs laid in depressions upon the petioles and underside of the leaves of the plant. Sometimes the egg stage is absent and the female which are enormously present in a colony are viviparous and bring forth large number of young ones parthenogenetically. These young nymphs mature very early and produce the next generation of young ones again viviparously and parthenogenetically. Thus they increase at an enormous rate producing sometimes more than 25 generations in a year. There can be upto 5 billion aphids/ha in a swarm.

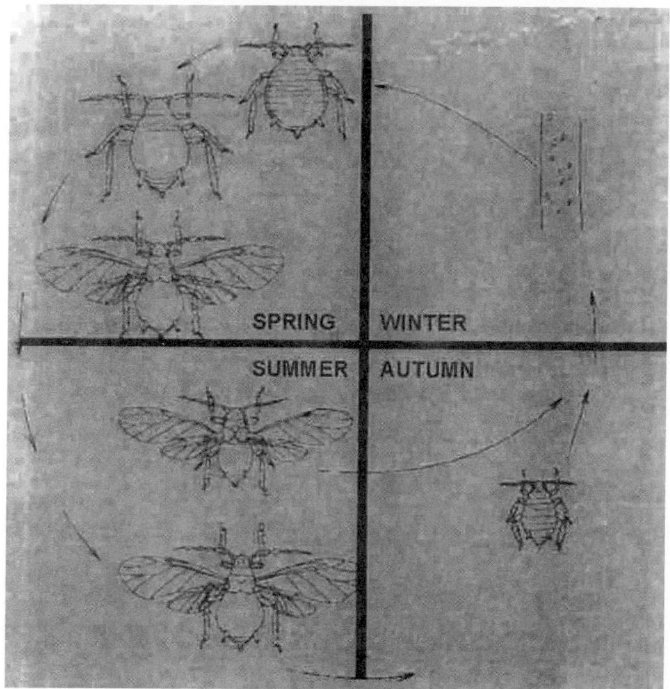

Fig. 80B. Life history of aphids. Eggs (top right) hatch to give wingless females which reproduce parthenogenetically (top left). After several generations, wingless females (bottom right) are produced in autumn which mate with winged males (upper figure bottom left) and lay winter resistant eggs (From Blaney, 1977).

In the case of hibernating Aphids, the eggs hatch into small nymphs under warm conditions. These grow quickly but never develop wings. Each nymph is the starting point of the great colony of Aphids and hence known as step mother. All the step mothers are females which have the remarkable ability to reproduce young, resembling them in being wingless; parthenogenetically and viviparously. These young in about a weeks time, in turn, reproduce parthenogenetically and viviparously the next generation of nymphs. In this way a succession of generation is produced.

Some time during the prolific production of aphids in the spring, all or a part of certain generations of females develop wings and these are called "spring migrants". These fly to the other host plants and establish themselves their starting a new colony parthenogenetically and viviparously. This usually happens in the summer month. With the approach of autumn a generation is produced by spring migrants, the individuals of which are all winged but made up of two sexes, male and female. Thus for the first time in an aphid colony males appear. The winged females of this generation reproduce parthenogenetically and viviparously a generation of nymphs which are all wingless true females in the sense that they can reproduce further only after mating with the winged males of the preceding generations. The winged females of autumn season are known as all migrants. The fertilized wingless true females lay one to 4 eggs each in the winter before they meet with their death. These eggs remain dormant during the cold season and hatch during spring into the step mothers which starts the life cycle again.

Control

Mustard aphid can be controlled by spraying the crop with any one of the following insecticides viz. nicotine sulphate 0.05%, diazinon/methyl parathion/phosphamidon 0.025%, dimethoate 0.03%, methyl demeton or malathion 0.1% or 0.04% monocrotophos 40 EC. Basal application of thimet 10% granules @10 kg/ha is also useful. In the flowering season, spraying should be carried out in the evening, when most of the pollinators including honey bees stop visiting the crop. When the crop is in the flowering stage, endosulfan and dimethoate which are comparatively safer to pollinators should be preferred.

3. ***Athalia lugens proxima*** (Klug.) (Fig. 81): This belongs to family Tenthredinidae of the order Hymenoptera. This is commonly

known as the "mustard sawfly". It feeds on various cruciferous plants like mustard, cabbage, cauliflower, turnip, radish etc. Damage is done by the larvae which are dark green, measuring about 16-18 mm in length. The adult stage is a rather thick set insect with a mixture of yellow-brown markings on the body and a reddish-brown colour on the wings.

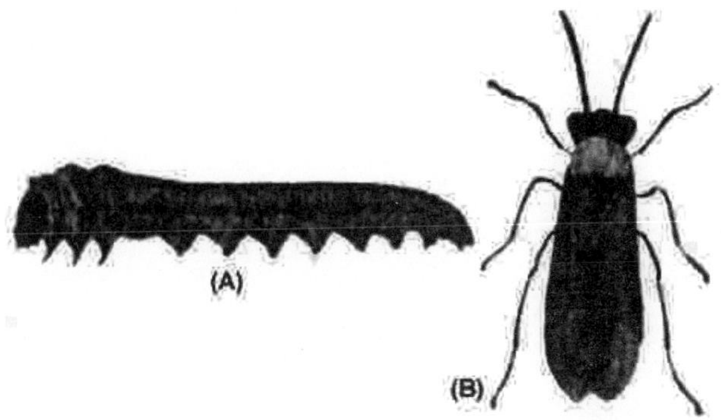

Fig. 81. Mustard sawfly - **A.** Larva, **B.** Adult

Life history

The female lays 30-35 eggs singly in slits made with the saw like ovipositor along the leaf margins on the underside. The eggs hatch in about 4-8 days and the larvae feed exposed in groups of 3-6 on the leaves during morning and the evening. They pass through seven stages and are fully grown in 16-35 days. The full grown larva measures 16-18 mm in length which enters the soil and pupates there in a waterproof, oval, silken cocoon and emerges as adult in about 11-37 days. Life cycle is completed in about 31-74 days depending on climatic conditions.

Control

Spray the crop with metasystox 25 EC @1 litre/ha or dimethoate (Rogor) @1 litre/ha in 1000 litres of water or malathion 50 EC or endosulfan 35 EC @ 1-1.25 litres/ha in 1000 litres of water.

4. ***Bagrada cruciferarum**** Kirk. (Fig. 82): This pest belongs to family Pentatomidae of the order Heteroptera. It is a serious pest of

---

* Same as *Bagrada picta* Fabr.

many of the cruciferous plants as well as those of the cucurbitaceous species, the seriousness being sporadic. All the stages in the insects life cycle cause damage to the plants. The individuals suck the sap from the leaves which consequently turn yellow and drop off exposing the plant to bacterial and fungal infection.

The adult is black in colour painted with irregular red markings, hence called painted bug. The scutellum is very large. The adults are extremely tenaceous and can withstand the absence of food for a fairly long period.

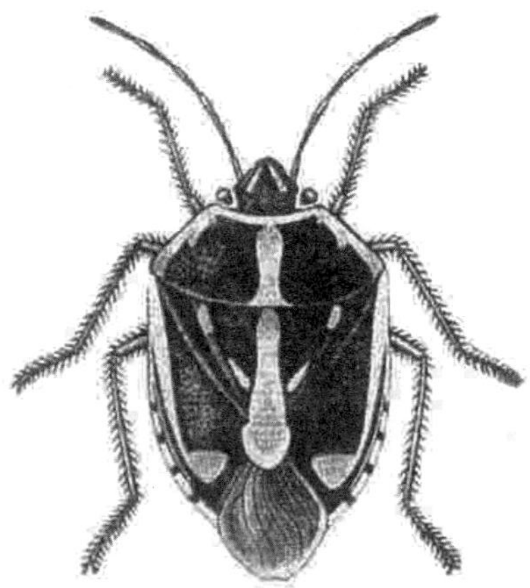

Fig. 82. Painted bug, *Bagrada cruciferarum*

Life history

The female lays about 230 eggs singly or in groups on the leaves, stalks, pods etc. The eggs which gradually become red in colour hatch in about a weeks time into young greenish red nymphs. They immediately start feeding on the plant and gradually change colour. At the same time they moult also, the number of moults varying often. The nymph transforms into adult in about 3 weeks time. The activity of pest ranges from October to March with 6-7 broods in a year.

Control

As in the case of majority of other pests both mechanical control and precautionary measures are often employed. The pest can be controlled by spraying the crop with any one of these insecticides: diazinon 0.02% or malathion 0.05% or dimethoate 0.03%.

5. ***Acherontia styx*** West. (Fig. 83): This insect belongs to family Sphingidae of the order Lepidoptera and is commonly referred as "Til Hawk Moth" "Sphinx Moth" or "Death's Head Moth". The larvae feed on the leaves and defoliate the *til* plants (*Sesamum indicum*). Besides *til*, it has also been found to damage potato, brinjal, lab-lab and the ornamental plants like balsam. The moth is harmful to apiarists because it sucks honey from the honey-combs, although at times it is stung to death by the bees. The adult of *Acherontia styx* is a large reddish – brown moth, with a wing span of at times more than 10 cm. The fore wings possess a mixture of dark- brown and grey ornamentation with dark blackish-wavy markings and a prominent yellow spot on each wing. The hind wings have an ochre back ground on which there are two broad dark brown wavy cross stripes. The head and thorax are also dark brown and on the latter there is a prominent Death's Head mark. The abdomen is ochracious in general hue with drak-brown cross stripes.

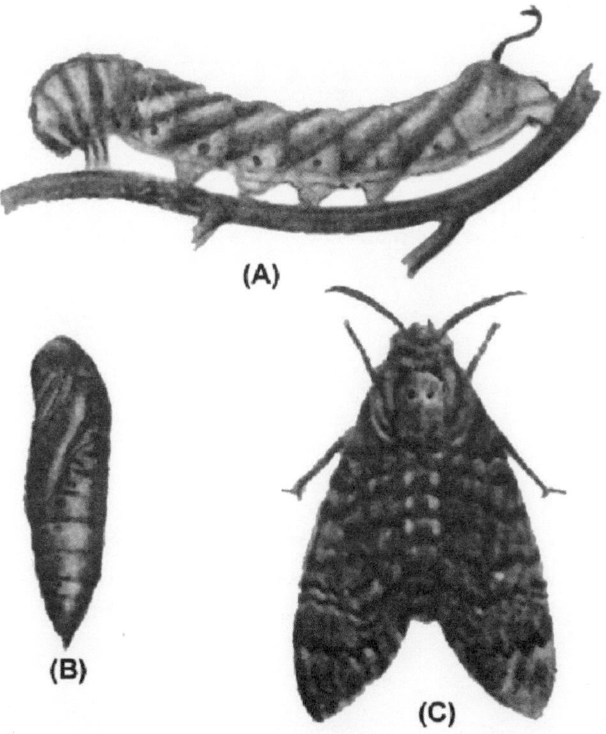

Fig. 83. Til - Hawk moth - **A.** Caterpillar, **B.** Pupa, **C.** Moth (From Pradhan, 1969).

Life history

The moths lay eggs singly on the under surface of the leaves of the food plant and the eggs are fairly large, their colour is at first greenish white but later on it turns yellow. Hatching takes place within a few days and the just hatched larva is pale-yellowish in colour. Soon it begins to feed on the leaves and develops broad green stripes running obliquely along the dorso-lateral surface. The full grown caterpillars which attain a length upto 5 cm or more and also a thickness of more than 1 cm are in habit of retracting and drawing in the head and some anterior segmants of the body into a sort of Sphinx- like posture, this along with a horn-like projection from the hinder end of the abdomen enhances the monstrosity of these larvae which otherwise have quite a plump body with a pleasent mixture of soft colours. The larval period is un-usually long and at times it lasts 2 months or more. The pupa is also almost equally long lived. The winter is passed in the pupal stage in the soil. There are 3 generations in a year.

Control

The Hawk moth larvae can be controlled by spraying the crop with endosulfan @ 1.25 litres/ha in 1000 litres of water or quinalphos 25EC @ 500 ml/ha in 1000 litres of water. Hand picking is also a good method to suppress the larvae.

33

PESTS OF PULSE CROPS

1. *Agrotis ypsilon* * Rott. (Fig. 84) : It is commonly known as "greasy gram cutworm". This belongs to family Noctuidae of the order Lepidoptera. The attack is extremely serious during winter when young seedlings, roots and leaves are nibbled at night and the stems cut at the level of the ground by the caterpillars. Besides gram, it is also a pest of potato, tobacco, peas, cotton, tomato, wheat, maize, sugarcane, cucurbits and other winter crops. The moth is dark brown with a grey body, grey fore wings with dark brownish black markings. The hind wings are almost white basally but with a dark terminal fringe particularly in the males. They measure about 40-50 mm along the wings. The moth is nocturnal.

Life history

The female moth lays eggs in clusters consisting of about 30 eggs in each underside the leaves. Several thousand eggs can be laid by a single female moth. The eggs are white, globular and ribbed 0.5 mm in diameter. The larva hatches in about 2-9 days and begins to feed on egg shell which forms the first meal of the larva. The larvae are also nocturnal and they remain underground in day time. The full grown caterpillar measures about 35-45 mm long, is dull grey or brownish in colour. Larval development takes about 28-34 days. The pupae are dark brown, 20 mm long. Pupation takes 10 to 30 days or more depending on the temperature and then moth emerges and starts new life cycle. The life cycle from egg to adult takes about 32 days at 30°C and 67 days at 20°C.

---

* Same as *Agrotis ipsilon* (Hufn.)

Fig. 84. Greasy gram cutworm, *Agrotis ypsilon* **A.** Larva, **B.** Adult

Control

Removal of weeds from the field helps to eliminate oviposition sites. Fenitrothion has generally given best control of many cut worm species.

The pest can be controlled by application of endosulfan 5% dust @25 kg/ha or 0.05% spray.

2. *Agrotis flammatra* Schiff. (Fig. 85): The gram cutworm belongs to family Noctuidae of the order Lepidoptera. It is one of the most important pests of gram in India. It is a polyphagous pest and besides gram, it feeds on a wide variety of plants belonging to different families. The adult moth is a stout, greyish brown in colour measuring about 5-6 cm in wing expanse. The fore wings of the moth have characteristic markings and smoky patches, the proximal areas being pale or yellowish in colour. Each wing is provided with a semi-circular spot below the pale or yellowish area and a greyish-brown kidney shaped spot towards the apical area. The damage is caused only by the caterpillars.

Fig. 85. Gram cutworm, *Agrotis flammatra*

Life history

The moths lay yellowish white eggs on the under side of the leaves, on shoots, stems or in the soil within 5-10 days of their life which is about 7-13 days only. A female moth lays about 950 or more eggs in her life time. Hatching takes place in about 4-14 days depending on the temperature. The larvae are full grown in about 4-9 weeks and measure 40-50 mm in length, dark grey or light green in colour. They pupate in the earthen cells constructed by them in the soil. The pupal stage lasts about 10-15 days in favourable weather but during winter it extends upto 5 weeks or more and after which the moth emerges. The whole life cycle is completed in about 6-13 weeks, depending on the climatic conditions. There are usually 2 generations in a year.

Control

The pest can be controlled with endosulfan 4% dust. Cutworms can also be controlled by spraying of 35 EC endosulfan in the concentration of 2 ml/litres of water at 15 days intervals. Two or three sprayings are sufficient.

3. *Heliothis armigera* * (Hubner) (Fig. 86): It belongs to family Noctuidae of the order Lepidoptera. It is commonly known as "gram pod borer", and is one of the most serious pest of gram. The adult is stout built Noctuid moth with a wing span of 30-40 mm. The colour of moth varies from olive grey to red brown. The forewing has a dark speck and a dark area near the outer margin. The hind wing has a distinct smoky dark outer border. Damage is caused by the caterpillars.

Life history

Eggs are laid singly on the tender parts of its food plant. The eggs are spherical, sculptured and with greenish shine when laid and measure about 0.5 mm in diameter, as the embryo develop the eggs turn brown. Depending on temperature the egg period lasts from 4-7 days. Each female may lays 1000 or more eggs.

The young larvae begin to feed on tender portions of the leaves and shoots. Later, the caterpillars feed on the developing seeds after

---

* Same as *Helicoverpa armigera* (Hb.)

cutting hole in the pod with head inside but with the posterior part of the body outside the pod. They also burrow into the large pod and eat developing seeds. Their colour and size vary considerably. The full grown larva measures about 35-40 mm in length. The larval period lasts about 2-3 weeks. Full grown larva burrows into the soil and pupates there. The pupa is shining dark brown in colour measuring about 12-16 mm in length. The moth emerges in 10-14 days after pupation. Egg laying starts about 4 days after emergence and may continue for about 9 weeks or more. There are several generations in a year.

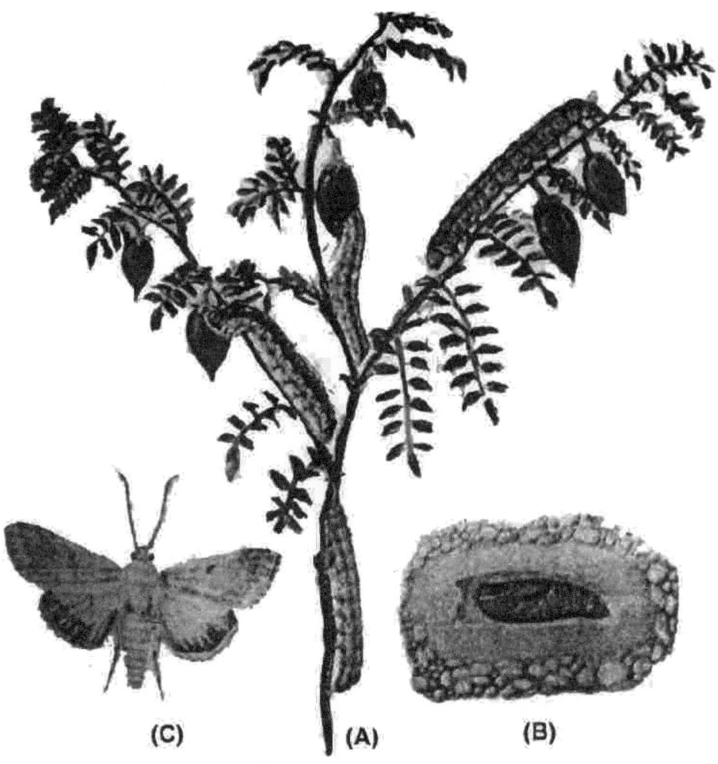

Fig. 86. Gram pod-borer, *Heliothis* **A.** Caterpillars on gram plant, eating into pods, **B.** Pupa in underground cell, **C.** Moth.

Control

It can be controlled effectively by spraying the crop with 0.1% carbaryl or 1.5 litres of endosulfan 35 EC or 750 ml monocrotophos 40

EC/ha in 1000 litres of water. The caterpillars can also be collected and destroyed by simple hand picking. Crop rotation is also extremely useful.

4. *Agromyza obtusa* * Mallas (Fig. 87): It is a small metallic black fly. It is known as "Red gram pod fly" which belongs to family Agromyzidae of the order Diptera. The damage is caused by the tiny maggots which bore into the pods and feed on the red gram seeds. It is distributed throughout the red gram grown area of the India.

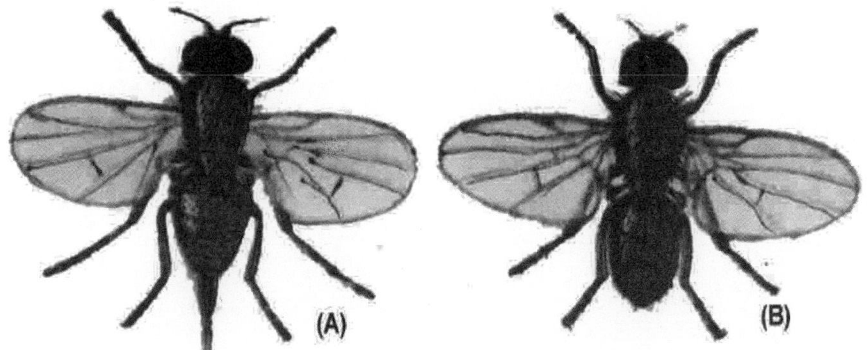

Fig. 87. Red gram pod fly - **A.** Female, **B.** Male (From Pradhan, 1969, 1983)

Life history

The fly thrusts its minute eggs into the shell of the tender pod. The maggots hatch in about 3-4 days, and after feeding under the epidermis, for some time they enters the seed. They become full grown in about 5-10 days. Pupation take place within the damaged pods. The pupal period lasts for 7-11 or more days depending on the climatic conditions. The adult flies emerge by cutting holes. The entire life cycle is completed in about 9-30 days depending on the prevailing temperature. There are several generations in a year.

Control

The pest can be controlled by spraying the crop with 1.5 litres of endosulfan 35 EC or 750 ml monocrotophos 40 EC in 1000 litres of water/ha or dusting with 5% carbaryl or 5% endosulfan.

---

*Same as *Melanagromyza obtusa* (Mallas).

5. *Prodenia litura** Fabr. (Fig. 88): This pest belongs to family Noctuidae of the order Lepidoptera. The larvae feed on a variety of plants and cause a lot of damage to various crops.

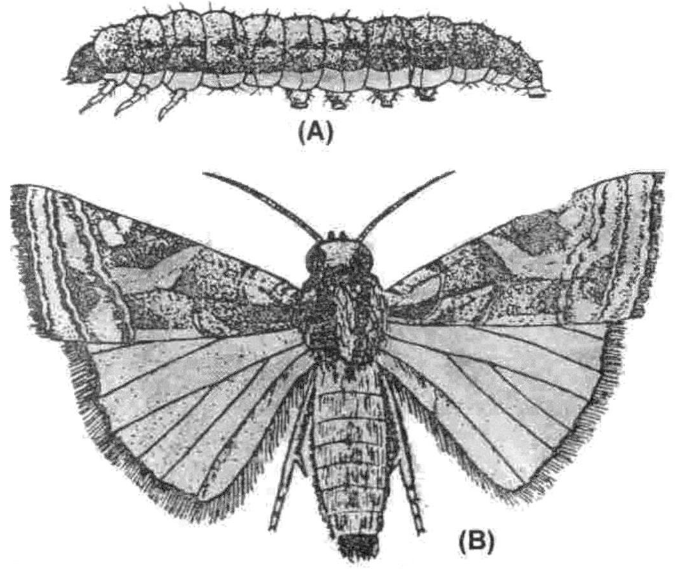

Fig. 88. *Prodenia litura* **A.** Larva, **B.** Moth

Life history

Each female lays spherical eggs about 0.3 mm in diameter in groups or clusters on underside the leaves in batches of 100 to 300. One female lays about 1500-to 2000 eggs. Hatching takes place in about 3-6 days, but in winters it can be upto 26 days, the newly hatched larvae are gregarious but later they disperse. The mature larvae are greenish-brown in colour with heavely dark markings. Mature larvae measures about 35 to 40 mm. Pupation takes place in the soil in an earthen cell or under fallen leaves. The total pupal period ranges from 7-15 or more days depending on temperature. The moths are robust with white markings on the fore wings. The hind wings are white and the margins of the wings are brown in colour. The total life cycle takes about 35-60 days depending on prevailing temperature.

---

*Same as *Spodoptera litura* Fabr.

Control

Cultural methods of control such as ploughing and burning of crop stabbles, flooding the infested fields and removal of weeds, all helps to lower the pest population. The pest can be controlled by spraying the crop with diazinon 0.02% or endosulfan 0.05%. Dusting the crop with carbaryl is also very effective. The infested crop should be sprayed with malathion 0.05% or carbaryl 0.15%.

34

PESTS OF SORGHUM

1. ***Atherigona soccata*** (Rond.): The adult is commonly known as ``sorghum shoot fly'', which belongs to family Anthomyidae of the order Diptera. In addition to sorghum it is also a pest of maize and wheat. It damages the seedlings and earlier crops. The maggots bore into the stem and feed on the basal parts of the young shoots causing a typical dipterous dead heart. The maggots cause a considerable loss in the yield of grains. The adult looks like a small grey housefly, about 3 mm long. These flies have long adult life of about 1 month or more. They feed on honey dew and other sweet material available in nature.

Life history

About 40 eggs are laid singly on the lower surface of the leaves. The eggs measure about 1.5 mm in length, are elongated and flattened, somewhat boat shaped and are provided with two wing like lateral projections. The eggs hatch in about 2-3 days. The maggots are extremely small which move down the leaves and pass between the leaf sheath and stem and bore into the stems of the seedling at about the level of the growing point. They feed on the developing shoots for about 7-10 days. Pupation takes place in the soil which lasts for about a week after which the adult fly emerges and starts laying eggs on other seedlings. The total life cycle takes about 15-32 days. There are several generations in a year.

Control

It can be controlled by planting the resistant varieties and by spraying 0.2% carbaryl or 0.05% endosulfan. Application of 10% phorate (thimet) granules in soil before sowing @ 15 kg/ha should be done. Two sprays of 0.01% methyl demeton (1 ml chemical in a litre of water) solution @560 litres/ha at 15 days interval after sowing controls the pest.

2. *Calocoris angustatus* Leth. (Fig. 89): The earhead bug belongs to the family Miridae of the order Heteroptera. The adult is a small greenish yellow bug of about 5-8 mm in length and a little more than 1 mm in breadth. This is one of the most serious pests of sorghum. The nymphs and adults suck out the juice from just developing sorghum grains which are in their milky stage. A continuous feeding by the nymphs and adults of these bugs results in the shrivelling of grains. When heavy infestation occurs the ears become black in the initial stage and finally they dry up hence the grains are not produced at all.

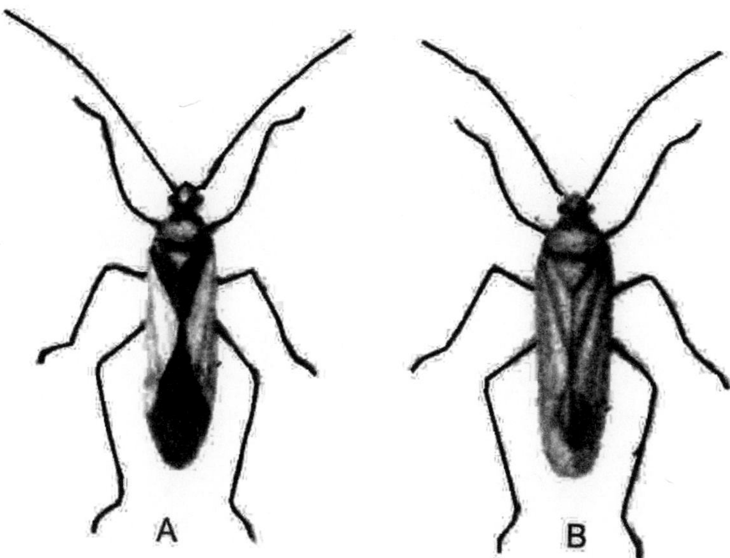

Fig. 89. Earhead bug, *Calocoris angustatus* **A.** Male adult, **B.** Female adult (From Pradhan, 1969)

Life history

A female bug lays as many as 150-200 eggs in between the anthers of the Sorghum. The eggs are extremely small, cigar shaped about 1.5 mm long. These eggs hatch out mostly in less than a week i.e. in about 6 days. As each insect lays between 150-200 eggs, the population of the pest increases by 75 to 100 times within about a week since the crop begins to attract the adults from the surrounding uncultivated grasses. The nymphs just after emergence start feeding on developing grains in milky stage. The nymphs pass through 5-instars and develop into adults in about 3 weeks. The life cycle is completed in one month under favourable conditions. There are several generations in the season.

Control

Nymphs and adults can be controlled by spraying 0.05% malathion or 0.05% endosulfan or 0.1% carbaryl (W.P.) or 4% carbaryl dust.

35

PESTS OF CEREAL CROPS

1. ***Chilo zonellus**** Swinh. (Fig. 90A) : This belongs to the family Pyralididae of the order Lepidoptera. It is a major pest of maize and jowar. However with respect to maize it causes damage only to the young plants while all stages of jowar are severely attacked. The period of infestation comes from the middle of March and continue upto September with a peak during August. Normally the attack on the plant starts from the top with the caterpillars boring into the shoots at the upper end. However, there are instances where the caterpillar enters the shoot from the side, punctures even stems as high as 8 cm. from the ground. In either case they ingest the sap from the plants and bring about its withering which ultimately results in the formation of dead hearts. The withered plants have reduced leaves and stem. An attack by this pest from the top end of the plant is indicated by the presence of numerous punctures on the leaves forming the whorl.

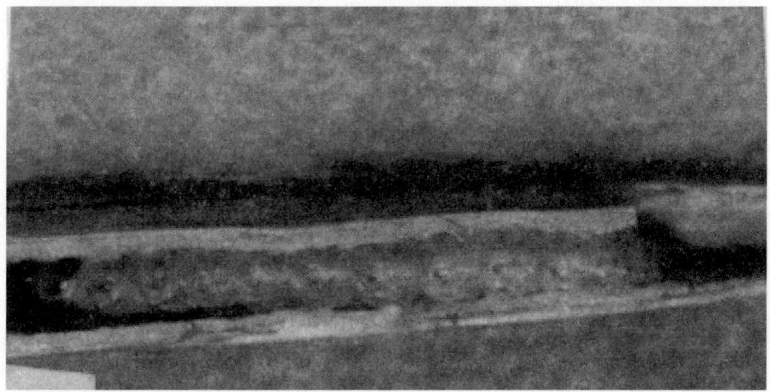

Fig. 90A. Jowar stem borer, *Chilo zonellus*

---

* Same as *C. partellus* (Swinh.)

The adult insect is a yellowish brown moth 20-30 mm across the wing with deep brown spots, long legs and two projecting palps in front of the head which give the appearance of a beak. Sexual dimorphism is characteristically exhibited in this species. The hind wings of male have a smoke like or pale straw colour, while those of females are bright whitish. Further, the abdominal tip of males is smooth and bluntly pointed while those of females is broad and covered with a thick tuft of hairs. The adults are short lived.

Life history

The female lays greenish or creamy white, flattened, oval about 0.8 mm long eggs at night on the lower surface of the leaf near mid rib in two rows overlapping each other. The eggs are normally laid from April to May. Each row contains about 50-100 eggs which in course of time change their colour first to orange and then to yellow. Hatching occurs in about 7-10 days liberating young caterpillars. These immediately feed on leaves, on which they have hatched and then gradually penetrate into the leaf at the mid rib region to start their journey towards the main stem for the principal attack. Often their course from the leaf to the stem is interspersed with small punctures on the leaf for the purpose of ventilation. Thus they reach the shoots tunnelling and boring their way from the leaf. Soon the main stem looses its strength and becomes a dead heart. The caterpillar thus draws its nourishment from the shoots and become fully matured in about 3-4 weeks. At this stage it prepares itself for pupation. It immediately makes an exit hole in the shoot and undergoes pupation within the infested stem in a chamber specially lined with thread like secretion. The pupal period lasts from 7-10 days. The pupae are shining brown in colour and possess six spines arranged in two rows at the tip of the abdomen. At the end of pupal period fresh moth emerges through the exit hole. The entire life cycle lasts for about 5-6 weeks. Generally there are two broods in a year.

Control

Many precautionary measures such as thorough examination of field before planting, removal of stubble trash and dead hearts from the plough field are employed. Besides these certain electrical devices such as attraction of moths by light traps are also adopted. Often biological control measures have been found to be successful. These include the introduction of egg parasites such as *Trichogramma minutum* and *Bracon chinensis*. Spraying of 1.5 litres thiodan 35 EC or 2 kg (50% soluble) carbaryl in 1000 litres of water/ha. 15-20 days after sowing, helps in controlling the pest.

2. ***Leptocorisa varicornis*** Fabr. (Fig. 90B): This belongs to the family Coreidae of the order Heteroptera. It is a common pest of rice and is a major one. Both the nymphs and adults destroy the ripening grains of the plants usually during August to October, extracting their milky juice with the consequence that the ears turn white forming shrivelled grains and preventing their maturation. A serious attack results in the appearance of empty, dry, brown ear heads on the plant. Yield losses of 10-40% are common and in severe infestations the entire crop may be destroyed. Plants such as those of jowar, maize, bajra, etc. form the secondary host of this insect.

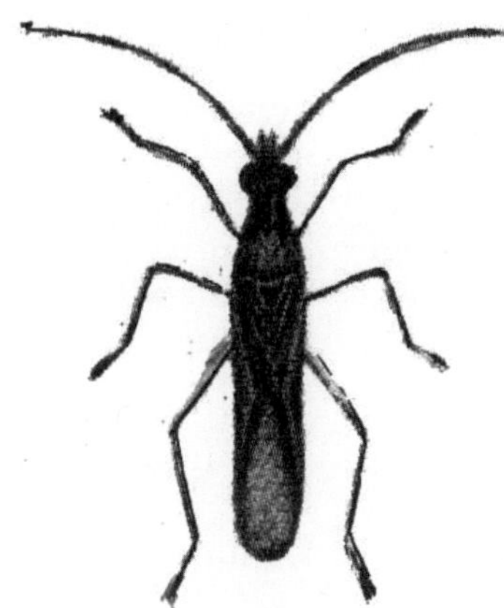

Fig. 90B. Gundhi bug, *Leptocorisa varicornis*

The adult insect is slender, green or light brown in colour. It emits strong unpleasant aromatic odour and hence the name `Gundhi bug'. Both male and female individuals are identical in appearance except that the abdominal apex of the latter is more swollen than that of the former.

Life history

The female deposits eggs in one to three rows on the leaves of the paddy plant, each row containing 5-20 eggs. The eggs are black, oval and flattened but concave from above. Each egg has a dark brown shell with a white spot indicating the point of rupture during hatching. The eggs are fixed to the leaves by a gummy substance. The eggs hatch into young, pale green nymphs in about 6-8 days. These nymphs are devoid of wings but green in colour in the beginning but possess long leg and antennae. Slowly the wing also develop and the nymph moults 5-times within a period of 17-27 days before attaining the adult condition. An important feature of the final moult is the transfer of functional odouriferous glands from the abdomen to the thorax and the formation of three tarsal joints. When the plant is at a tender stage both nymphs and adults congregate in large number around ripening ears, tender shoots and leaves. The entire life cycle is completed in about 4-5 weeks and in a year there are generally 5 broods during the paddy season. Multiplication normally occurs from July to November, hibernation from December to February and their population on grasses from March to June.

Control

Besides many precautionary steps like extermination of wild grasses from the sowing fields etc. and certain mechanical (electrical) measures such as light trap, arrangement for catching the insects, the general tendency has been to employ either biological control or chemical control. Biological control measures warrant the introduction of certain beetles like *Cicindela - sixpunctata* or certain species of Pentatomid and Reduvid bugs which prey on the adults.

Control by chemical measures is through the application of insecticides. Dusting of crop with carbaryl (Sevin) 5% or malathion 5% dust @ 20 kg/ha in the morning hours or spraying the crop with 0.05% or 0.07% endosulfan is useful. Dusting is usually done on the earheads, especially when they are in the milky stage and the bugs appear in large numbers.

3. ***Hieroglyphus banian*** Fabr. (Fig. 90C) : This belongs to the family Acrididae of the order Orthoptera. It is a serious though sporadic pest of eastern U.P. damage to the paddy is caused both by the young and adult stages of the insects which feed upon the leaves and soft ear grain of the plants. The insects also attack other plants like sugarcane, jowar and different types of grasses.

The adult is a green, dry grass coloured medium sized individual with long slender legs. The young nymphs are very much like the adults except that they are smaller in size and possess a brownish yellow colour and allied species of this genus *H. nigrorepletus* Bol. is a minor pest of jowar.

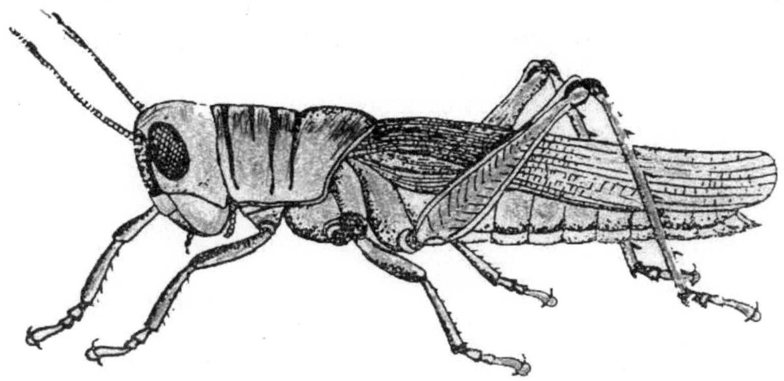

Fig. 90C. Rice grasshopper, *Hieroglyphus banian.*

Life history

The female lays yellow eggs resembling rice grains in clusters usually during October month. The eggs are closely packed in hard cement which forms a pod like cylindrical structure termed ``egg pod". The eggs remain burried under soil during hibernation stage i.e. upto June.

The eggs hatch during the monsoon season into young nymphs which immediately starts feeding on nearby grasses and later on migrate to the main crops. These nymphs become full grown in about 8-10 weeks during which period they moult 5-times (sometime 6 times) at an interval of 10-15 days each. The young nymphs known as hoppers resemble very much the adult except for the fact that these hoppers are devoid of wings. The nymphal period on the whole extends from 10-12 weeks after which a fully developed adult is formed. The entire life cycle lasts for about 17 weeks. Generally there is one brood in a year which appear during the monsoon rain.

Control

As in the case of other pests, certain measures of success is achieved by adopting physical measures such as hand collecting and destroying the hoppers, and arrangement of light traps etc. However, insecticidal treatment has been found to be very effective by dusting 12.5 kg. 10% carbaryl (sevin) dust/ha on the first appearance of the pest. The pest can also be controlled by spraying the infested crops with endosulfan 1250 ml in about 1000 litres of water per hectare.

4. ***Sesamia inferens*** Walk. (Fig. 91) : This belongs to the family Noctuidae of the order Lepidoptera. It is a serious pest of maize. Its larvae bore down the stem making long tunnels throughout the whole length. The central shoot is completely destroyed and dead hearts are formed. It feeds on wheat, barley, paddy, sugarcane and jowar etc.

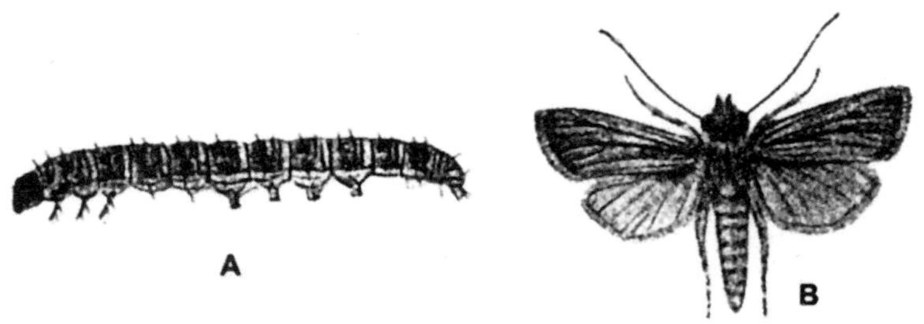

Fig. 91. *Sesamia inferens* :**A.** Caterpillar, **B.** Moth (From Pruthi, 1969)

Life history

The pest breeds actively from March-April to November on rice and then migrates to the wheat crop. The moths are nocturnal and lay about 150 eggs on leaves or on the ground. The eggs hatch in about 6-8 days and the young caterpillars bore in to the epidermal layers of the leaf sheaths. Later on, they bore into the stem as a result of which the growing shoots dries up producing dead hearts. When the attacked plants dry the larvae move on to adjoining plants, they are full fed in 3-4 weeks and pupate inside the stem or in between the stem and leaves. The pupal stage lasts about a week and the life cycle is completed in about 6-7 weeks. There are 4-5 generations of the pest in a year.

Control

Spraying endosulfan 0.07%, methyl parathion 0.08%, phosphamidon 0.05% and diazinon 0.06% is effective in checking the infestation of this insect.

5. ***Mythimna separata*** Walk. (Fig. 92) : It is commonly known as ``Armyworm", belongs to the family Noctuidae of the order Lepidoptera. It is a pest of wheat, maize, jowar, bajra, sugarcane etc. The caterpillars are voracious, eating entire leaves and the whole plant usually during nights and migrate from one field to another field.

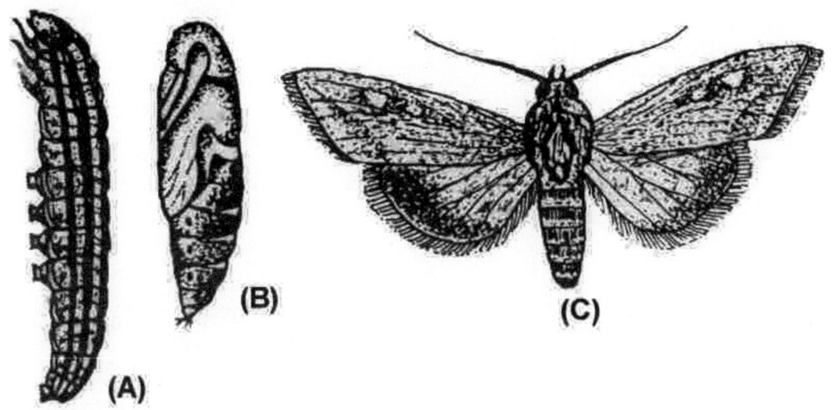

Fig. 92. Armyworm, **A.** Caterpillar, **B.** Pupa, **C.** Adult

Life history

Eggs are laid either singly or in clusters of about 100 inside the rolled leaves or between the leaf sheath and stems. They are subspherical, greenish white, turning yellow and take about 4-13 days or more to hatch, which depends on the climatic conditions. Older caterpillars are gregarious and have voracious appetites. The caterpillars vary considerably in colour, from green to pink but have four distinct longitudinal black stripes and reach a mature size at 35-40 mm after about 14-22 days in spring or 80- 100 days in winter. The pupa is dark brown in colour 15-19 mm in length. Pupation takes place 4-5 cm deep in the soil. Pupal stage is completed in about 7-22 or even more days. The length of pupal stage depends on temperature. The adults are reddish or grey moths with a central pale spot in each forewing. Wing span is from 35-50 mm. The whole life cycle takes about 25-64 days (31 days on an average) with 5 generations per year.

Control

Cultural methods such as ploughing, stubble burning, flooding of the infested fields, removal of grass and alternative host plants from around the fields, all help to reduce the pest population. Dusting contact insecticides are very effective. Spraying the crop with 0.05% monocrotophos or 0.07% endosulfan at the time of heavy infestation reduces the pest population.

6. ***Hispa armigera*** Oliv. (Fig. 93): This insect belongs to family Hispidae of the order Coleoptera. The grubs are leaf-miners. This is also a pest of paddy which has been found widely distributed in this country and abroad. The damage is caused by the grubs and adults both which feed on the green delicate parts of the paddy leaves. The adult beetles are blue-black in appearance.

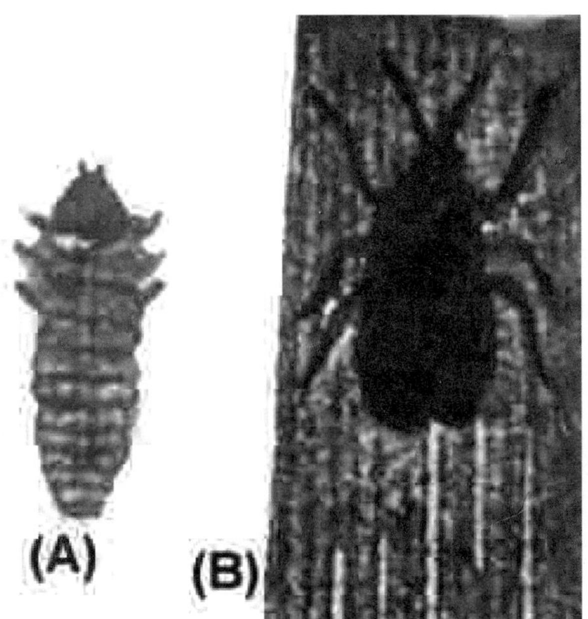

Fig. 93. *Hispa armigera* **A.** Grub, **B.** Adult.

Life history

Each female individual lays about 50 to 60 eggs inside the leaves. The tender parts of the leaves are preferred for laying the eggs. After about 5-days, yellowish grub emerges which feed inside the leaf-mine for about 10 days. Pupation takes place inside the mine and adult emerges in about 4 to 6 days. There are several generations in a year.

Control

Spraying the crop with 0.05% monocrotophos, as well as 0.7% endosulfan at the time of severe infestation also controls the pest-population.

7. *Holotrichia consanguinea* (Blanc.) (Fig. 94): The larva of this beetle is commonly known as ``White grub". It belongs to the family Melolonthidae of the order Coleoptera. The grub feeds on the fibrous roots of maize, sorghum, brinjal, sugarcane etc. and is found in the upper layer of the soil about 5-10 cm. The full grown grubs measure about 35 mm long, is white in colour with brown head. The beetles are dull brown and about 18 mm in length. While the grubs feed on roots, the adult beetle feeds on the plant leaves.

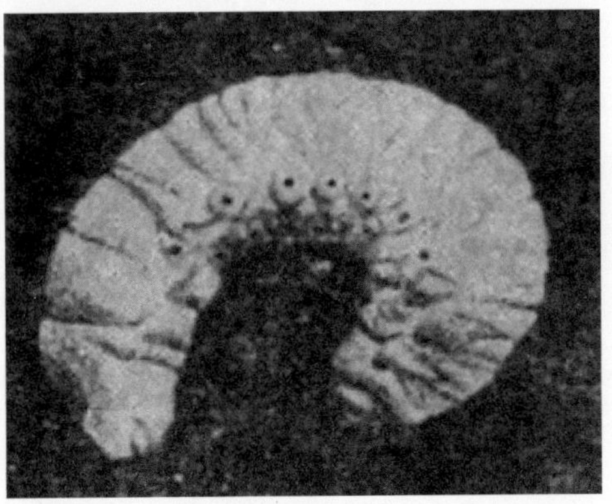

Fig. 94. A white grub, *Holotrichia* sp.*

Life history

The adult beetles lay eggs singly upto a depth of 10 cm which hatch in about 7-10 days. The development completes in about 10 weeks. After the monsoon the grubs pupate in the soil. The pupal stage lasts for about 15 days. The beetles remain in soil and come out only at night for feeding. The adults remain in soil and undergo diapause for about 8 months. They remain active only in rainy season i.e. June. There is only one generation in a year.

---

* The other common species of *Holotrichia* are: *H. serrata* Fab. *H. insularis* Brenske, *H. conferta* Sharp.

Control

The white grubs can be checked by applying carbaryl @2-2.5 kg/ha in the field. The soil can be treated with phorate granules and quinalphos to control this pest.

36

POLYPHAGOUS PESTS

1. TERMITES (Order : Isoptera) (Figs. 95, 96): These are commonly called "white-ants", the name being used because though not related to ants, they live in colonies and in many of their ways resembles the ants. Their colour varies from whitish pale to brown. Termites form the most important insect pest as they cause very great damage. The termites are polyphagous. Their food is principally composed of cellulose which they obtain from living or dead vegetation, books and other articles of wood, fabric etc.

Maize, jowar, bajra, pulses, paddy, wheat, sugarcane, potato, tea, coffee, coconut, jute, sunnhemp, fruit trees, forest trees etc. are badly damaged by termites. They cause considerable damage to the dry and dead wood, furniture, made out of the dry wood, books and almost all articles containing cellulose.

Life history

The termites have two main forms viz. reproductive and sterile. The colony consists of one male or king, one female or queen and a variable but generally large number of workers and soldiers (Figs. 95 A-D, 96). The queen is a giant and has no other work than of continuously laying millions of yellowish or cream coloured kidney shaped small eggs. They are deposited in particular cells for hatching. The nymphs resemble miniature adults. The nymphs undergo several moults producing workers and soldiers. Periodically they produce winged forms or primary reproductives, which swarm out and repeat the cycle. With the first few showers of rain there particularly appear a few winged male and female forms. They leave the termitarium when mature, fly about, copulate, shed their wings become kings and queens, burrow in soil and establish new colonies.

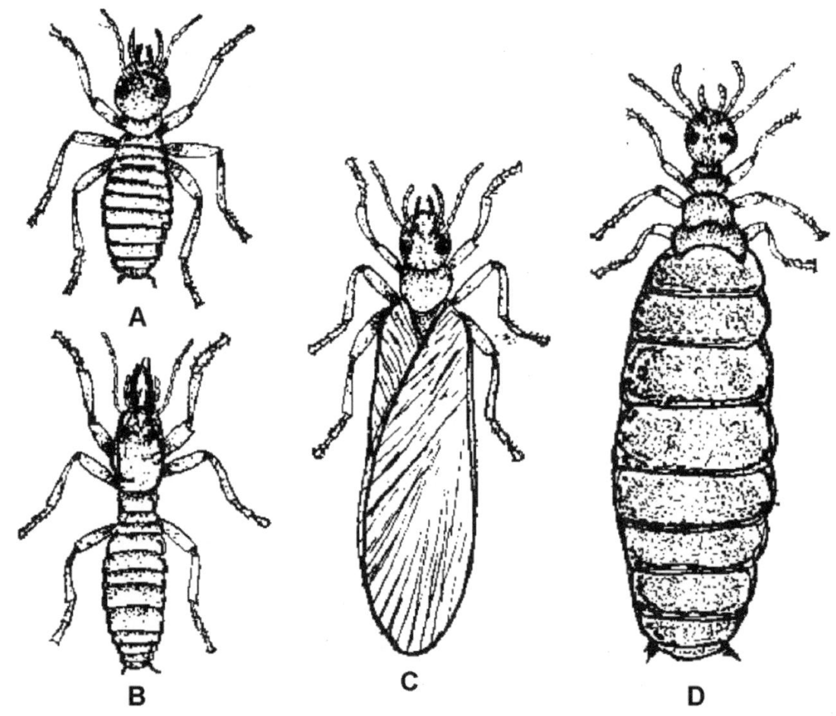

Fig. 95. Castes of termite, **A.** Worker, **B.** Soldier, **C.** Winged male, **D.** Queen.

Workers

These are small white insects most commonly seen doing damage. They can be easily recognized by their vertically carried head and small broad jaws. These are injurious form and constitute 80-90% of the whole population.

Soldiers

These are a little larger than workers and carry long pointed jaws placed horizontally in front of the head. They are about 2 to 3% of the population.

Winged adults - They are bigger than the soldiers and possess four equally long wings, which when folded lie flat on the back and are readily cast off after nuptial flights.

King - This is a still bigger wingless male which remains with queen after nuptial flight and is the partner to form the colony. It lives in the nest and repeatedly copulates. There is generally only one king in a colony.

Queen - This is the biggest wingless member of the colony about 10 cm. in length and is the cofounder of the colony. The head and thorax are small when compared with the abdomen. Thousands of eggs are laid per day for several years.

Supplementary reproductives - These are special females whose duty is to reproduce in emergency if the queen happens to die. These are fed on special food to activate their reproductive organs.

Control - The affected parts in the trees should be sprayed with sanitary fluid or osfoetoid solution banding the trees by coal tar, renewed half yearly.

Protection of wooden structures can best be obtained by taking care that no wood comes in contact with the ground and that light and dryness are ensured as much as possible. Timber thoroughly treated with coal tar, creosote are repellant to the insects.

When termites are already in the wood, all infested parts will generally need to be removed and the nest should be located if possible and destroyed by using kerosene or by fumigation. Treatment of the soil near a building with a mixture of trichlorobenzine one part, fuel oil or creosote two parts aids for a few years in keeping termites away.

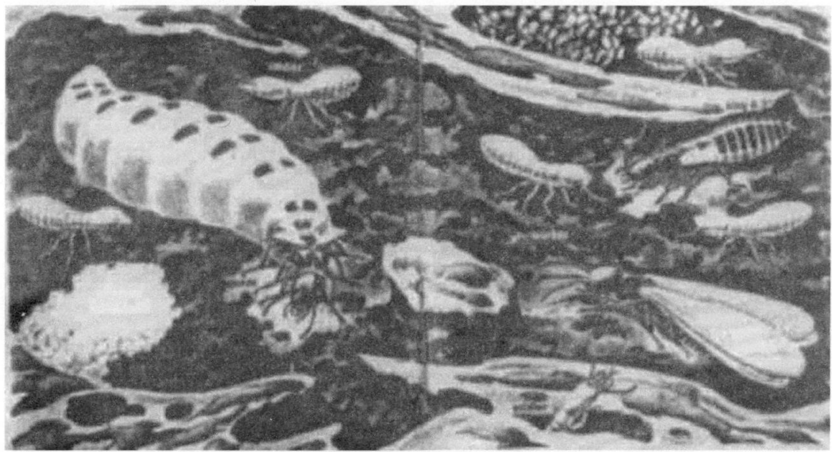

Fig. 96. Colony of termite

Termites penetrate through lime mortar but not cement provided there are no cracks. A plinth 8-15 cm. thick of cement before erecting walls will bring about desired safety. The floor of the house may be treated with chlorpyriphos solution.

Soil treatment for eradication or as a protective barrier must be thoroughly applied. Around the foundation, the normal procedure is to trench the soil and flood or rod-inject the formulation into the soil. The rate for shallow foundation is 9 litres per lineal 1.5 m. or deep foundation 18 litres per lineal 1.5 m.

All wood that must have contact with the soil should be termite proof by pressure impregnation with coal tar, creosote, zinc chloride, mercuric chloride, sodium fluosilicate, pentachlorophenol, and its sodium and copper salts or other tested preservatives after the wood is cut. When building, no bit of waste wood such as grade stakes and concrete forms should be left burried in the soil near the structure. Stumps and waste wood should be removed from the premises.

Soil treatment with the dust of endosulfan 4% @ 25-30% kg/ha be done uniformly after the last ploughing and before sowing the seeds.

2. *Schistocerca gregaria* Forskall (Fig. 97): Commonly known as `desert locust', belongs to family Acrididae of the order Orthoptera, is a polyphagous pest and virtually all crops are at risk so far as locusts are concerned but there is some preference of cereals. The leaves and soft shoots are eaten from margin inwards leaving irregularly shaped feeding marks. Swarm damage usually results in complete defoliation of crops. A swarm of desert locust may contain about 50 billion insects.

Life history

The life history of a locust consist of egg stage, hopper stage and the adult stage. The eggs are laid in a hole made by thrusting the ovipositor 10-15 cm deep in sand embedded in a frothy mass which hardens to form a tubular egg pod. Each egg looks like a grain of rice which is yellowish in colour and 7 to 8 mm in length. Each egg pod contains 70-150 eggs and each female can lay 4-5 pods of eggs. Egg development takes 2-4 weeks according to temperature.

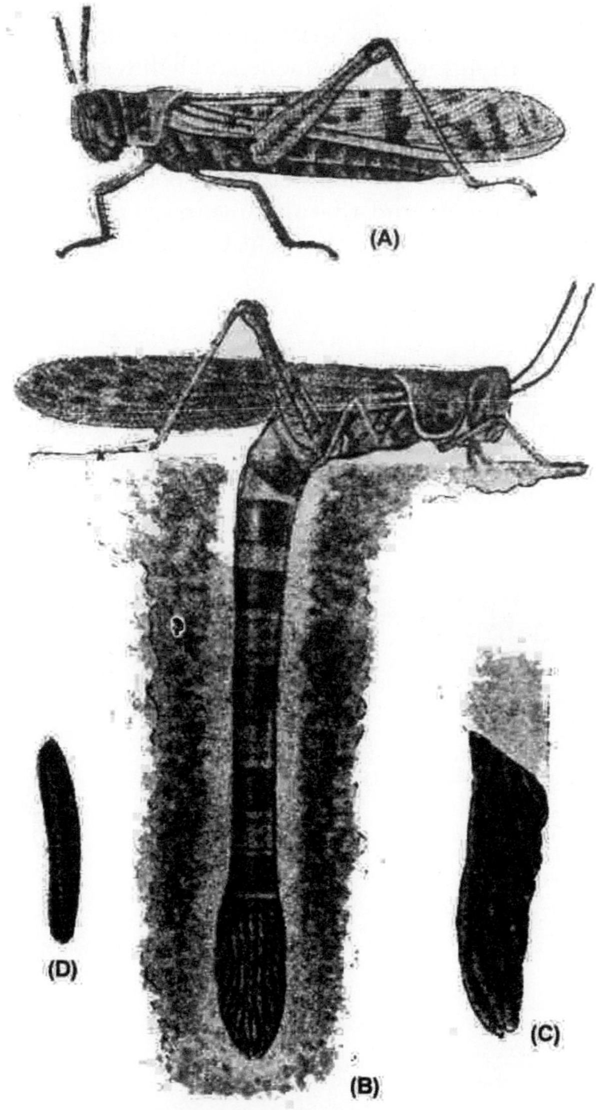

Fig. 97. Desert locust, *Schistocerca gregaria.* **A.** Adult, gregarious phase, **B.** Egg-laying, **C.** Egg cluster, **D.** Egg (magnified) (From Pruthi, 1969)

The first larva is rather vermiform and it has to wriggle up through the egg pod to the sand surface. On reaching the surface it moults and becomes a hopper over a period of 3-10 weeks depending on the temperature. The hoppers develop through 5 instars before they become adult. The hopper exist in two distinct phases, solitary and gregarious with difference in colour and behaviour.

The adults are large; male 40-50 mm and female 50-60 mm pale yellow or brownish. The elytra are greenish yellow, translucent with many brown spots. There is a characteristic peg like process between the bases of the forelegs and the under surface of the thorax is not smooth. There are several generations per year.

Control

The hoppers may be attracted through the use of baits as well as dusting, ground spraying and aerial spraying with carbaryl and endosulfan. Against aerial swarms malathion is widely used. Barrier spraying with persistent insecticides can be very successfully used against large swarms.

(3) *Amsacta species* (Fig. 98): The red hairy caterpillar commonly known as `Katra' belongs to family Arctiidae of the order Lepidoptera, is a polyphagous insect which feeds on a variety of plants like jowar, bajra, maize etc.

Fig. 98. Red hairy caterpillar, **A.** Adult, **B.** Caterpillar (From Pradhan, 1969, 1983).

The fully mature caterpillar measures about 4-6 cm. in length. The body of the caterpillar is covered with reddish brown or brownish orange hairs. The moths are robust having white wings with black spots. The outer margins of the fore wings, the anterior margin of thorax and the abdomen is scarlet red. There are black bands and dots on the abdomen.

Life History

The caterpillars of *Amsacta* are active from late June to September. The insect is normally univoltine and occasionally bivoltine; when there are two generations a year, the first generation is non-diapausing and the pupae of the second generation undergoes diapause.

The moths of the red-hairy caterpillars appear in large number after the first shower of monsoon. They lay eggs in clusters on the cultivated plants as well as on weeds. A single female may lay upto 1500 or more eggs. The larvae hatch out in about 3-5 days and begin to feed. They feed continuously, grow and migrate from one field to another field. The full grown caterpillars go down into the soil for pupation. In the soil they remain for about 8-10 months i.e. whole of the winter and summer and emerge as moths after the first few good showers of rain. The duration of pupal period largely depends on the arrival of monsoon.

Control

The moths can be attracted by light traps. The clusters of eggs are quite large in size and that can be easily destroyed by hand picking. The young caterpillars of the *Amsacta* can be killed by dusting the cultivated crops with malathion 5% @30 kg/ha. For full grown caterpillars spraying endosulfan 35 EC @1.5 litres in 1000 litres of water per hectare controls the pest.

4. ***Chrotogonus trachypterus*** Blanch. (Fig. 99): The surface grasshopper belongs to family Acrididae of the order Orthoptera. It is world wide in distribution. The grasshoppers are polyphagous and besides cotton and wheat, they also feed on a number of other cultivated crops. Both the nymphs as well as the adults feed on leaves by cutting germinating plants of cotton, wheat etc. particularly in areas near by the waste lands.

Life history

Chrotogonus remains active throughout the year but its activity is reduced during winter when only eggs and adults are found. A female may lay 36-434 eggs in 2-15 egg pods at a depth of 3-6 cm in the soil. Each egg pod consists of 16-46 eggs. In the summer, eggs hatch in 19-33 days where as in winter they hatch in about 5 months. The nymphs feed on grasses or other soft plants, moult 5-7 times and are full fed in 40-70 days. The adults are stoutly built, about 20mm long and 8 mm broad with a rough greyish appearance. There are two longtudinal rows of black dots on the under side of the body which is otherwise white. The body of the female is larger and flat than the males; the females live for about 3 months and the male almost half as long. There are two generations in a year.

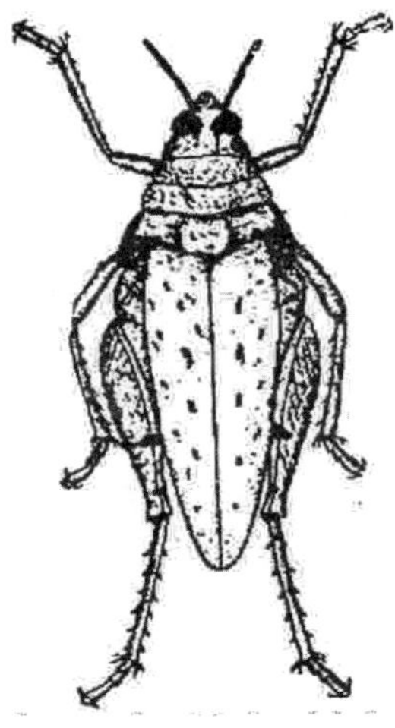

Fig. 99. Surface grasshopper, *Chrotogonus trachypterus*

Control

The pest can be controlled by dusting the crops with malathion 5% or 12.5 Kg. 5% carbaryl (sevin) dust per hectare on first appearance of the grasshoppers.

37

HOUSEHOLD INSECTS

COCKROACHES (Order : Dictyoptera) (Fig. 100): The most common household insect pests are the cockroaches. All cockroaches belong to family Blattidae of the order Dictyoptera. Cockroaches are generally brown or dark brown in colour. They are broad and flattened, with the head bent under the body so that the mouth opens backward and the eyes look downward. The antennae are long slender and of many segments. Wings are usually developed in adults. The mouth parts are strong. The legs long and in most species bear numerous spines. Cockroaches are active at night, hiding in dark places such as cracks and crevices during day light and running very rapidly when disturbed.

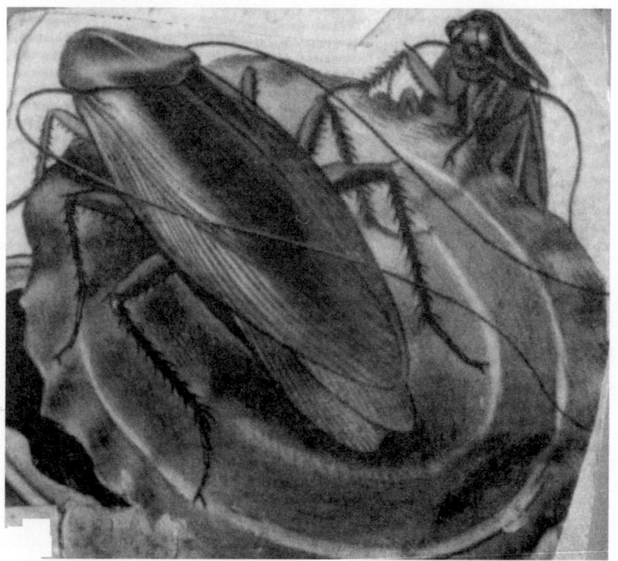

Fig. 100A. Cockroach, *Periplaneta americana*

The cockroaches consume foods and food materials freely, gnaw woollen goods, leather, and anything that has paste on it, and thus often injure book bindings. In fact, they are practically omnivorous. Besides eating, they leave a disagreeable odour which spoils food where they have been. They lay their eggs in pockets or capsules or ootheca, the number of capsules vary with species and the outside of ootheca is horny in nature. The nymphs are active, feed freely and moult several times. Several types of cockroaches are found in buildings, the most common are:

1. ***Periplaneta americana*** Linn. (Fig. 100A) : This is the largest of the house cockroaches, about 4 cm. in length. It is dark reddish brown with a more or less definite yellow band around the margin of the pronotum. Both males and females are winged bearing long active antennae.

Life history

As many as 16 nymphs can hatch from an ootheca, incubation taking less than two months to complete. The nymphs are wingless but otherwise resemble the adults except for size. A generation requires ten to twenty months to develop which is also influenced by temperature. Adults are long lived and females may produce as many as 50 capsules in her life time.

2. ***Blatta orientalis*** Linn. (Fig. 100B). This is more commonly called `black beetle'. It is almost black. The female is oval and has only wing stubs. Its length is about 3 cm. The males have developed wings nearly covering the abdomen, are more slender and about 2.5 cm. long.

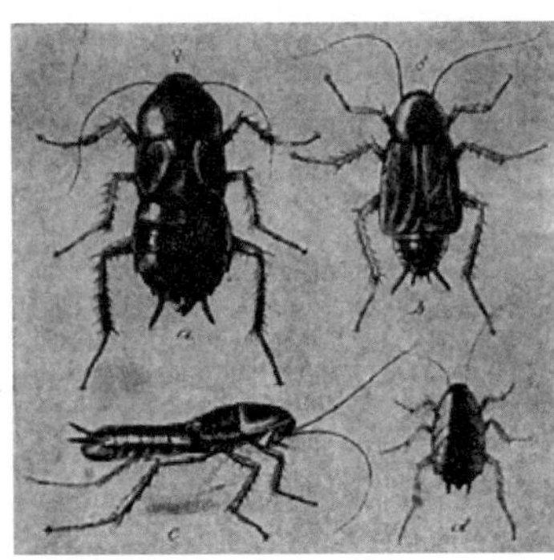

Fig. 100B. Oriental cockroach, **a.** adult female, **b.** adult male, **c.** and **d.** nymphs (From Pfadt, 1962).

Life history

The egg capsules, containing a maximum 16 eggs are dropped as soon as formed and often are glued to objects. The time required for its young to develop is about the same as for the *Periplaneta americana*.

3. ***Supella supellectilium*** (Serville) (Fig. 100C) - These are small cockroaches and when full grown these are about 1 cm. long, bearing light tan band across the base of the wings and a broken band across the centre of the wings. The male has wings covering the tip of its slender abdomen, while the female is broad with stubby wings, egg capsules are small and tan in colour.

Life history

The capsules of this species can have as many as 18 eggs each and hatching occurs in about 2 months. There are two generations in a year.

Fig. 100C. Nymphs of the brown-banded cockroach *Supella supellectilium* natural size (From Pfadt, 1962).

4. ***Blatella germanica*** Linn. (Fig. 100D): It is pale brown in colour with two parallel, dark stripes on the pronotum. The adults are about 1.5 cm. long, boat shaped in appearance and possess fully developed wings covering the abdomen. The antennae are longer than

the body. The egg capsules or ootheca are slender and light brown in colour.

Fig. 100D. German cockroach life stages, winged adults, nymphs, and egg cases. Note female carrying egg case (From Pfadt, 1962).

Life history

Each egg capsule contains upto 48 eggs. The female carries the capsule throughout most of the three week incubation period and produces about five capsules during her life.

Control

Dusting of 5% malathion in dark or concealed places where these insects are found is the most effective control method. Those which are dusted directly, are killed by contact with powder. The cockroaches can also be killed by spraying with 0.2% malathion or 0.5-1% diazinon or by dusting with 5% carbaryl or 2% diazinon.

5. ***Musca domestica*** Linn. (Fig. 101A) : The house fly is a cosmopolitan species, belonging to family Muscidae of the order Diptera. The eggs are pearly white in colour, elongate and about 1 mm long. The larva is a maggot, blunt at the posterior end, tapering at a point anteriorly. The cuticle is transparent except towards the end of the larval period when the maggot assumes a creamy or yellowish appearance. A full grown maggot is about 12 mm or more in length. The puparium which is shorter and more robust than the

larva is about 6 mm long, when mature, is dark brown in colour. The adult fly is dark, mottled grey and 6 to 7 mm long.

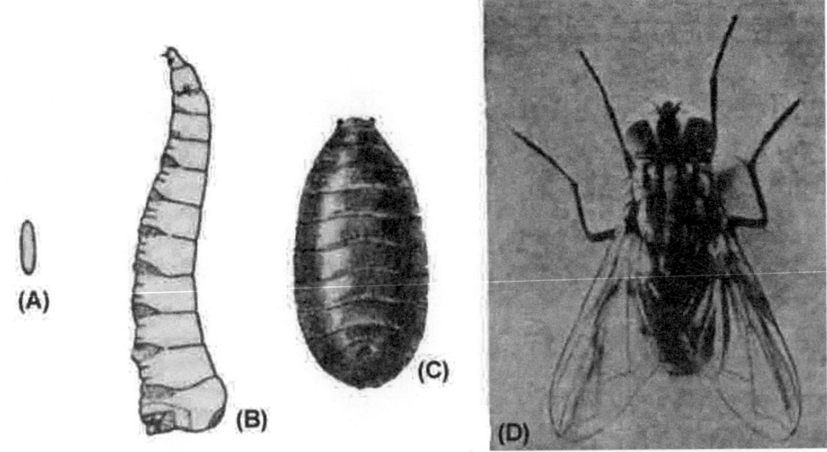

Fig. 101A. Life history of the house fly, *Musca domestica* **A.** Egg; **B.** Larva, **C.** Puparium; **D.** Adult

Life history

As stated above, the small white eggs are laid in considerable numbers in decaying organic matter. A single gravid female fly may lay from 100 to 150 eggs at a time and upto 600 or more eggs in two months of active adult life, though in mid summer longevity is thought to be under a month. Development of the egg is rapid. The larva hatches within a day in normal summer temperatures. The larva feeds on the organic matter in which it hatches and grows rapidly. It goes through three instars in 5 to 15 days before pupation. The pupa remains within the shed integument of the last larval instar which is referred to as puparium. The pupal stage lasts from four to ten days. Two to twelve days after emergence the adults are ready to mate and within two to three additional days, the gravid female is ready to lay her first batch of eggs.

Control

The house flies breed in dung, all kinds of manure and decaying meat and fermenting grass and straw. Some of these materials can be eliminated by burning or burrying. Others can be treated with poison the fly larvae.

The use of insecticides in the form of baits e.g. dipterex, diazinon and malathion used as baits mixed with molasses and water. They are also used as dry baits impregnated on sugar or other materials.

Spraying with malathion or diazinon 2% is effective in killing the house flies.

6. ***Cimex*** species (Fig. 101B): The bed bug belongs to family Cimicidae of the order Heteroptera. It is a small flat insect, reddish brown in colour about 5 mm. long when adult. It is wingless, only tiny stubbs of wings remaining to show that it has been derived from winged ancestors. It produces a very noticeable odour. It is nocturnal, hiding during the day in any cracks and crevices it may find, in the bed stead, behind loose wall paper and elsewhere.

Life history

A female lays about two hundred eggs, these hatching in a week to a much longer period depending upon the temperature. The nymphs are yellowish white at first while turning brown gradually with the increasing age. Under favourable conditions, about seven weeks are required to produce the adult bug.

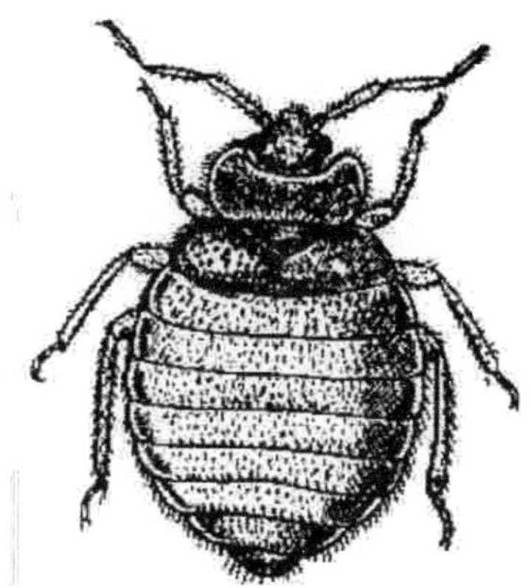

Fig. 101B. Bed bug, *Cimex* sp.

Control

Bed bugs can be readily controlled by residual spraying of houses, barns, barracks etc. with malathion 1% or 0.5% lindane. These materials may be applied as wettables, emulsions, or solutions and will give good control for several months or more. If bugs are located in the room or furniture, they may be killed by spraying all hiding places with deodorized kerosene spray containing 57% EC of malathion, lindane, etc. They can also be killed by pouring boiling water on the infected furniture. Where spray treatments are not desirable, infestations may be eliminated by fumigation with hydrogen cyanide. The rooms and apartments can be more safely fumigated with methyl bromide, methyl formate or ethylene oxide mixed with carbon dioxide.

7. ***Pediculus humanus*** Linn. (Fig. 101C): Commonly known as "Human head louse" belongs to family Pediculidae of the order Anoplura. Human lice are greyish, flattened wingless insects 1.5 to 3 mm. long with short five segmented antennae, simple eyes, heavy clinging legs for grasping hairs and piercing mouth parts.

Life history

The head louse lives on the skin and among the hairs of the head and glues its eggs to hairs of the head, back of the ears and at the back of the neck. The eggs are elongate oval in form about 1 mm. long. They hatch in about a week under favourable conditions and the very tiny slender lice at once begin feeding which continues from two to six or more times a day for the next 1 to 4 weeks growing meantime and moulting three times before they become adult. The adults live for about one month. The female lays an average 8 to 10 eggs a day, until a total of 50-100 are deposited by the head louse.

Fig. 101C. Head louse, *Pediculus humanus.*

Control

For the control of head or body louse application of malathion 5% dust twice at 10 day intervals is effective. Treatment with 0.2% lindane in hair oil has given good results.

8. ***Tineola bisselliella*** (Hummel) (Fig. 102): Commonly called Clothes moth belongs to family Tinaeidae of the order Lepidoptera. The buff coloured moth is small about 1 cm. across the wings. The larvae of the moth feed on wool, hairs, feathers, furs, upholstered furniture, and nearly all animal products like bristles, dried hair and leather.

Life history

The adult moths are harmless and take no food of any sort. They lay their eggs singly on the products on which the larvae feed. Each female laying from 100-150 eggs. The eggs are small, 0.5 mm. long and white in colour which is very distinct when deposited on black material. They hatch in about 5 days. The larvae are white and vary in size. The newly emerged larva is about 1.5 mm. while the mature larva is about 8 mm. long. The length of larval period varies from 6 weeks to several years. The larvae spins tough cocoon, within the cocoon pupation takes place. In one to four weeks adult moth emerges. The entire life cycle may be passed in about 2 months or may be prolonged over 4 years.

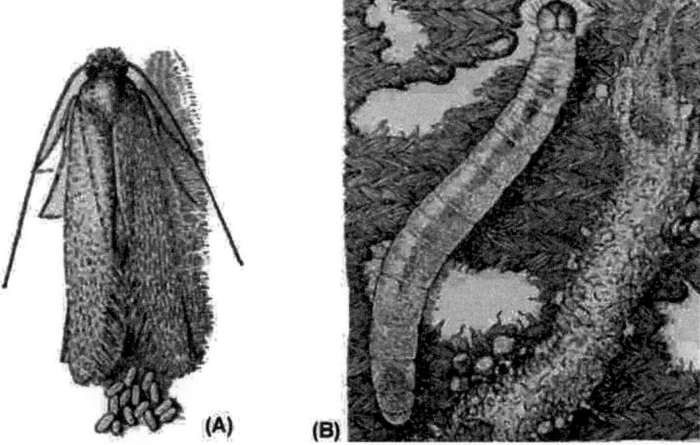

Fig. 102. The clothes moth. **A.** Adult and Eggs; **B.** Larva feeding on woolen cloth and silken tube of larva (From Pfadt 1962).

Control

Use of naphthalene flakes at the rate of 125 gm. to 500 gm to 0.5 cum. of space is useful if the boxes are tightly closed. Paradichloro benzene and naphthalene mixed in equal parts is still better. Camphor can also be used for this purpose but it is extremely volatile. Carbon disulphide is also useful but it is oily and gives an offensive odour. Dry cleaning with petrol keeps the wool safe for a sufficiently long time.

Frequent dusting and brushing are very important control measures. Before storage, clothes may be kept in paper bags with pasted flaps to prevent re-infestation. Stuffed animals can be saved by spraying them with petrol. A mixture of ethylene-dichloride (3 parts by volume) and carbon tetrachloride (1 part) if sprayed is known to kill all stages of clothes moths.

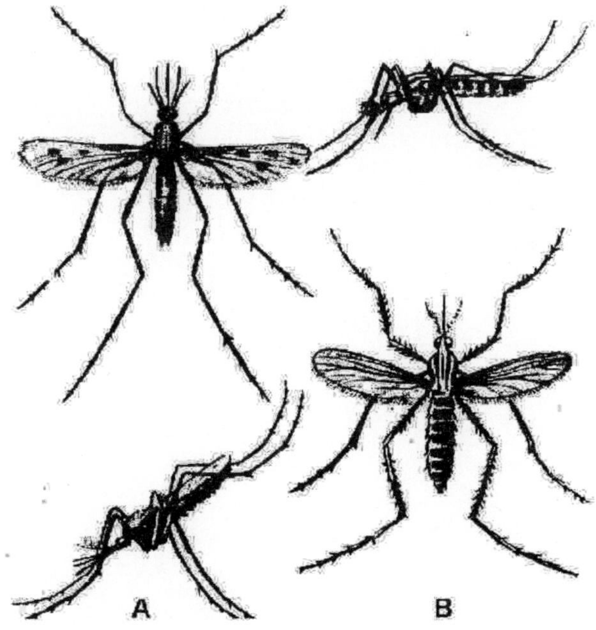

Fig. 103A. *Anopheles* mosquito, **B.** *Aedes* mosquito

9. Mosquitoes (Fig. 103 A-C) The mosquitoes belong to family Culicidae of the order Diptera. This large group of small but imporant insect has been amply studied as part of public health campaigns against malaria and yellow fever. The conquest of malaria is a scientific milestone. The common carriers of the disease, the *Anopheles* mosquitoes, are recognized by the ``three-pronged" beak of the female and by the tilted position they assume when resting. The *Aedes* mosquitoes, one of which carries yellow fever, are more like the common house and swamp mosquito *Culex* in appearance. The disease carrier is limited to tropics and subtropics. In most species the females feed upon blood. Male mosquitoes are harmless. Their beaks are not fitted for piercing. Females bite and hence can transmit disease. They are the only ones which buzz. There are many kinds of mosquitoes but the larvae of all live in stagnant water.

Life history

The eggs may be laid singly in small clusters or in masses often called "rafts" on the surface of standing water. The number of eggs laid by one female varies in different species but averages several hundred. The larvae are predaceous, feeding on tiny water animals including other species of mosquitoes. They moult four times and in a week transform into pupae. After a brief pupal stage the pupal skin splits along the middle of the back of the thorax through which the adult mosquito escapes (see Fig. 103C).

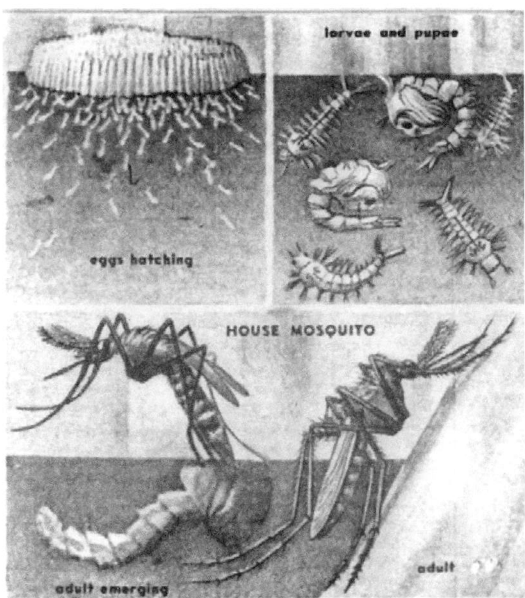

Fig. 103C. Life history of *Culex* mosquito

Control

There are many ways by which mosquitoes can be effectively controlled. Destruction of the larvae, pupae and eggs is the most direct way to control mosquitoes in large number. Where water cannot be drained off, cover it with a film of kerosene or a mixture of kerosene and crude oil. It will suffice to destroy eggs on the surface, larvae and pupae at or coming to the surface for air. Oil used should be sprayed on water. Sawdust soaked in kerosene has been found to give up the oil slowly and thereby preserve the film on the surface longer when this material is scattered along the edge of water. Swamps may be filled up, drains, thoroughly cleaned and stagnant water drained off. For protection against mosquito bites, use of net is recommended.

10. ***Lepisma saccharina*** Linn. (Fig. 103D) : It is commonly known as ``silver fish" which belongs to family Lepismatidae of the order Thysanura. It is silvery grey in colour. Usually less than 1 cm. long, very active and hard to catch. It prefers dark places and feeds on anything containing starch, the binding of a paper, wall paper and rayon and on damp wheat flour.

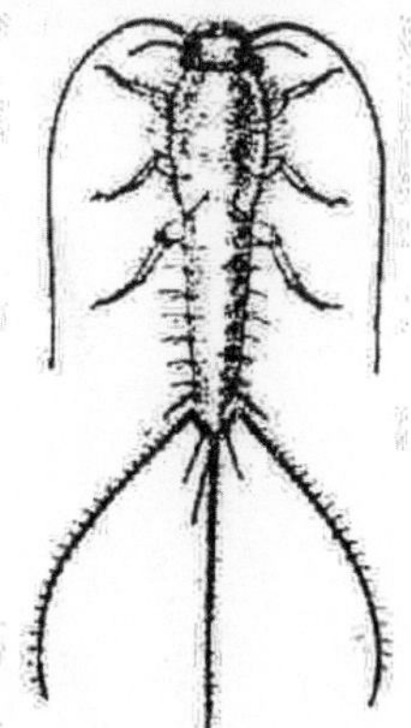

Fig. 103D. Silver fish, *Lepisma* sp.

Control

Dusting pyrethrum or 5% malathion where the insect exist, and poison baits using a starchy base and a stomach poison are often of value. Diazinon 0.5-1% emulsion or malathion 3% emulsion kills the pest.

11. ***Gryllodes sigillatus*** Walk. (Fig. 103E). All the crickets belong to family Gryllidae of the order Orthoptera. These insects are as common as the cockroaches and are also nocturnal in habit. They make a very shrill noise by their stridulatory apparatus. They live in various crevices in the walls, floor, ceilings, doors, furniture, and books etc.

Fig. 103E. House cricket, *Gryllodes sigillatus*

Life history

The house crickets have only one generation in a year. They lay yellowish, cylindrical eggs in the late summer and are found in clusters in soil, crevices and moist places. The nymphs hatch in about 8-11 days and are without wings. They are full grown in about 3-4 months.

Control

It is difficult to exterminate them but they can be checked to some extent by baiting, or spraying the house with suitable contact insectides. The crickets can be killed by spraying 50% WP of diazinon or by dusting malathion or carbaryl 5% at night, taking care that food materials are not contaminated by the insecticides.

12. **Black ants** (Fig. 103F) The ants are cosmopolitan in distribution and belong to family Formicidae of the order Hymenoptera. They eat away and contaminate the food.

The common ants are *Monomorium* sp. the little black ant, and *Camponotus* sp. the large black ant. The damage is done by workers.

Control

The simplest thing to do when ants appear in dwellings is to scatter dust in their paths or spray the infested area with a deodorized oil spray. Endosulfan used as 5% dust or as 2 to 5% solution in deodorized kerosene is the most effective material for the purpose as it does not repel the ants and remains effective for a month or more if properly applied. Dusting with lindane will destroy almost all the species of free living ants.

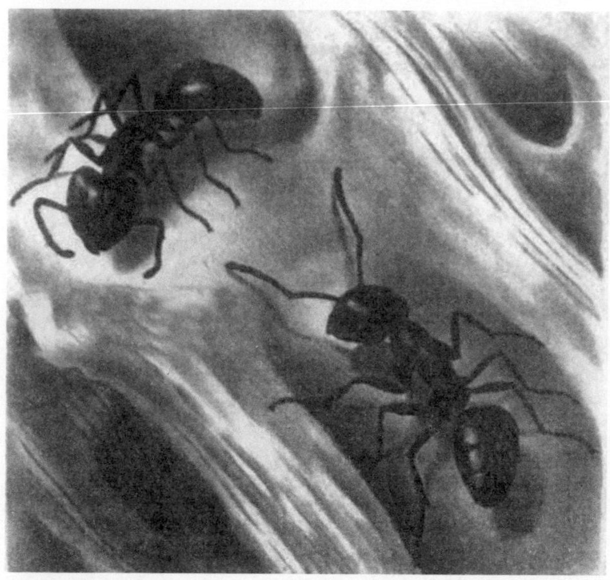

Fig. 103F. Black ants

Although it is difficult to locate nests if they are there they may be fumigated with a suitable fumigant. If the nests cannot be located, a poison bait should be used which are available in the market for the control of ants. Spraying diazinon 0.5% emulsion in the ant travel area is also very effective.

38

PESTS OF FRUITS AND VEGETABLES

1. ***Idiocerus atkinsoni**** (Leth.) (Fig. 104): This insect belongs to family Cicadellidae of the order Homoptera. It is a major pest of mango. The damage is caused by both nymphs and adults which suck the juice of the young shoots and flower buds of the entire inflorescence. Like aphids, the mango leaf hopper secretes honey dew which covers the branches of the trees and facilitates fungus growth.

The adult insect is small, wedge shaped with dark grey wings. The head is broad and prominent and possess three dark brown spots. The hind legs are thickly clothed with a number of bristles.

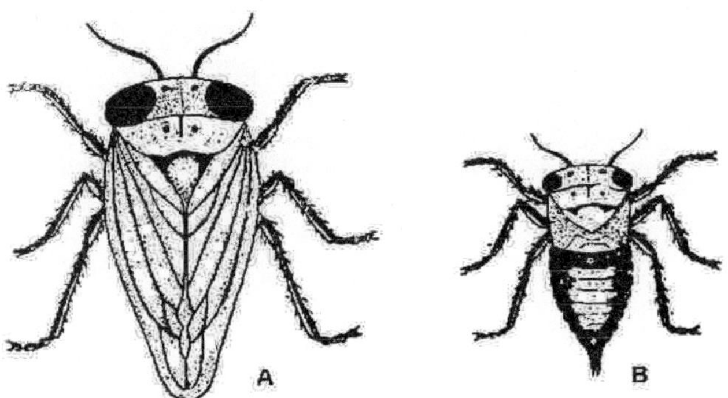

Fig. 104. *Idiocerus atkinsoni*, **A.** Adult, **B.** Nymph

Life history

The female lays eggs singly inside tissues of the leaves or flower buds in a streak made by ovipositor. A single female lays 100-200 eggs which are dull white in colour. The eggs hatch in about 8-10 days

---

* Same as *Amritodus atkinsoni* (Leth.)

releasing the nymphs which immediately ingest the juice from the plant. They resemble the adult in all respects except in that they do not possess wings and have fairly long legs. There are usually 4-5 moults undergone by the nymphs in about 15-20 days and the entire life cycle is completed in about a month or so. There are several generations produced in summer while the adults hibernate in winter.

Control

Chemical control has been primarily employed to check the population of the mango leaf hoppers. This pest can be effectively controlled by spraying the trees with a number of insecticides. Some of the important being carbaryl 0.1% or diazinon or endosulfan or dimethoate 0.03% or malathion 0.05%.

2. ***Ophideres species*** (Fig. 105): This belongs to family Noctuidae of the order Lepidoptera and is commonly known as `fruit piercing or fruit sucking moth'. It is a minor pest of mango, grapes, apple and citrus. The general body of *O. conjuncta* is faint orange brown. Its forewings are dark grey and the hind wings are orange red having two black curved patches. The larvae are semiloopers. A fully grown larva is about 50-60 mm in length.

Fig. 105. Fruit sucking moth

Life history

The moths are nocturnal. The female lays eggs on a number of wild plants and weeds growing near the citrus trees. Eggs are round translucent about 1 mm in diameter. They hatch in about 2 weeks and within 24 hours of emergence the growing larvae start feeding on the leaves of the plants. A larva passes through 5 instars in 4-5 weeks. The pupa is dark red or brown which lasts about 2 weeks. The moth on emergence fly to nearby orchards for feeding on fruit juice. The exact duration of the life of the moth is not known.

The moths damages the plant in the adult stages with the help of its strong piercing mouth parts. It punctures the fruits for sucking juice. Bacterial and fungal infections take place at the side of attack. The area surrounding the puncture become pale and the fruit drops off.

Control

Spraying trees with diazinon 0.02% is effective to control this pest. Poison baiting with 20 gm malathion 50% W.P. or 50 ml diazinon + 200 gm gur or molasses in 2 litres or water attracts and kill the moths.

3. ***Papilio demoleus*** Linn. (Fig. 106) : This belongs to family Papilionidae of the order Lepidoptera. This is not only a pest of citrus but of all the plants belonging to family Rutaceae. The young larvae feed on the tender leaves while the older ones on the mature leaves. As a result of heavy infestation the entire plant may be left without leaves and bears no fruits.

Life history

A female butterfly usually mates once and lays about 75-120 eggs within 2-5 days time. A comparatively smaller number of eggs are laid in summer. The freshly laid eggs are pale green which become dark grey before hatching. The hatching takes place in about 3-4 days in favourable conditions and 5-8 days under unfavourable conditions. The larval life lasts 8-16 days in favourable conditions and 4 weeks under unfavourable conditions. A full grown larva is bright yellowish green measuring about 40 mm long and 6.5 mm wide. The young larvae are provided with white black brown markings thus appearing like the droppings of the bird. The pupal period lasts about 8 days in favourable conditions and 9-12 days under unfavourable conditions. The butterfly is beautiful, measuring 94 mm from wing tip to wing tip. The head and thorax of the butterfly are black; and the underside of the abdomen is creamy yellow in colour, the wings being dull black which are ornamented with yellow markings. There are 3-4 generations in a year and the winter is passed in the pupal stage.

Fig. 106. Lemon butterfly - **A.** Full grown larva, **B.** Larva suspended for pupation, **C.** Pupa, **D.** Butterfly (From Pradhan, 1969)

Control

The larvae can be controlled by hand picking and by spraying the plants with 0.05% malathion or 0.05% endosulfan or 0.1% carbaryl.

4. ***Phyllocnistis citrella*** Staint. (Fig. 107). This insect belongs to family Gracillariidae of the order Lepidoptera. It is known as "citrus leafminer". This insect is widely distributed throughout the India and is a serious pest of citrus. It is also a pest of several other plants. The damage is caused by the larvae by making zigzag silvery mines in the young leaves. Due to the extensive mining by the pest,

the leaf suffers badly, gets deformed and irregularly curled up in shape, unhealthy in look and defective in its function and finally it dries and falls off. The attack is comparatively much greater in leaves of fresh growth, young seedlings and varieties having soft succulent leaves.

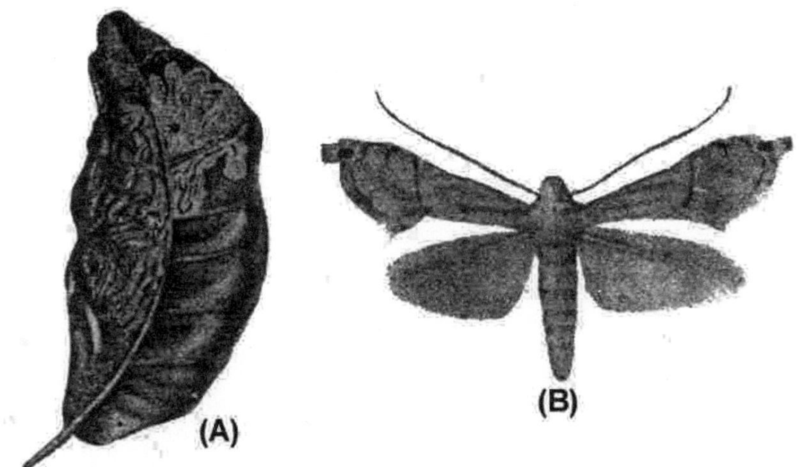

Fig. 107. Citrus leaf miner - **A.** Mined citrus leaf, **B.** Adult

Life history

The moths lay tiny flattend, transparent eggs on the under side of the young leaves. The eggs are laid singly which hatch in about 2-7 days. Soon after emergence, the legless larva enters the leaf tissue and begins to feed inside it and mines the leaf lamina. The larva is full fed in about 5-25 days. The full grown larva measures about 5mm in length and is yellowish green in colour. Pupation takes place within the galleries. The pupal stage lasts about 5-21 days after which the moth comes out. The adult stage is a tiny greyish moth with a wing span of 8-10 mm. The fore wings are white with two narrow grey stripes and hind wing has a pale grey fringe. The whole life cycle is completed in about 14-58 days depending on weather conditions. There are several generations in a year.

Control

The pest can be controlled by repeatedly spraying the trees with methyl demeton/dimethoate 0.03% at 15 days intervals. The collecting and burning the mined leaves is also useful to control this pest.

5. ***Sthenias grisator*** Fabr. (Fig. 108): The grapevine girdler as it is known belongs to family Cerambycidae of the order Coleoptera. It occurs throughout the grape growing areas of the country. Besides grapes, it is also a pest of almond, cashew, mulberry, mango, jack fruit and several garden plants. The damage is caused by girdling of the adult beetles which is followed by the drying up of the wood. The adult is a medium sized, stout beetle. The beetle is greyish brown in colour and is provided with white and brown irregular markings. The insect resembles a mottled bark. The pest prefers girdling of the green branches before laying eggs, which causes a lot of damage to the grapevines as a result of which the vines dry up above the level of girdling.

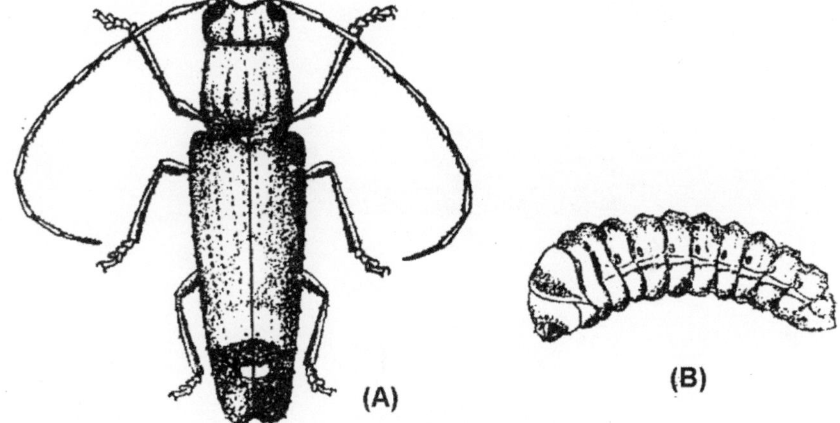

Fig. 108. Grapevine girdler. **A.** Adult, **B.** Larva (Grub)

Life history

The eggs are deposited at night underneath the bark in cuts made by the beetle on girdled vines. Eggs are laid in groups of 2-5 which are oval in shape and measure about 3-4 mm in length and 1 mm in width. The eggs hatch in about 7-8 days. The freshly hatched grubs are very small measuring about 2-4 mm in length and soon tunnel into the wood. The head is dark brown and the thorax is globular provided with chitinous spines. As with other cerambycids, the mouth parts specially a pair of mandibles are very prominent and powerful. The full grown grub is about 10-12 mm long, which pupates in the tunnel. The whole life cycle is completed in more than a year.

Control

Collection and destruction of the girdled branches is a simple method to control the pest. Hand picking adult beetles is also very effective. The pest can be controlled by spraying the grape vines at the ground level with malathion 0.05%.

6. ***Aulacophora foveicollis**** Lucas. (Fig. 109) : This belongs to family Chrysomelidae of the order Coleoptera. It is a serious pest of Cucurbitaceous plants like melon, pumpkin, cucumber etc. The attack caused by the feeding of adult insect or its larva on leaves, is indicated by the formation of irregular holes in the leaves. The larvae also damage the shoot, roots of the plants which ultimately withers and dies. The chief damage however is done to the fruits which are rendered useless for human consumption.

The adult is brilliant orange coloured beetle, with the body surface shiny and smooth. The underside is entirely black and is covered over with short, wide, soft hair like setae.

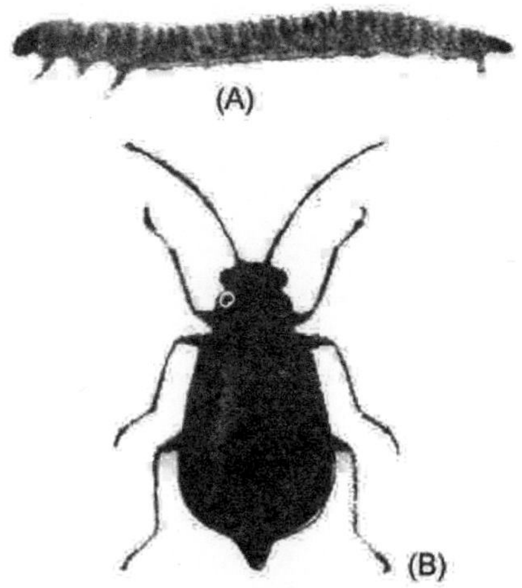

Fig. 109. Red pumpkin beetle, *Aulacophora foveicollis* **A.** Larva, **B.** Adult

Life history

The female lays about 60-300 orange or yellowish-brown eggs within a period of slightly more than a month, either singly or in

* Same as *Raphidopalpa foveicollis* (Lucas.)

clusters on the moist soil underneath the plants. These eggs under optimal humidity and temperature conditions, hatch in about 6-15 days into young larvae. Right from the start the larvae attack the roots of the plants and moult 4-times before changing into pupa. For every moult the larva goes into the soil and casts off its cuticle. After a larval period of 13-23 days they penetrate into the soil to a depth of 1cm. to 25 cm. where they construct an oval waterproof creamy white cocoon. The pupal period lasts for 7-17 days after which the adult escapes out of the soil. The entire life cycle ranges from 30-55 days. Normally there are 3-5 broods in a year. These pests are extremely active, from March to October although occasionally they are found damaging plants in winter months also.

Control

Besides certain precautionary measures such as adjustment in the time of sowing of plants, ploughing after harvest etc. it is the chemical control mostly which has been found to be effective. Using the carbaryl dust is very effective. Spraying the crop with dimethoate 0.02% or malathion 0.05% also controls the pest.

7. ***Dacus cucurbitae*** Coqu. (Fig. 110): This belongs to family Tephritidae of the order Diptera and is commonly known as "melon fruitfly". It is a serious pest of melons and other cucurbits like ribbed gourd, bitter gourd and gourd. Tomato is also attacked occasionally. It is distributed throughout India. The damage is caused by larvae which bore into the fruits and feed on the pulp. The infested fruits can be easily recognized due to deformation or rotting area round the seat of oviposition.

Life history

The female fly lays 5-27 eggs singly or in batches underneath the rind (skin of young fruits) which is punctured by the ovipositor. The eggs hatch in 2-5 days liberating small dirty white apodous larvae. The larvae are typical Dipterous maggots, 10-12 mm when fully mature. These feed continuously on the pulp of the fruit resulting in its complete destruction. They can jump to a height of 30 cm. which helps them in dispersal. They become fully grown within 8-15 days and then leave the fruit; after falling down to the ground, they burrow in the soil to a depth of 2-8 cm. and pupate there. In about 5-10 days adult flies emerge and infest the whole crop. Complete life cycle takes about 3 weeks. The adult fly is reddish brown measuring about 5 mm with black and white spots. The wings are transparent with brown bands and grey spots at the apex with a wing-span of 12-15 mm. There are several generations in a year.

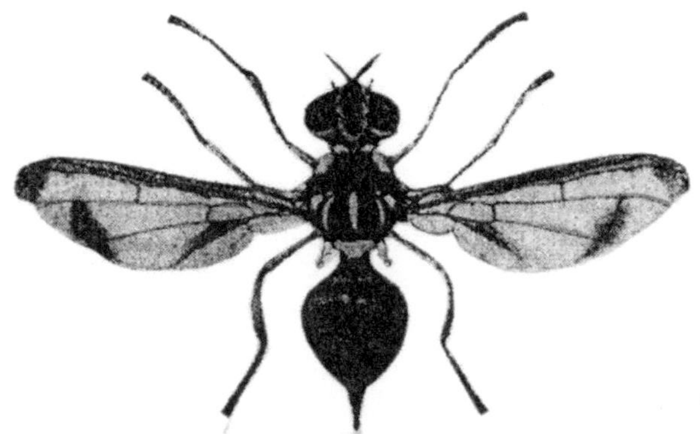

Fig. 110. Fruit fly, *Dacus* sp. (From Pradhan, 1969)

Control

All infested fruits should be collected and destroyed as soon as they have fallen to kill the maturing maggots inside. In addition, use of 0.03% dimethoate, 0.05% malathion in a sugar solution form of spray on the foliage, is also very effective. A poison bait (20 gm malathion 50% W.P. or 50 ml of diazinon + 200 gm of gur or molasses in 2 litres of water) kept in flat containers attracts and kill flies.

8. ***Plusia orichalcea*** Fabr. (Fig. 111A): This belongs to family Noctuidae of the order Lepidoptera. It is commonly known as ``cabbage semi-looper''. It is a minor pest of cabbage, cauliflower and other winter vegetables. The pest is polyphagous. The caterpillars are greenish yellow in colour, cause damage by biting round holes into the cabbage leaves. The adult pest is yellowish brown with large golden patches on each forewing. The moth measures about 42 mm across the spread wings.

Life history

The female lays eggs on leaves of the host plants and caterpillars feed actively on the leaves. The full grown larvae pupate in the debris lying on the ground. The moths are very active at dusk on garden flowers.

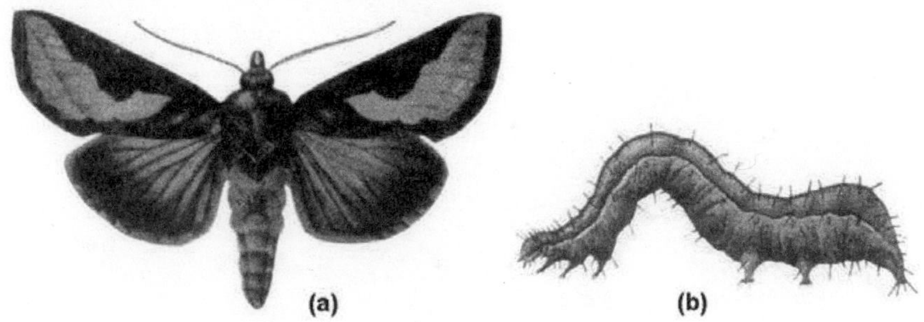

Fig. 111A. Cabbage semilooper, *Plusia orichalcea* **a.** Adult, **b.** Semilooper. (From Pruthi, 1969).

Control

The pest can be controlled by spraying the crop with malathion 0.05% upto three weeks before harvest.

9. ***Epilachna*** beetles (Fig. 111B) : These are commonly known as ``*Epilachna* Beetles', belong to family Coccinellidae of the order Coleoptera. Both the adults and larvae feed on the leaves and fruits of cucurbitaceous, solanaceous and crop plants. The leaves are eaten between the veins sometimes being completely stripped to the midrib. Stems are often gnawed and holes are eaten in the fruits.

Life history

The eggs are laid in clusters, varying in number, usually on the underside of the leaves and placed vertically. The eggs are pale, yellow, elongate oval with comb like hexagonal sculpturing and measure about 0.5 mm in length. Incubation takes about 4-5 days. The larvae are pale yellow, covered with delicate spines when first hatched. The young larvae start feeding soon after hatching making rows of small windows in the leaves. Full grown larvae are dark yellow, broad with a dark head and strong branched spines and 6-7 mm long. Larval development takes about 16 days. Pupation takes place on the leaves of the host plant and the pupa is dark yellow. The adult beetles are oval, 6-9 mm long, reddish to brownish yellow but colour is variable. Each elytra is marked with a series of black spots. *Epilachna* beetles are the only phytophagous representatives of the family Coccinellidae. The whole life cycle takes about 30-35 days under suitable conditions.

Fig. 111B. *Epilachna* beetles with eggs, Larva and Pupa. Note skeletonized leaves.

Control

The *Epilachna* beetles can be controlled by spraying them with malathion 0.05% or 0.1% carbaryl. On cucurbitaceous vegetables, only malathion 0.05% should be sprayed as the other insecticides are phytotoxic.

10. ***Leucinodes orbonalis*** Gue. (Fig. 111C-E) : This pest belongs to family Pyraustidae of the Order Lepidoptera. This is commonly known as ``brinjal fruit borer''. The pest is widely distributed in several countries. The young caterpillars are creamy white, the mature larvae are pinkish in colour. The moths are white having brown or black spots on the dorsal side of the thorax as well as abdomen. The wing margins are provided with fine bristles and the fore wings have numerous black or brown spots. The damage is

caused mainly by the caterpillars which attack the terminal part of the shoots, thus destroying the growing point. Fruit buds and fruits are badly damaged by borers specially in autumn.

Life history

Moths of the brinjal borer appear in the beginning of summer i.e. in late March. Each female lays about 100 or 120 eggs singly or in groups of 3 to 4 on almost all the parts of the plant i.e. leaves, stem, flowers and tender fruits. Hatching occurs in about 4 to 6 days. The newly emerged larvae bore the terminal part of the shoot, flowers, fruits etc. where the eggs are laid. There are 5 instars and the larval stage lasts 1 to 4 weeks. The mature larvae pupate inside the cocoon. Pupal duration is about 1 to 2 weeks. The whole life cycle is completed in 3 to 6 weeks. There are several generations in a year.

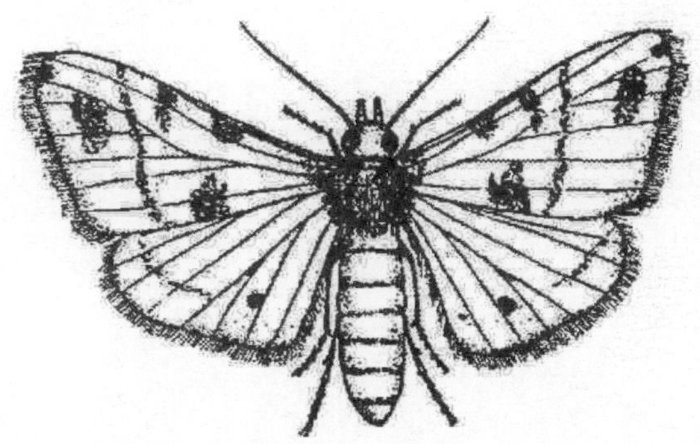

Fig. 111C. Brinjal fruit-borer, *Leucinodes orbonalis*, moth.

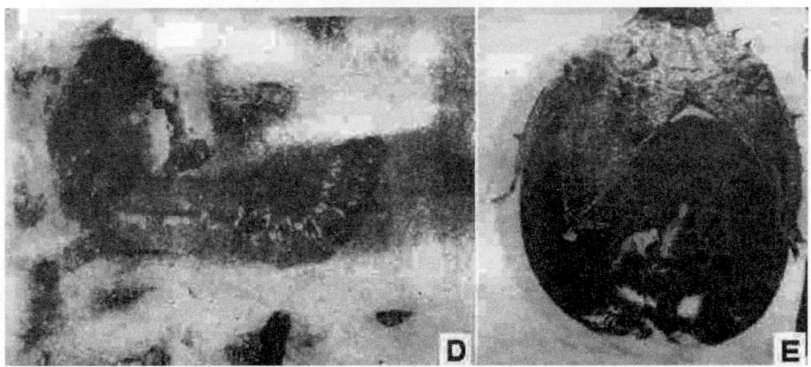

Fig. 111. D. Larva of the brinjal fruit-borer, **E.** Brinjal damage by fruit borer (From Atwal, 1976).

Control

The borers can be controlled by spraying malathion 0.04% or carbaryl 0.1%. The borers can be destroyed by removing the infected fruits and leaves.

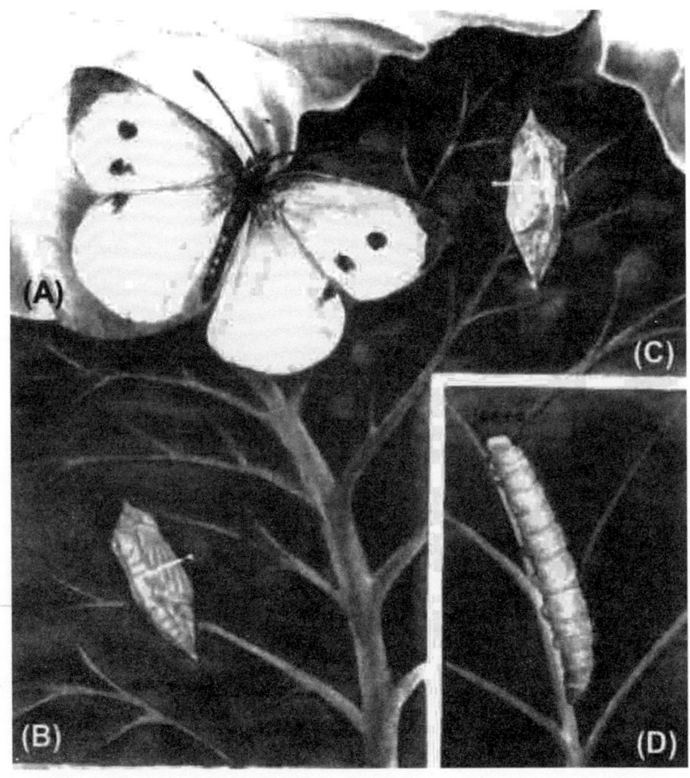

Fig. 112. Cabbage butterfly **A.** Female butterfly, **B-C.** Chrysalis (Pupa) **D.** Larva

11. ***Pieris brassicae*** Linn. (Fig. 112). The cabbage butterfly belongs to family Pieridae of the order Lepidoptera. It is a serious pest of almost all the vegetable plants of the family Cruciferae. It is world wide in distribution. The damage is caused only by the larvae which are known as "cabbage caterpillars". The butterflies are pale white with a smoky shade on the dorsal side of the body. The wings are pale white with a black patch on the apical angle of each fore wing and a black spot on the costal margin of each hind wing. The females measure 6.5 cm across the spread wings and have two distinct black circular dots on the dorsal side of each fore wing. The males are smaller than the females and have black spot on the lower side of each fore wing.

Life history

The butterflies are very active in the field and lay about 164 yellowish conical eggs in clusters of 50-90 on the upper or lower side of a leaf. The eggs hatch in about 11-17 days in November-February and 3-7 days in March-May. The caterpillars feed gregariously during early instars. The mature caterpillars measure 40-50 mm in length. The larvae are pale yellow when young and turn greenish yellow later on. The head is black and the whole body is marked with black spots. They pass through 5 stages and are full fed in 15-22 days during March-April and 30-40 days during November-February. Then they pupate in barns or on the trees. The pupal stage lasts 8-15 days in March-April and 20-28 days in November-February. The butterflies live for 3-13 days and breeds 4 times in a year.

Control

The pest can be controlled by spraying the crop with malathion 0.05% up to three weeks, before harvest. During the later stages of growth of the plants 0.2% carbaryl is very useful and safe.

39

BENEFICIAL INSECTS

The advantages of beneficial insects are manifold. For instance, the curious fact of honey production, silk production and shellac production are of great significance. Therefore, the present chapter deals with the economic importance of honey bees, lac insects and silkworms.

THE HONEY BEES

Honey was one of the man's first food and the first available sweet. It has been highly valued by Indians from time immemorial. It has been defined as an aromatic viscid sweet material derived from the nectaries of the plants through the collection of honey-bees and modified and stored by them as a denser liquid. Here some detailed account of honey production is given.

Honeybees, utilize the sun and polarized light for orientation during foraging activities. Once a food source is located, workers return to the colony and communicate the direction (in relation to the sun) and distance to the food by a series of dances (each species or strain of honeybee has slightly different dances and interpretations). Figure 133 illustrates two such dances. A strong, vigorous dance creates excitement because the food source is near and abundant; a weak dance elicits little response, and hive energy is therefore, conserved. This communication by dances is unusual and is not found in most bees, although some stingless bees may leave a scent trail to assist others in locating food.

Several hundred thousand flowers may be visited by a single bee each day to collect pollen or nectar. Usually, a bee forages for either nectar or pollen, but a number of worker bees (about 17%) take both on the same trip. Pollen adheres to the body hair and is then combed off and packed into special baskets on the outer surface of the hind tibia. Nectar is sucked into the crop, or "honey sac". When the bee returns to the hive, the pollen ball is scraped off each hind leg and is packed into a wax cell while nectar is either fed to others or regurgitated into cells for storage.

Thus honey, the product of the honeybee. *Apis* sp. represents conentrated and partly hydrolyzed nectar. The principal sugar in nectar is sucrose which is broken into glucose and fructose by salivary enzymes as the bee returns to the hive. This fluid is regurgitated (vide supra) from the crop into comb cells where the solution is concentrated by bees fanning their wings and evaporating water from mixture. Once the mixture is concentrated to approximately 80 per cent sugar, the cells are sealed with wax (Fig. 113 F) for future use by the bees.

It is interesting to note that it requires about 40,000 to 80,000 trips of a honeybee, and visits to many times this number of flowers, to find and assemble nectar enough to make about 500 gm. of honey. Each trip is about 2 or 3 km. Thus for a single bee to collect nectar enough for 1/2 kg of honey would mean travelling at least double the distance around the world.

Chemical composition of honey

Chemically, honey is a viscous water solution of sugars, containing 13 to 20% water, 40 to 50% fructose, 32 to 37% glucose, 2% sucrose, traces of maltose, 1 to 12% dextrins and gums; and traces of minerals, free organic acids, vitamin A and B-complexes, enzymes, plant-pigments, pollen grains and bee wax.

The value of the honey lies in its simple sugars found in large amount in it which are directly assimilable and are a ready source of bodily energy. The colour of honey and taste are no guides to the purity and quality of honey. Honey from soapnut and white clover, is light (almost water white) in colour and mild in flavour, that from mustard looks yellow and has a little pronounced flavour and that from shisham (*Dalbergia*) is dark amber and strong in flavour. Barberry and buckwheat honey are very rich dark in colour and have a strong taste.

Honey as food – It has been estimated that 200 gm of honey is as nourishing as 1.135 kg of milk or 1.658 kg of cream cheese or 340 kg of meat or 425 kg of boneless codfish and 8 oranges or 10 eggs. It is a rich energy giving food and with milk forms a perfect food. Honey is used in Indian homes in a variety of ways. Honey can be considered as a complete food. It has also been referred as `Ambrasia' or ``the food of Gods".

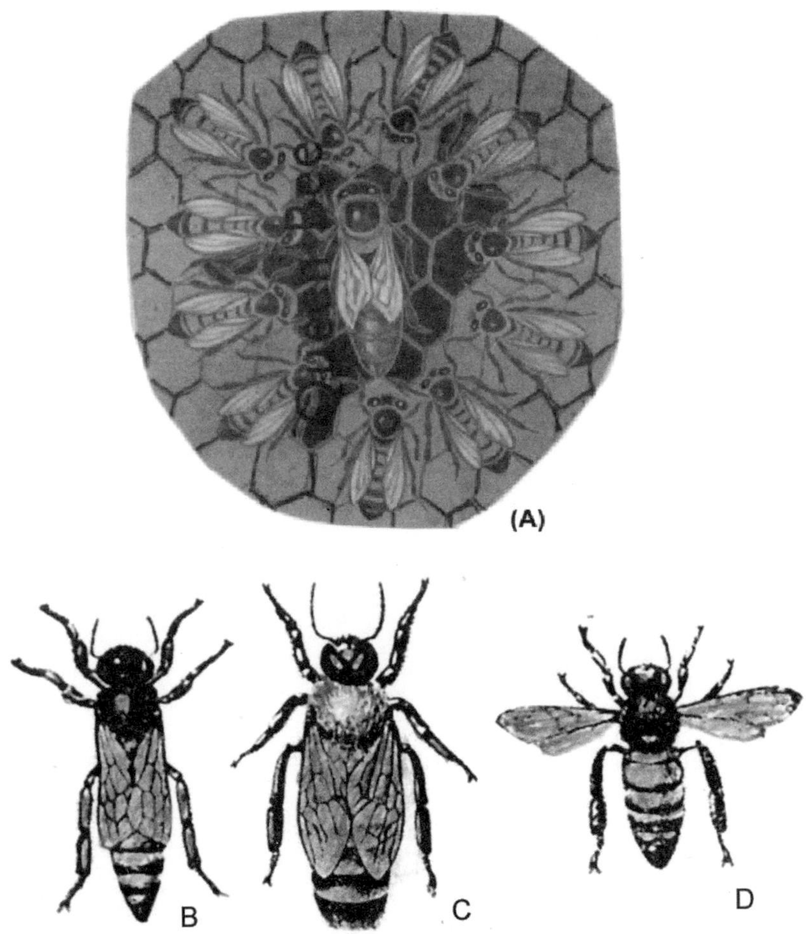

Fig. 113. Castes of Honey bee, **A.** Queen in the hive, **B.** Queen, **C.** Drone **D.** Worker.

Honey as tonic - As a tonic, it contains 365 UG Vitamin B-1 thiamin, 268 UG Vitamin G (Riboflavin), 18 mg vitamin C (Ascorbic acid), 254 UG Pantothenic acid, or 0.60 mg Nicotinic acid. It also contains a good amount of leulose (fruit sugar), dextrose (glucose), sucrose, dextrins, gums, moisture and various salts.

Honey as medicine - Just as lactose is used as the carrier of almost all Homeopathic medicines, similarly honey is put to the same use in the Ayurvedic and Unani system of medicine. It acts as laxative, a blood purifier, a preventive against cold, cough and fever and a curative for sores, eyes, for ulcer of the tongue, sore of throat, and burns etc.

Bee wax

The worker bees secrete wax from glands situated in the abdomen. The secretion is exuded between the segments of the underside of the abdomen and scales of wax can be seen as a result of hardening of this secretion. These scales are detached from the body by the setae of the tarsae and passed onwards to the mouth wherein they are chewed and made plastic to be used in building the comb walls. Thus it is an important by-product of the bee keeping industry. Bees wax is a yellowish to grey brown solid and has an agreeable honey like odour. It is somewhat brittle when cold and when broken presents dull, granular, noncrystalline fractures.

Uses of wax

Bees wax is used in the manufacture of more than 300 items. The largest consumer is the cosmetic industry closely followed by the Catholic Church which needs it for candles and beekeepers for comb foundation. Bee wax goes into the making of face creams (cold, cleansing) shaving creams, ointments, lotions, cosmetics (lipstics, pomads), polishes for boot, furniture and floor, lubricants, pharmaceutical preparations, modelling and plastic work, water proofing compounds, different kinds of inks, paints, varnishes, carbon paper, crayons etc.

THE BEES

There are several kinds of social bees which store honey. The most important are the stingless bees and the honeybees. The stingless bees are tropical occurring in Central and South America, Africa, Asia and Australia. These bees nest in hollow trees, caves or other cavities and store considerable honey which is sought mainly by man and monkeys. In India four forms of bees are commonly found belonging to family Apidae of the order Hymenoptera viz. *Apis mellifera*, the famous honeybee; *Apis dorsata* the giant bee of India, *Apis florea* and *Apis indica*.

Castes of honeybee (Fig. 113 A-D)

A good colony of honey bees (e.g. *A. mellifera* or *A. indica*) have about 40-50 thousand individuals consist mainly of three castes viz. queen, drones and workers. The number of workers in one colony exceeds 90% of the total population.

Queen (Fig. 113 A,B) There is only one queen in each hive. This is the mature female from which the whole colony develops. The only duty of the queen is to lay eggs and increase the population of the colony. The number of eggs deposited per day ranges from several hundreds to 2000 depending upon the favourability of the season. A queen lives from 2-5 years during which the total number of eggs laid may reach as high as 1.5 million. The fertile eggs produce workers and the unfertilized ones hatch into drones.

Drone (Fig. 113 C) The male insect is known as drone. The sting and wax glands are absent but male reproductive organs fully developed. It is reared from an unfertilized egg. It depend upon the workers for food and it cannot feed itself. The sole duty of drone is to fertilize the virgin female before they leave the colony for nuptial flight.

Workers (Fig. 113 D)

These are the main spring of the colony. One beehive may contain about 50,000 worker bees. The duties of workers are gathering of honey, pollen, building the nests, nourishing the broods, attending the queen and drones, keeping hive clean and warm, ripening the honey and defending the colony from any outside attack. They are sterile females, much smaller in size than queen and drone with their reproductive organ undeveloped. They have well developed special structures for various functions: a long proboscis for sucking honey, a pollen basket for collecting pollen, wax glands for building the comb, strong wings for fanning the honey and powerful stings to defend the colony against intruders.

Life history

The queen lays one egg in one brood cell. The eggs are pinkish, elongated cylindrical, attached to the bottom of the comb cell. Larval honeybees develop in wax cells (Fig. 113E) where they are fed worker jelly, a secretion that has low amounts of pharyngeal gland material mixed with mandibular gland secretions and honey. After 3 days, pharyngeal secretions are withdrawn and beebread; honey mixed with pollen, is fed to the larvae. This switch in diet and a reduced number of feedings seem to produce stunted, sterile females, the workers. Worker development requires approximately 21 days. After emerging, honeybee workers pass through a series of behavioural steps during their short existence of approximately 6 weeks, progressing from no apparent work initially to feeding the young to producing wax (from special abdominal glands) to foraging for nectar and pollen.

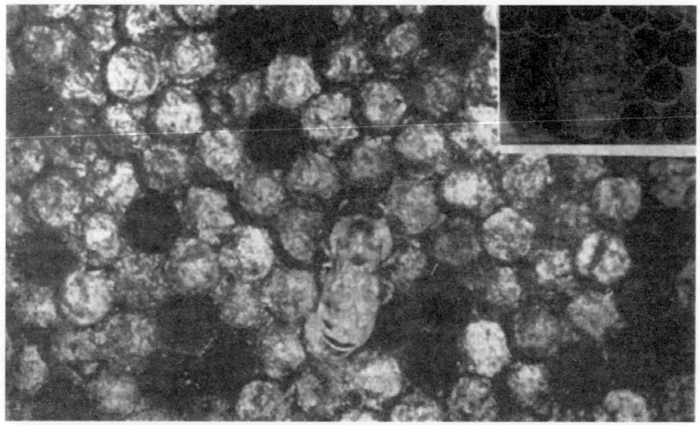

Fig. 113E. Capped brood cells in honey bee hive (From Elzinga, 1978). Inset, a sealed honey bee queen cell (Blaney, 1977).

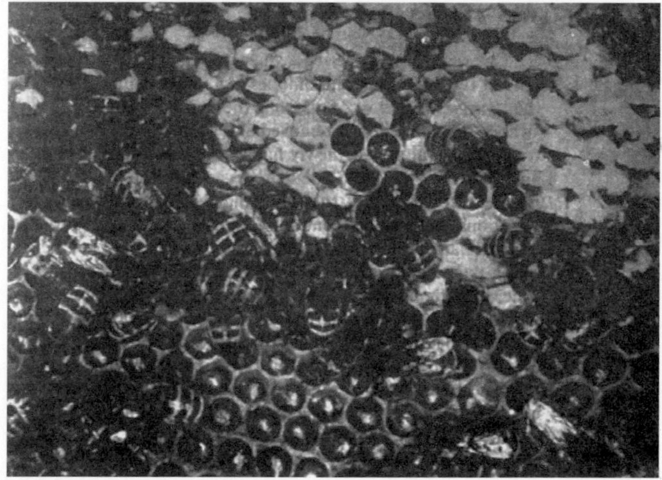

Fig. 113F. Honeybees (*Apis mellifera*) and honey production. Some cells are capped while others are in the process of being filled (From Elzinga, 1978)

Normally, in higher bees, there is only one queen per colony because of at least two inhibiting pheromones in a material produced by the queen called the *`queen substance'*. The queen substance secreted by the mandibular glands of the queen honeybee has been identified as 9-oxodecenoic acid; used by the workers to recognize the continued presence of the queen, and by the queen during the mating flight as a sexual attractant for the drones. The worker honey bees are non-reproductive females and the queen substance is not secreted by their mandibular glands. Following the death or departure of the old queen in a swarm, this inhibition is removed and the surviving workers are stimulated to transfer several eggs or newly hatched larvae to prepared queen cells. This is apparently to ``remind" the workers to feed the female larvae in a different manner. Royal jelly, a substance that has high amounts of pharyngeal secretions, must be fed continually to female larvae to produce queens. After approximately 16 days of development, the first queen emerges and the other queen pupae are systematically destroyed. If two queens emerge at the same time, a battle ensues until only one survives. She then takes a nuptial flight with drones, mates, and returns to propogate the colony. Approximately 1,000 eggs must be oviposited per day to maintain a large colony.

Thus in short it can be stated that when the larvae fed throughout with "royal jelly" from the pharyngeal glands of nurses become queens, those fed after about the middle of the third day with honey and pollen become workers, while intermediates appear if the diet of royal jelly is resumed to late stages of larval growth. There is no particular change in the composition of the diet of worker larvae or queen larvae, until about the third day all the larvae receive massive feeding; thereafter the worker larvae are under nourished, these are sterilized by partial inanition whereas the queen larvae suffer no under feeding (Wigglesworth, 1965).

In the more advanced honeybees or stingless bees of the tropics, caste distinctions (Fig. 113 A-D) are more obvious. New colonies originate when a queen leaves with a swarm of up to 30,000 workers (queens cannot survive alone). Scouts locate a new site where a nest is started. New colonies grow rapidly, reaching approximately from 50,000 to 100,000 (Elzinga, 1978).

THE LAC INSECT

The lac producing insect, *Laccifer lacca* (Fig. 114 A-D) is a scale insect belonging to the family Lacciferidae of the order Homoptera. It is cultivated mainly in India and Burma, which produce over 90% of the world supply of lac (shellac when purified).

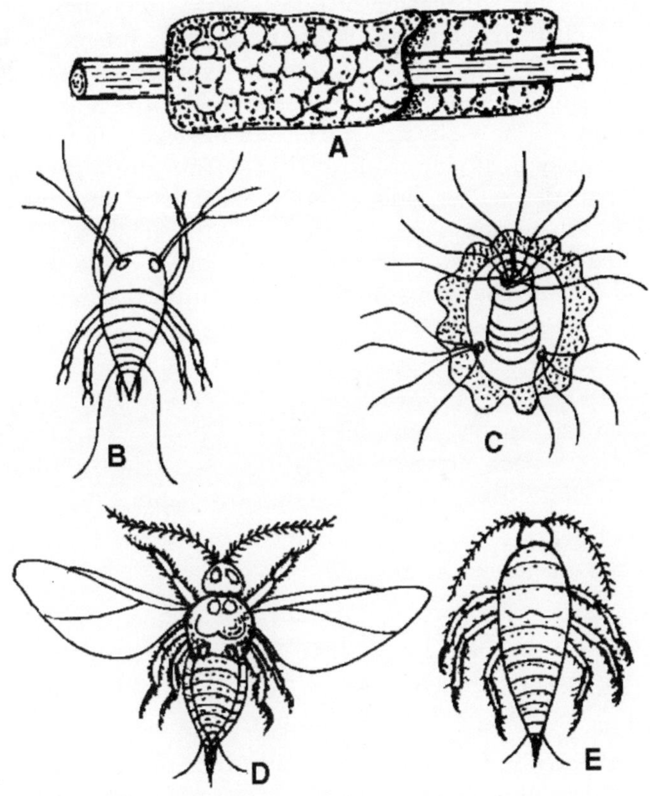

Fig. 114. Lac insect **A.** A twig with lac, **B.** Nymph, **C.** Female insect, **D.** Male (Winged), **E.** Male (Wingless)

Lac is a resinous secretion of this insect. The insect feeds on certain tropical trees and is so injurious to its hosts that rather exacting methods must be followed in order that adequate crops of lac be obtained. The chief host trees are Kusum - *(Schleichera trijuga),* Palas (*Butea monosperma*), Ber (*Ziziphus jujuba*), Peepal (*Ficus religiosa*), Khair (*Acacia catechu*) etc. A brief account of the life history of the lac insect will indicate how the lac is produced.

Life history

The female insects secrete maximum amount of lac causing a considerable increase in size of lac crusts or cells. In the older lac crusts or cells the eggs are laid. The eggs hatch after laying in the form of tiny, soft bodied, crimson coloured nymphs which are also called crawlers. For cultivation of lac, selected twigs bearing mature females are tied to the host trees at the proper times. These female lac insects produce enormous number of nymphs as many as 1000. These nymphs after emergence scatter over the trees and eventually settle down on the new shoots. Here they insert their proboscis deep into the phloem and xylem and suck out the plant juices. The lac produced from internal organs, soon appears as a shining layer over the bodies of the nymph. As the insect grows it secretes more and more lac until it is almost concealed in a globular cell and the cells of the crowded insects coalasce and form a continuous encrustation (Fig. 114 A). The males numbering usually about 30%, do not produce much lac. They are not essential for normal females, and crops of lac can be secured parthenogenetically.

The life cycle of the lac insect takes about six months. There are three crops in a year however, for the insect has different growth cycles on different host trees. These crops are as follows:

1. **Katki crop,** trees infected in June to July and crop harvested in October to November i.e. in the month of Kartik.

2. **Baisakhi crop** - trees infected in October and November and reaped in April to May i.e. Baisakh onwards.

3. **Kusumi crop** - harvested in December to January and from infection made in June to July and January to February respectively.

The harvesting of the crop, the planting, pruning, the preparation of the trees for new crops, the method of infection and defence against insect enemies are all under investigation and much progress has been made in past several decades. India produces the maximum amount of lac about 90% which is obtained from Bihar, Orissa, M.P., U.P. etc. After removal from the trees the lac is variously treated to produce crushed lac, seed-lac and shellac. In these processes it is necessary to get rid of the dye (formerly used for colouring the silk and wool but now replaced by synthetic dyes) and that is done by crushing, washing and drying. This produces the seed lac. The seed lac is melted, filtered by squeezing while hot through cotton bags to hot plates and put through water cooled rollers, from which it is removed in the form of flakes or sheets. The flake-lac when dissolved produces orange ``shellac" which is in fact is prepared by boiling seed lac with yellow arsenic. The shellac is the purified form of lac.

Uses of lac: Lac is used in preparing a large number of articles mainly by jwellers for filling hollow gold ornaments, bangles and bracelets etc. It is a very essential ingredient of paints and varnishes, and polishes. Lithographic ink, sealing wax. Shellac is used in gramophone records, buttons, pottery, artificial leathers etc.

THE SILK WORMS

The rearing of the silkworm is undoubtedly one of the large animal industries in the world. The insect was under domestication in China long before the earliest historical records. The beautiful silk of the orient were sought and greatly prized by the ancient Greeks, Romans, long before they knew that such wonders were manufactured from a lowly insect. The secrets of the culture and of the weaving of the silk were carefully guarded by the Chinese for more than 2000 years. For centuries, the Greeks and later the Romans endeavoured to learn the secrets of the sericulture. Two monks who had lived in China and had become familiar with sericulture are said to have informed the Emperor Justinian I, of their knowledge. The two monks were sent back and smuggled the eggs of the silkworm and seeds of the mulberry tree to Constantinpole about 500 AD. Silk culture had been introduced into Japan by similar underhanded means about 300 AD.

Silk: It is a secretion in the form of fine threads, produced by several kinds of caterpillars in preparing cocoons for their pupae. The long silk glands which are modified salivary glands secrete a thick pasty substance which is passed out through a pair of fine ducts that open on the labium. The secretion is spun by the caterpillar into fine threads which hardens on exposure to air to form fairly strong and pliable silk threads. It has been reported that the mulberry silkworm prepares silk filaments several thousand meters in length at the rate of 15 cm per minute.

Types of silk

(1) *MULBERRY SILK* (Fig. 115)

This silk is supposed to be superior in quality to all other types due to its shining and creamy white colour. It is secreted by the caterpillars of *Bombyx mori* which belong to family Bombycidae of the order Lepidoptera. The caterpillars feed on mulberry leaves.

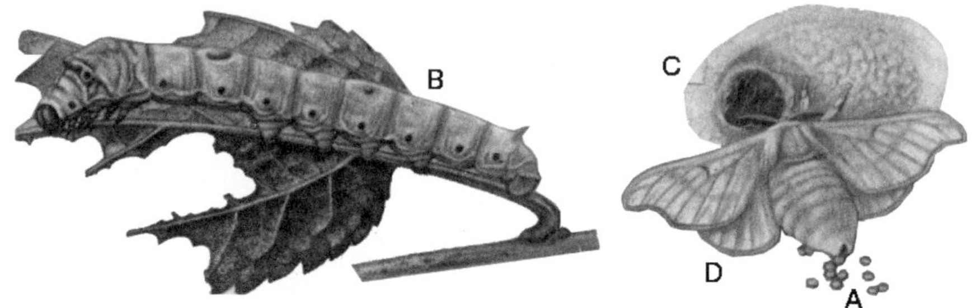

Fig. 115. Mulberry silkworm, *Bombyx mori,* **A.** Eggs, **B.** Mature larva, **C.** Cocoon, **D.** Adult female

(2) ERISILK (Fig. 116)

It is produced by the caterpillars of *Philosamia ricini* (Saturniidae : Lepidoptera) which feeds on castor leaves *Ricinus communis*. Its colour is also creamy white but is less shining than the mulberry silk. The cocoons of *P. ricini* has loose texture and the silk is not reelable and is of poor quality.

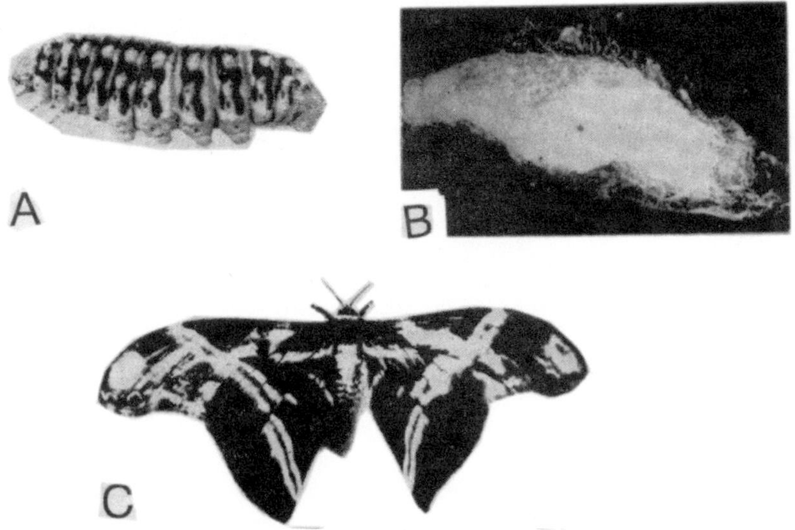

Fig. 116. Eri silkworm, *Philosamia ricini.* **A.** Late larva, **B.** Cocoon, **C.** Adult moth (Awasthi, 1985)

(3) TUSSER SILK (Fig. 117)

It is secreted by caterpillars of several species of *Antheraea* which belongs to family Saturniidae of the order Lepidoptera. This silk is coppery colour. The larvae feed on the leaves of Arjun (*Terminalia arjuna*), Sal (*Shorea robusta*), Ber (*Ziziphus jujuba*), Jamun (*Eugenia jambolana*) etc.

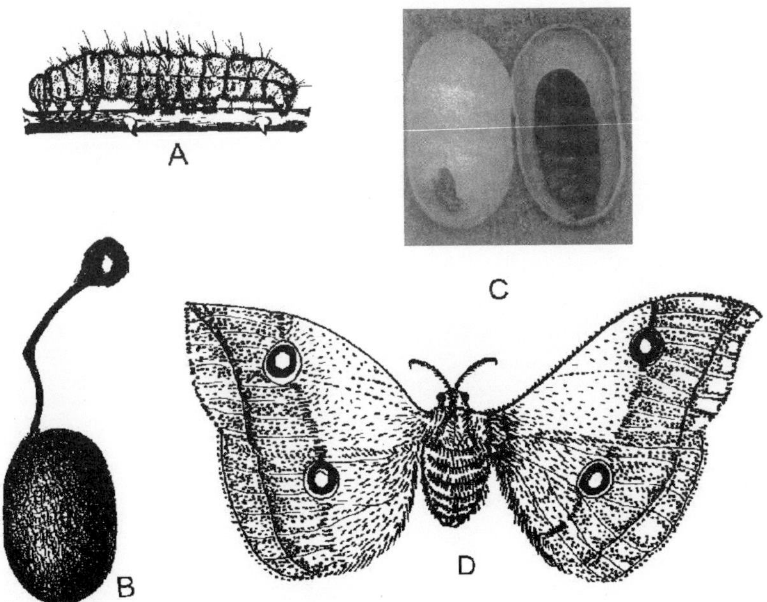

Fig. 117. Tusser silkworm, **A.** Larva, **B.** Cocoon, **C.** Cocoon cut open to show the pupa inside, **D.** Adult moth.

(4) MUGA SILK

It is secreted by the caterpillars of *Antheraea assamensis* which feeds on *Machilus bombycina* (Som), and Champa, *Michelia champaka* etc.

Life history

The mulberry and erisilk is manufactured from domesticated silk worms, whereas the Tussar and Muga silkworms are wild in nature. Life cycle of all the four silkworms is very common as they lay eggs from which caterpillar hatches and only the duration varies. The caterpillars eat, grow and finally metamorphose into cocoons within which pupation takes places. The male and female moths after emergence from the cocoons, immediately start mating and laying eggs and thus beginning a new life cycle.

40

PESTS OF VETERINARY IMPORTANCE

The livestock are sensitive to the attack of insect pests. Their heavy infestation causes stunted growth and sometimes death. A considerable loss to the tissues and various body parts is done by different types of myiasis. The constant irritation of biting flies make livestock rest less, and withdraw a large amount of blood, produces anaemia. Cattle grubs migrate in the body tissues make holes in the skin of the back and spoil meat. Screw-worms infest wounds and feed on living tissues, often causing death of untreated animals. Certain flies e.g. horse flies, stable flies, blow flies etc transmit the bacilli of anthrax, infectious keratitis, mastitis and "swollen joints". In the following pages account of some of the insect pests of veterinary importance is given :

1. ***Hypoderma species*** (Fig. 118): The larvae of this genus are commonly known as "cattle grubs" and the adults as "heel flies or warble flies". This fly belongs to family Hypodermatidae of the order Diptera. Although the normal hosts are sheep, cattle, goat and deer, they occasionally parasitise horses and humans. Two species infest cattle, *H. lineatum* and *H. bovis*. The mature larvae are easily distinguished by examination of the spiny armature. *H. bovis* having the last two segments completely devoid of spines, H. *lineatum* having only the last one smooth, there are other distinguishing features also. The mature larva of *H. bovis* measures from 27-28 mm in length and that of the *H. lineatum* about 25 mm. in length.

Life history

The eggs of both the species are laid on the hairs of the cattle, *H. lineatum* attaching 5 to 15 to a single hair, *H. bovis* attaches only one egg to a hair. As many as 800 eggs may be laid by a female of either species. In both the cases, eggs hatch within a week and the tiny armoured larvae crawl down the hair of the host and bore either directly into the skin or into the hair follicles. In doing so, they cause a lot of irritation. The first stage larva then migrates through the body of host finally lodging beneath the skin of the back. After about 4 months, the larva of *H. bovis* reaches the spinal cord where it burrows between the periosteum and dura mater for a period of time before leaving the spinal canal and cutting through muscles and connective tissue to the back; *H. lineatum* rests for a while in the wall of the oesophagus probably without having entered the spinal canal. After reaching the skin, the larva cuts a small opening to the surface.

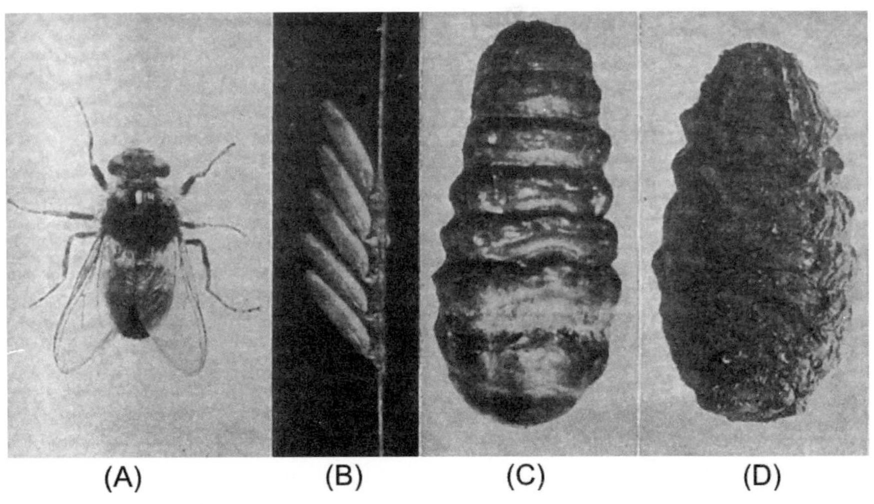

Fig. 118. Life stages of the common cattle grub, *Hypoderma lineatum*
A. Adult or heel fly, **B.** Eggs attached to hair; **C.** nearly full-grown larva; **D.** Pupa (From Pfadt, 1962)

At the end of the developmental period in the warble which requires 5 to 8 weeks for *H. lineatum* and upto 11 weeks for *H. bovis*, the grub works its way out and falls to the ground. There it crawls and enters the loose earth or debris, where pupation takes place. The pupae are dark-brown or black in colour and in 4 to 5 weeks emerge as warble flies. The complete life cycle takes about one year. The injury done by the larvae is primarily due to irritation caused by their migrations in the body of the animal and by their emergence from beneath the skin. The escape of the larva from the warble leaves an open wound that persists for a long time and is subject to the bacterial infection and other tormenting insects. The economic losses include (1) reduction in the milk production of the cattle from 10% to 20% of the normal yield, (2) loss of flesh resulting from the wild efforts of the animals to escape from the flies, (3) flesh becomes greenish-yellow, jelly like in appearance and is unfit for consumption at the parts where the grubs are located, (4) injury to the hide which becomes ``grubby", full of holes where the grubs have emerged.

Control

The removal of larvae by squeezing the infected parts is a simple mechanical method. In addition, use of suitable insecticide in the form of water soluble spray is used for controlling the pest. The body of the animals is treated 40 days after the appearance of the warbles. The spray is repeated every month till the grubs are seen on the body of the animal.

(2) *Xenopsylla* species (Fig. 119A): The *Xenopsylla* belongs to the family Pulicidae of the order Siphonaptera. The common rat flea is known as *X. cheopis*. This species is cosmopolitan occurring just about wherever *Rattus rattus* is found. An analysis of the major urban centres shows *Rattus rattus* to be most important and *R. narvegicus* to be of minor significance e.g. in Calcutta *X. astia* is the predominating rat flea whereas in Bombay *X. cheopis* predominates. However in Calcutta, wards with plague have rats more heavily infected with *X. cheopis*. In Uttar Pradesh it was thought that only *Rattus rattus* and its fleas were implicated in the genesis of rural plague in man but a reservoir was found in a relatively resistant and abundant field rodent, the gerbil, *Tatera indica*. The domestic rat acts as a liaison agent between wild rodents and man with the flea, *Cerotophyllus punjabensis*.

Fig. 119 A. Rat flea

Life history of a typical flea

The eggs of a flea are fairly large, 0.5 mm long, glistening white and rounded at both ends. The number of eggs deposited in each laying is 3 to 8. The eggs are deposited over the hairs of the host. The larvae are very active, slender 13 segmented, yellowish with bristles. The mouth parts are of biting type. Flea larvae feed on a variety of material found in den and associated areas of host. Under favourable conditions the larval period ranges from 9 to 15 days but in unfavourable conditions it may extend 200 days. The mature larva spins a cocoon and pupates inside. The cocoon is whitish in appearance and loosely spined. Pupal period ranges from 7 days to a year. The life cycle accordingly may vary from 18 days to 20 months or even more depending on temperature and other conditions. The rat flea *X. cheopis* transmits a number of diseses and carries germs to the host causing plague.

Control

The fleas can be controlled successfully by dusting the Rat runs and harbourages with insecticides like methyl parathion, lindane, malathion etc. and by the control of rodents.

(3) *Stomoxys* species (Fig. 119B): Flies of this genus belong to family Muscidae of the order Diptera. These flies are commonly known as ``Stable flies''. Of them *S. calcitrans* is the most familiar and widely distributed member of its genus. On account of similarity in colour the stable fly is often mistaken for the housefly, *Musca domestica.* However, the former is more robust and has broader abdomen. In colour also the stable fly is brownish-grey with a greenish yellow sheen, the outer of the four longitudinal thoracic stripes are broken and the abdomen is more checkered. The wings in resting condition are widely spread apart at the tips and are distinctly iridescent.

Life history

The eggs of stable fly are laid on manure or dung of horse, cattle and sheep, but it does not breed commonly in excrement under field condition unless this is mixed with decaying vegetable matters. The eggs are about 1 mm long curved on one side and straight as well as grooved on the other i.e. opposite side. The eggs are laid in pairs. Egg deposition ranges from 23 to 100, usually between 25 to 50. There are four or five layings but there may be as many as 20 layings. Thus, a female fly can produce as many as 600 eggs. Incubation period varies from 2 to 5 days. The newly hatched larvae burring themselves in their food at once, thus protecting from dryness. They are full grown in about 14 to 26 days, under different temperatures.

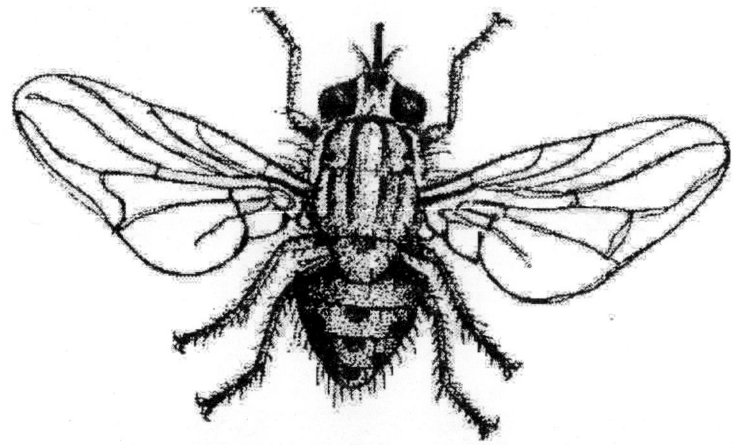

Fig. 119 B. Stable fly. *Stomaxys* sp.

The larvae crawl into drier parts of the medium and pupate. The puparia are 6 to 7 mm long. The pupal period also varies from 5 to 26 days depending on temperature. The adult fly emerges, unfold its wings and is ready to fly. It transmits virus of infectious anaemia of horses, the *Antherax bacillus* and several pathogenic *Trypanosomes*. The *T. evansi* is the causitive of "Surra" a fatal disease to horses and mules and often with serious effects on camels and dogs. The fly is infective for only a brief period of time.

Control

Spraying the domestic animals with 0.05% pyrethrum mixed with 0.5% piperonyl butaoxide solution once or twice in a week is very effective. Dusting the animals with 0.5% malathion is also very useful.

4. The Tabanid flies (Fig. 120 A-D): Species of *Tabanus* and *Chrysops* are commonly known as ``Horse flies" and ``Deer flies". They belong to family Tabanidae of the order Diptera. The size of the flies of various species ranges from 10 to 25 mm long and they are robust with large occasionally bright-coloured or iridescent eyes, mouth parts are of cutting and sponging type or piercing and sucking type. The different species of *Tabanus* and *Chrysops* attack different body parts of their hosts for feeding.

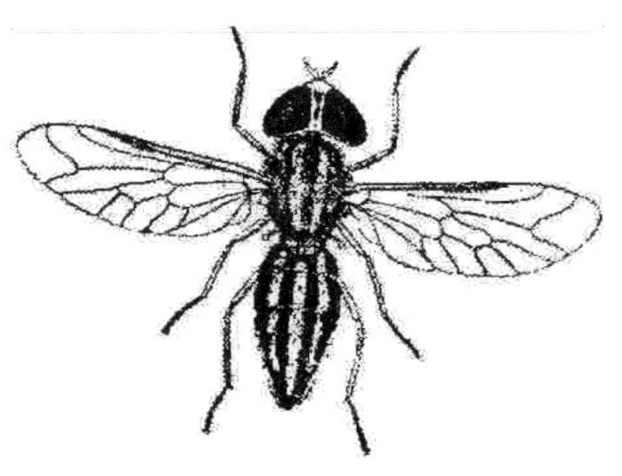

Fig. 120. A. Striped horse fly. *Tabanus* sp.

Fig. 120. B. Black horse fly *Tabanus atratus*

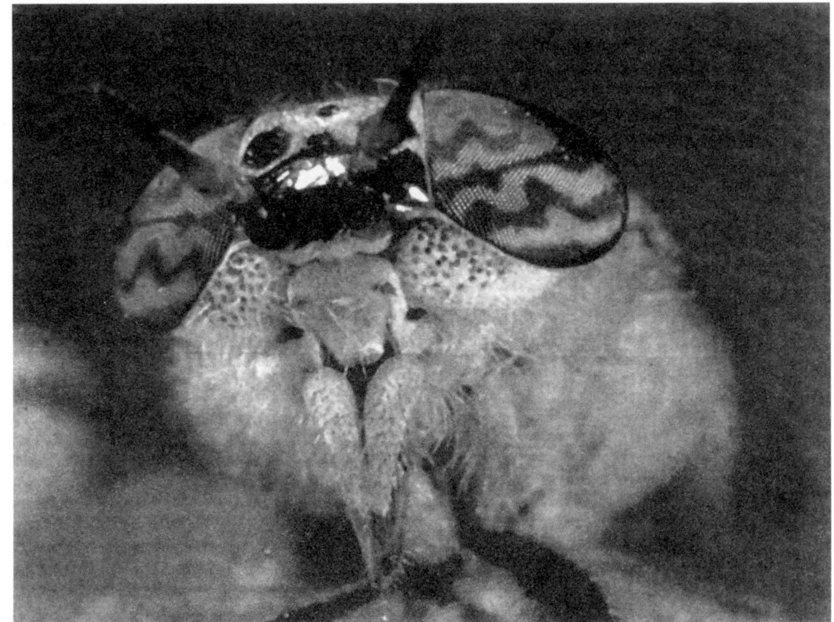

Fig. 120. C. The head of a Horse fly showing the remarkable colour patterns typical of the compound eyes of this family (Tabanidae). This iridescence is due to the form of the commatidia and disappears on death (From Blaney, 1977).

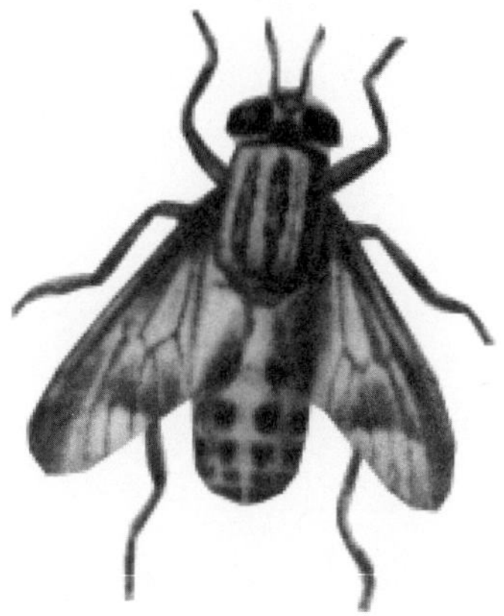

Fig. 120. D. Deer fly. *Chrysops* sp.

Life history

A female fly deposits 130 to 450 eggs in groups on the wood pieces near water. The newly emerged larvae feed on mosquitoes and other aquatic insects. The mature larvae hibernate near water in winter and in spring pupation takes place in dry soil. In summer, the flies emerge and start breeding. There are several generations in a year. The flies specially females are harmful. They cut out the skin of the animals and feed on the blood thus making large wounds on the body causing decrease in the quantity of milk. As is the case with *Stomoxys,* the species of *Tabanus* and *Chrysops* also transmits ``Surra'' anthrax and other diseases in several domestic animals.

Control

The water bodies should be properly treated with kerosine oil. Like Stable flies, spraying with pyrethrum mixed with piperonyl butaoxide, three to four times in a week on the animals is very effective to control this pest.

(5) THE CATTLE LICE (Fig. 121) : Lice are important pests of cattle in all parts of the world. Some of them are *Haematopinus eurysternus, H. quadripertusus, Linognathus vituli* and *Solenopotes capillatus.* All these belong to family Haematopinidae of the order Anoplura. Several species of cattle lice are found and they have preference for particular breeds and ages of cattle. Intense irritation set up by lice causes cattle to rub and scratch. Among grossly infested animals large areas of skin become raw, bruised and denuded of hairs due to constant rubbing against posts and fences. Because of irritation and loss of blood cattle neither thrive nor gain weight normally. Heavy infestations decreases in milk yield. Cattle lice are small wingless insects. The various species of the cattle lice measures from 2 mm to 4.5 mm long.

Fig. 121. Common species of cattle lice, **A.** Short-nosed cattle louse *Haematopinus eurysternus* (Nitzsch), **B.** Long-nosed cattle louse *Linognathus vituli* (Linnaeus), **C.** Little blue cattle louse *Solenopotes capillatus* Enderlein.

Life history

Female lice glue their eggs close to the skin on hairs. Sometimes a single hair will bear 2, 3 or more eggs. Incubation period of lice is about 8 days, and development from egg to adult takes about 25-30 days. Each female during its life time lays 30 to 35 eggs. Lice population fluctuates with the seasons. They are lowest in the summer and reach their peak in winter and early spring.

Control

For control of lice on dairy cattle, only the application of rotenone or pyrethrins plus piperonyl butaxide is permissible. Other recommended sprays include 0.03% lindane, and 0.5% malathion or 1/2 to 1 kg of 5% rotenone powder per 450 litres of water, a recommended dust. Two treatments of a two or three week interval are generally needed to control infestation.

6. ***Hippobosca species*** (Fig. 122) : The *Hippobosca* species belong to family Hippoboscidae of the order Diptera. These are commonly known as ``Louseflies". The larvae of the flies are retained in the body of the female, being nourished by special glands within the mother until for pupation is reached. They are then extruded and pupation takes place, hence the term `pupipara'. The adult flies are flattened, leathery in appearance, the legs are short and strong and antennae are inserted in pits near the border of mouth. The wings are always well developed in the genus and are functional throughout the adult life. The species of the genus are ectoparasite on mammals, *Hippobosca equina* found on horses, *H. variegata* occurs on cattles, *H. capensis* attack on dogs and *H. camelina* is a parasite of camels.

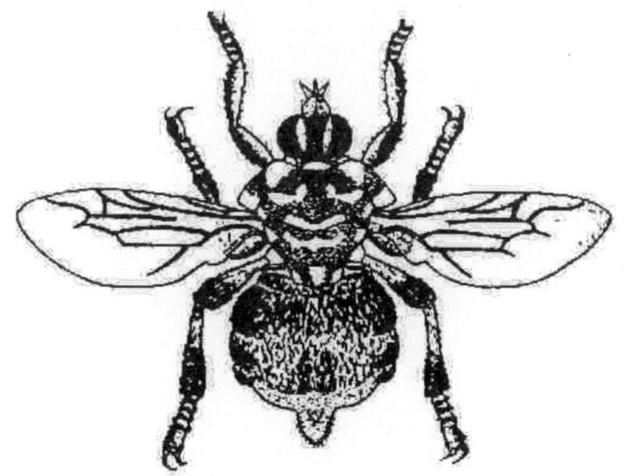

Fig. 122. Louse fly, *Hippobosca* sp.

Control

A number of insecticides have been recommended for livestock such as lindane, malathion, methoxychlor and are widely used to control the pest. Rotenone and pyrethrum are the insecticide of plant origin used on dairy animals. Several other chemicals are also applied directly to livestock including synergists as piperonyl butoxide and repellent as butaxy-polypropylene glycol.

7. ***Blow flies*** (Fig. 123): These flies belong to family Calliphoridae of the order Diptera. They have bright metallic appearance being shining blue or green and are known as "blue or green bottle flies". The flies measure about 1 to 1.5 cm long. Three genera of the family Calliphoridae viz. *Calliphora, Chrysomia* and *Lucilia* include the largest number of flies that feed on carrion. The blow flies are widely distributed in various parts of the world.

Life history

Eggs are laid on dead animals, garbage or in open wounds of wild or domestic animals. Eggs hatch very soon, within 24 hours after being laid. The larvae feed mainly on decaying flesh; screw-worm invades living tissues. The larval and pupal stages are completed in about 10-15 days. Some species are parasites of man and domestic animals causing myiasis. There are several generations in a season.

Fig. 123. Blue bottle fly, *Calliphora* sp. **A.** Larva, **B.** Pupa, **C.** Adult

Control

The larvae can be destroyed *in situ* by spraying the wounds with chloroform-water to induce them to leave the tissue or by removing them with forceps. The wounds should be dressed with some repellent e.g. pinetar, larvicides can be used on wounds to control this pest.

41

PESTS OF FORESTRY

Beeson (1941) has described in considerable detail the ecology and control of the forest insects of India. His work deals with several hundred forest insects belonging to different orders that are destructive to the living forest trees, wood and the various materials prepared out of the wood. In the following pages, description of a few insect pests of forestry is given.

1. *Hoplocerambyx spinicornis* Newman : The sal borer, *H. spinicornis*, belongs to family Cerambycidae of the order Coleoptora. The beetle is about 5 cm uniformly very dark brown, the elytra varying in colour from piceous black to reddish brown, antennal segments with spines at the corners. The beetles begin to appear each year as soon as the monsoon rain starts. If several centimetres of rain falls early in June the beetles appear a few days afterwards and if the monsoon does not start until July, no beetles will be seen until after the rain has fallen. They do not all emerge at once, each rain-storm brings out a fresh lot of beetles and this goes on for about 8-weeks when probably 95% of the total will have emerged; about 60 cm of rain after June 1st is sufficient to cause a total emergence. A beetle lives for about 3 or 4 weeks so some of the earliest will have died before the latest have emerged. The beetles are active by day light particularly during the warmest part of the day. For food they gnaw the bark of sal particularly the inner and living bark when it is exposed by some accident to the tree. Fresh sap from the bast and sapwood of *Shorea robusta* is highly attractive. A beetle that is gorged on sap appears intoxicated and often is unable to stand or fly.

Life history

The female begins to lay eggs 7 to 9 days after the fertilization. Each beetle will lay 100 to 300 eggs. The maximum number of eggs laid by one female is 468, 250 eggs is the average laying of one female and 180 eggs may be laid in one day. The maximum oviposition period is 30 days. The eggs do not all survive to beetles, a big sal tree will rarely produce more than 300 beetles, although early in the attack there may have been a thousand grubs in the bark.

Eggs are laid singly in crack holes and overhanging flake of bark. The normal incubation period varies from 3 to 7 days. Under very dry conditions the egg does not hatch out but begins to shrivel within 10 to 14 hours and is completly desiccated in 24 to 36 hours.

The grubs that hatch from the eggs feed under the bark and then in the sapwood and finally bore into the heart-wood. The grubs become full grown by November. During winter and hot weather they stay in tunnels in the heart wood shut in by calcium carbonate and wood fibres closely packed. The insects turn into beetles in May-June and wait until monsoon arrives.

2. *Xylotrechus buqueti*: A large number of forest trees are damaged by this beetle. This belongs to family Cerambycidae of the order Coleoptera. Beetle measures about 6-12 mm, dark brown in colour with greyish or tawny-yellow pubescence with a black patch on the pronotum and three curved transverse bands on the elytra.

Life history

Eggs are laid in bark in groups of 2 to 5 from which the larval galleries radiate horizontally just grooving the inner bark or laying entirely in the sapwood. The larvae make very irregular tunnels which frequently cross each other. The pupal chamber is made at a depth of about 2 cm and is not constant in shape or orientation, varying from vertical to radial. The beetle emerges by the most direct route, excavating a short horizontal tunnel that takes off from the upper end of the pupal chamber. The exit hole is circular.

The life cycle may be semi-annual or longer. Emergence occurs almost throughout the year but predominantly in March-June. The emergence period of one infestation may extend for 5 months.

3. *Batocera horsfieldi* (Fig. 124A): This beetle belongs to family Cerambycidae of the order Coleoptera. The beetle is about 5-6 cm long, black in colour with fine ashy or yellowish grey pubescence, pronotum with 2 elongate white or yellowish spots, elytra with numerous shining black tubercles at the base and several rounded or broken elongate white marks extending to truncate apex. Scutellum white or yellowish. It is a borer of living trees in the Himalayan forests.

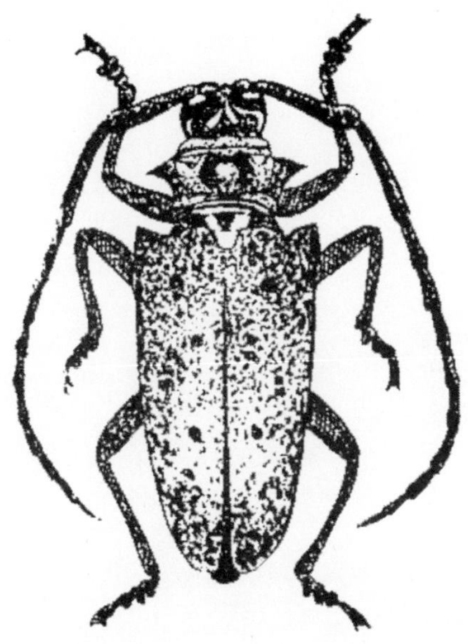

Fig. 124A. *Batocera horsfieldi*

Life history

The egg is laid on the bark of the living tree usually within the basal 1.25 or 1.5 m of the bole but a tree with heavy attack may have tunnels as high as 3 m from the ground. The larva on hatching bores down to the surface of the sapwood and excavates a shallow more or less circular patch from which a relatively enormous tunnel is later bored upwards into the centre of the heart wood. The tunnel gradually expands in diameter and reaches a length of 45 cm or more and a width of 5 cm. The upper end of the tunnel is widened into a pupal chamber and is not very far from the bark. The full grown larva is nearly 15 cm long. Beetle escapes by boring an exit tunnel through the intervening about 3 cm or so of wood and leaves a circular exit hole of about 2.5 cm in diameter. Length of the life cycle is unknown. It may take two or three seasons. Beetles occur in September-October and possibly emerge earlier.

4. ***Batocera rufomaculata*** Dejean: It is a borer of a very large number of trees. The beetle belongs to family Cerambycidae of the order Coleoptera. Beetle is about 4 to 6 cm long, dark with a fine

greyish vestiture, pronotum with 2 kidney shaped orange yellow spots, scutellum white elytra in basal third with numerous black tubercles and several yellowish spots variable in number and shape.

Life history

Each female lays about 200 eggs placed one by one in incisions cut in the bark with the mandibles. Oviposition normally takes place on dead trees but also on the branches and trunks of living trees that are not in good health and on the roots of trees exposed to erosion. Each egg is brownish white cylindrical 6 x 2 mm long with narrow rounded ends. Newly hatched larva is about 10 mm long. It feeds at first in the meristem and later penetrate deeper. The tunnels in the depth of the wood are irregular and 2 or more centimetres in cross section. The pupal chamber is 4-6 cm cavity in a widened extension of part of the larval gallery and is surrounded by long interlaced fibres. The larva is bright yellow without any food in the alimentary canal. The pupal period lasts about 3 to 4 weeks and is followed by an immature beetle stage of variable duration. The beetle emerges by a short tunnel running direct to the outside and ending in a circular exit hole.

The emergence period is from March to August, about 50% beetles emerge in May and 30% in June. The life cycle is thus annual.

5. *Dinoderus species* (Fig. 124 B) : There are four species of *Dinoderus* viz. *D. brevis, D. minutus*, *D. ocellaris* and *D. pilifrons*. These beetles belong to family Bostrychidae of the order Coleoptera. They are commonly known as ``Bamboo borers". The habits of all four species are similar. The beetle bores into the cut bamboo at a spot where the external rind has been severed or removed e.g. by trimming off the side branches or smoothing nodes, by splits or abrasions. It also bores into the exposed transverse sections of cut ends and into the internal walls of the terminal internodes of the hollow bamboo. The dust derived from the construction of the tunnel is all ejected.

Life history

Pairing takes place inside the tunnel and eggs are laid in such pores as are exposed in the walls. The larva on emergence bores a tunnel upwards or downwards coaxially with the fibres. In a crowded infestation the course of the tunnel is very irregular crossing other tunnels and reversing its direction. In a hollow bamboo the larval tunnels tend to be concentrated in the innermost zone of the internodal wall avoiding the outer zone where the fibro-vascular tissue is dense and nutritious parenchyma is scarce or absent. The larva tears off the small particles of tissue and taste them rejecting the horny fibres. The whole of the chewed material is ingested and is passed through the alimentary canal without visible change other than the loss of starch grains. The number of larval instars has not been determined, there are at least three or probably four. Pupation takes place in a cell at the end of the larval tunnel and the pupal stage is of very short duration. The immature beetle spends a variable amount of time in further feeding before escaping. The adults are dark brown in colour. There are three generations in a year. The first generation starting in March is completed in 11 to 12 weeks. The second generation starting in June is also completed in 11 or 12 weeks with a pupal period of 4 days. The third generation starting at the end of September over winters as larvae and pupates in March (pupal stage 4 to 5 days) thus taking 22 to 24 weeks from boring in of beetle to emergence of brood. Under most favourable conditions this period (i.e. time spent in tunnelling, sexual maturation, pairing, oviposition and incubation etc. which takes 4 to 6 weeks) may be reduced to 2 or 3 weeks and sufficient time would then be available for 4 generations in one year at the quickest rate of development.

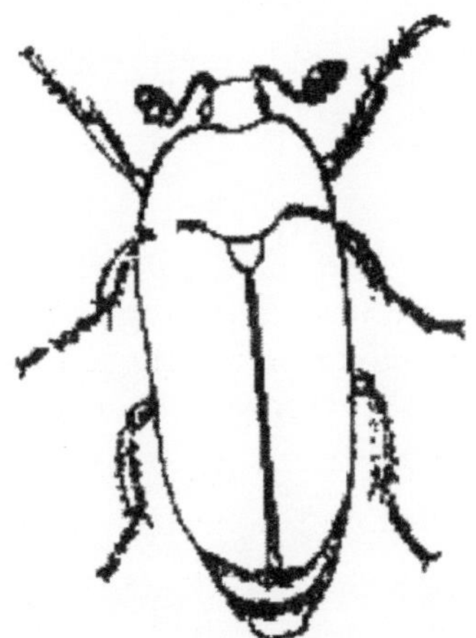

Fig. 124B. Bamboo borer, *Dinoderus* sp.

6. *Sinoxylon species* (Fig. 124C): There are several species of *Sinoxylon* all of which belong to family Bostrychidae of the order Coleoptera. There are about two dozen or more species. Some of them are *S. anale, S. atratum, S. beesoni, S. capillatum, S. crassum, S. cucumella, S. dichroum, S. indicum, S. lycturum, S. olerae* etc. The *Sinoxylon anale* is common and found as frequently within the forest as in timber depots, sawmills and factories. The larvae of *Sinoxylon* are wood borers. The adult beetles sometimes bore into green shoots and twigs for the purpose of feeding or hibernation making axial tunnels. Beetles bore into the stems of sickly or dying poles with more success. They cause damage to more than 50 species of the forest trees. The damage is caused by both larvae as well as adult beetles.

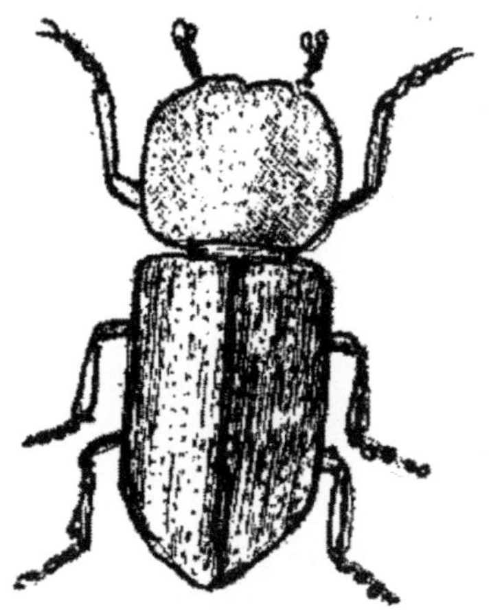

Fig. 124C. *Sinoxylon* sp.

Life history

The adults are dark brown in colour. The eggs are laid in the tunnels made by the females in the tree, the larvae on emergence feed on the soft part of heart wood of the tree and they pupate within the wood tunnel. The adult emerges by boring the wood.

The life cycle varies between very wide limits; its minimum may be round about 3 months and the maximum recorded is over 4 years. The duration of life cycle depends on the plant species. There are several generations in a year.

Lyctus Beetles

The *Lyctus* beetles comprises several species of the genera *Lyctoderma, Lyctoxylon, Lyctus* etc. The species of *Lyctus* are known as `Powder post beetles'. From the manner in which the larvae reduce attacked wood to a fine flour like powder are of considerable economic importance as pests of sawmills, and wood using factories and of furniture and wooden fitting of buildings.

7. *Lyctus africanus* Lesne. : The *L. africanus* belongs to family Bostrychidae of the order Coleoptera. Almost all variety of forest trees are damaged by this species. Hence it is described here. *L. africanus* attacks logs, branchwood and stump of felled trees in the forest and practically every kind of manufactured wooden article that contains sapwood e.g. packing cases, boards, wooden framework and fittings in buildings, furniture tool handles, plywood, woody seeds and roots, bamboo and cane etc.

Life history

The beetle *L. africanus* is about 2-2.4 mm in length. Large number of beetles emerge suddenly in warm weather i.e. the phase of abundant emergence usually commences in the first half of May and lasts for 3 to 4 weeks and then fall to relatively low figures. As remarked by Beeson (1941) that the period of *L. africanus*, after the first swarm in May is so variable that it is difficult to say what is the normal behaviour of the species. *L. africanus* may complete three generations in a year.

As with other bostrychid beetles, a portion of the population developing from one brood of eggs matures at the rate necessary to produce 3 generations a year and remainder mature at slower rates equivalent to 2 to 3 or more short life cycles. Larval development may be prolonged for several years at least seven years and probably more.

8. *Cryptorrhynchus rufescens* : This beetle belongs to family Curculionidae of the order Coleoptera. It is a borer of the bark and sapwood of the *Pinus longifolia* and *P. khasya*. The beetle attacks living Chir saplings, poles and trees that are sickly, burnt, injured by fire or other accidents, diseased owing to fungus attack. The weevil larvae bore in the diseased trees and in the relatively healthy tissues nearby in the bark and cambial region and thereby hasten the girdling of the stem which causes the tree to die.

Life history

The eggs are laid in deep crevices or cracks in the bark of the tree or in pits gnawed for the purpose. The larva is a bark-borer making very irregular tunnels without definite direction but running for nearly 25 cm. These are circular in cross section and packed with red dust. For pupation a shallow chamber is excavated which may be mainly in the sapwood or wholly in the thick inner bark. The pupal chamber is oval and lined with coarse interlaced wood fibres. On maturing in the cocoon of wood fibres the beetle gnaws a straight exit tunnel to outside. There are 3 generations in a year. The beetles of the overwintering generation emerge from April onwards. Eggs laid in April give rise to beetles at the end of June. A second generation may be completed between July and September and the third over winters in any stage larva, pupa or beetle.

Control measures

1. For bamboo borers

The best method of protecting bamboo during seasoning is by immersion in water which causes unpleasant smell and changes the composition of other substances on which bamboo borers feed. The soaking should last for 2 to 3 months in still water or less running water. Adequate immunity from borers attack is obtained by soaking in a 50:50 mixture of cold creosote and heavy fuel oil for 48 hours.

2. For long horn beetles

1. Removal of bark from logs shortly after felling prevents egg laying by all Cerambycidae. This is one of the standard measure for protecting sapwood and heartwood.

2. The damage to cultivated mango and other trees by *Batocera rufomaculata* can be prevented the stem with stout paper coated with coal-tar or with wire gauge 1.5 mm mesh or spraying bark with a strong repellent like Bordeaux mixture during the oviposition period.

3. For *Sinoxylon* and *Lyctus* beetles

1. Seasoning: Poles which are seasoned to practically air dryness in the sapwood are not thereafter attacked by powder post beetles.

2. Cold water: Pols and logs already infested with bostrychid borers can be sterilized by submersion in cold water. 80 to 100% of borers in the larval and adult stages are killed by one week's immersion.

3. Hot water: Immersion of poles and logs in hot water, will kill bostrychid borers in the sapwood to an extent depending on the temperature of water, the period of immersion and the thickness of bamboo.

4. Use of paints on wooden materials prevents egg laying. A similar effect has polishing the material with varnish.

42

PESTS OF PLANTATION CROPS

The plantation crops viz. tea (*Camellia* spp.), coffee (*Caffea* spp.) coconut-palm (*Cocos nucifera*) etc. are very important and useful from economic point of view. They are perennial and provide ideal situation for intensive pest infestation. There are a large number of insect pests of plantation crops. In the following pages only a brief account of some of the most destructive insect pests of tea, coffee and coconut palm plantation crops, is given.

PESTS OF TEA

(1) *Andraca bipunctata* Walk. (Fig. 125): This belongs to family Notodontidae of the order Lepidoptera. The larvae are known as "Bunch caterpillars" because they are gregarious in nature. The moths measure about 40 to 50 mm across the wings. They are brown in colour. Besides the wavy cross lines, the forewings have two white spots near to the outer margin. The hind wings are rather pale in the anterior region and brown in the posterior region.

Life history : Each female moth lays about 500 yellowish eggs in linear clusters on the underside of the tea leaves. Hatching of eggs takes place in about 7 to 10 days. The just hatched larvae are yellowish in colour. They eat up their egg-shell before starting feeding on leaf tissues. There are 5 instars and the total time taken for development varies from 3 to 4 weeks. The fully mature larva constructs a cocoon in between the dry leaves of the tea. The full grown larvae are 50 to 65 mm in length. The pupa is about 25 mm in size and reddish brown in colour. The pupal period lasts 2 to 4 weeks. There are generally 4 generations in a year in the north eastern region of India. The larvae feed on the tea leaves thus causing a considerable loss to the tea crop.

Fig. 125. Bunch caterpillar, *Andraca bipunctata* **A.** Clusters of caterpillars, **B.** A Caterpillar.

Control

In addition to hand picking and destroying, insecticidal methods are usually applied for the control of tea pests. Spraying of 0.05% malathion or 0.05% endosulfan or 0.02% carbaryl was found useful to kill the larvae.

(2) *Helopeltis theivora* Wat. (Fig. 126) : This bug belongs to family Miridae of the order Heteroptera. This is commonly known as the ``Tea mosquito bug". The female bug is orange and the male bug is black in colour. Since the bugs are provided with piercing and sucking type of mouth parts, they cause considerable damage in nymphal as well as adult stages. The nymphs and adults both suck the sap from leaves, buds and delicate parts of the tea stems. There are two species of this bug the *H. antonii* is found in South India and the *H. theivora* in eastern part of India.

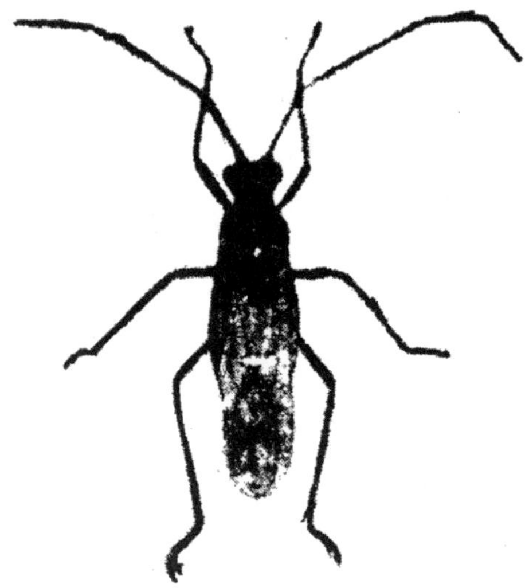

Fig. 126. Tea mosquito bug, *Helopeltis theivora*

Life History

About 500 eggs are laid by each female bug on the buds of tea. The eggs are like rice-grains when hatch in about 4 to 30 days depending on the temperature. The nymphs have 5 instars and the bugs complete their development in 2 weeks in warm season. In cold climatic condition, complete development of the nymph from 1st to 5th instar takes about 8 to 10 weeks. The adult bug is slender, small measuring about 6 to 8 mm in length. They are nocturnal and remain hidden under the tea leaves during the day time. The damage is caused to the young tea leaves by making punctures and sucking the sap. When large number of punctures are made, the leaves of the tea appear black and fall off. This bug, as with other bugs, prefers humid and warm climate. The tea bushes provide such suitable conditions for this bug.

Control

The incidence of the pest can be suppressed by collecting and destroying the nymphs as well as adults with hand nets in the morning and evening. The use of 0.05% malathion, as in other cases has also been found very effective to kill the bugs.

PESTS OF COFFEE

(1) *Xylotrechus quadripes* Chev. (Fig. 127) : This beetle is known as "Coffee stem borer" and belongs to family Cerambycidae of the order Coleoptera. It is the larval stage that bores into the main stem some times down to the tap root and primary branches. It is common in the coffee growing areas of the South India and Assam. The adult is blackish-brown beetle a little more than 1 cm in length.

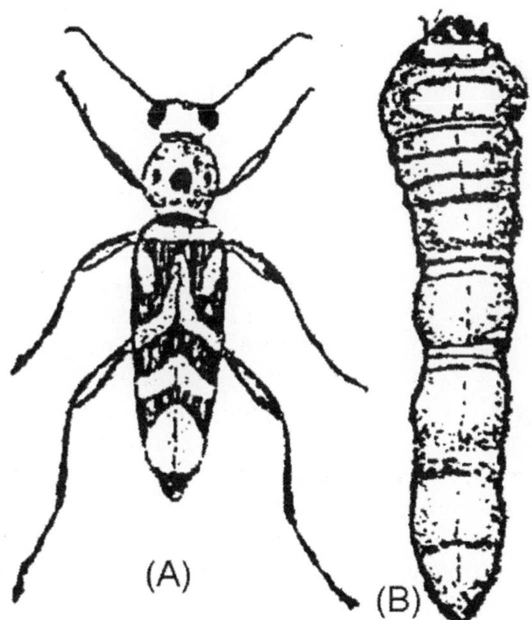

Fig. 127. Coffee stem borer, *Xylotrechus quadripes* **A.** Beetle; **B.** Borer (Grub)

Life History

A female beetle lays as many as 100 eggs in about 4 weeks time. The larvae which are white or grey-white in colour, hatch in about 10 days and start feeding as well as boring the stem of the coffee. The larvae feed and make tunnels in the stem. The fully developed larvae pupate inside the tunnels. The pupal period lasts about a month after which the adult beetle emerges and starts the next generation. There is only one generation in a year. The main damage is caused by the larvae which damage coffee plants by making tunnels as a result of which the branches break.

Control

The larvae can be picked and destroyed and the infected parts of the plants can be removed to prevent further damage.

(2) *Xylosandrus compactus* Eichh. : This beetle belongs to family Scolytidae of the order Coleoptera. The beetle in appearance is brown in colour and cylindrical in shape. It has been found that the larvae and adults both form small holes in the bark hence the name "Coffee shot-hole borers." The lower side of the plant parts i.e. stems and bark is preferred. The attack of this beetle is seen by the wilting, which is due to large number of holes.

Life history

The eggs are laid in tunnels of old branches. The number of eggs laid by a female individual is up to 50 or more The eggs hatch in about 7 days under favourable condition. The larvae after hatching inside the tunnel feed on various fungi growing in the tunnels on the faeces of the beetles. The larval period lasts about 3 weeks. The cocoons are made by the full grown larvae near to the exit hole of the future adult. The pupal development is completed in about 10 days. Thus, the total time taken to complete the life cycle varies from 35 to 40 days. Depending on the suitable conditions, there are several generations in one year. The damage is caused by the larvae and adults both by forming tunnels which are responsible for wilting of the coffee plants as well as falling of the leaves. The affected stems are completely damaged by the attack of micro-organisms.

Control

Hand picking and destroying the larvae and adults is the simplest method of control of this pest. Use of certain chemicals in the form of spray like malathion-0.05% and methyl parathion 2% dust kills the insects.

PESTS OF COCONUT PALM

(1) *Oryctes rhinoceros* Linn. (Fig. 128): The adult beetle is known as "Rhinoceros beetle" because it has a pointed horn on the head. This beetle belongs to family Scarabaeidae of the order Coleoptera. The body of the beetle is robust and elongate measuring about 5 cm. Unlike other beetles the wings are well developed and it is a good flier. The adult beetle is one of the most destructive pests which feeds on the crown of the coconut trees. The larvae are harmless.

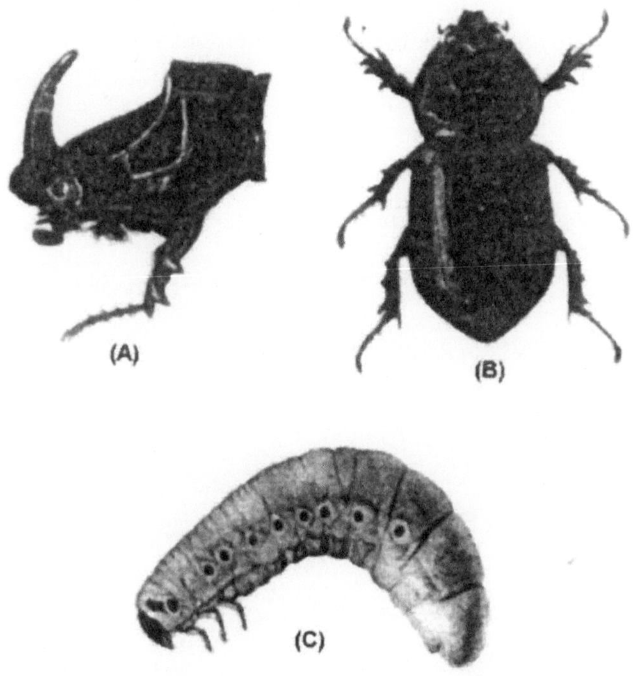

Fig. 128. Rhinoceros beetle, **A.** Side view of male, **B.** Dorsal view of male, **C.** Full grown larva.

Life history

Each female beetle lays about 100 to 150 oval eggs, white in colour, which are laid 10-15 cm below the soil level in decaying vegetable matter viz. decaying leaves, logs and stem of palms, rotten vegetables, old cattle dung, wet saw dust etc. The larvae emerge in about 10 to 15 days. The larval development takes about 100 to 180 days and passes through 3-instars. The full grown larva has an average length of about 75 mm. The body is covered with short brownish hairs. Pupation takes place in the earthen chamber. The pupa is brown in colour and its size varies from 50 to 70 mm. The pupal period lasts for 2 to 4 weeks and even after emergence from the pupal shell the beetle remains in the cocoon for about 11-20 days before coming out of the soil. The adult beetle emerges in about 10 to

25 days. The beetles are nocturnal. There is only one generation in a year. As a result of the attack, the growing point is cut off and the tree dies. The young trees are more seriously attacked by this beetle. In favourable climate the total life cycle from deposition of eggs to emergence of the beetle lasts for 4 to 9 months Egg laying occurs practically throughout the year. The incidence of attack is highest in the dry season i.e. December to May.

Control

The eggs, larvae and pupae can be destroyed by using carbaryl or lindane which kills them at the depth of about 55 to 60 cm of their breading places. The crownless trunks as well as dead trees can be cut and dried. In IPM, green muscardine fungus *Metarhizium anisopliae* and *Baculovirus oryctes* is widely recommended.

(2) *Nephantis serinopa* Meyr. (Fig. 129): This moth belongs to family Cryptophagidae of the order Lepidoptera. Only the caterpillar is extremely harmful to coconut trees and is known as ``Black headed coconut caterpillar". The larval stage is a voracious feeder on coconut leaves and the damage done to the foliage is so severe that the vitality of the whole tree is lowered and the yield is very adversely affected.

Life History

Each moth lays about 100 to 125 eggs in clusters on the tips of the older leaves. The eggs are pale yellow or cream coloured but later become pink. The larva emerges in about 5 days which is extremely small about 1.5 mm. A full grown larva measures about 12 to 15 mm before pupation. The caterpillar feeds on the green matter and makes silken tunnel. About 35 to 40 days are required for full development; the larva later on transforms itself into a brownish pupa within the tunnel. In about 10 to 12 days, a moth emerges which is white in colour. The old moth is ash-grey in colour measuring about 8 to 15 mm. The damage is caused by caterpillars which construct a large number of tunnels when feeding on leaves thus causing leaves to dry. Heavy infestation causes a considerable loss of yield of coconuts.

Control

The control of this pest by releasing the hymenopterous and dipterous parasites is a simple method to check the attack of the pest on coconut trees but when parasites are used, the chemicals must be carefully managed so that the parasites must not be destroyed.

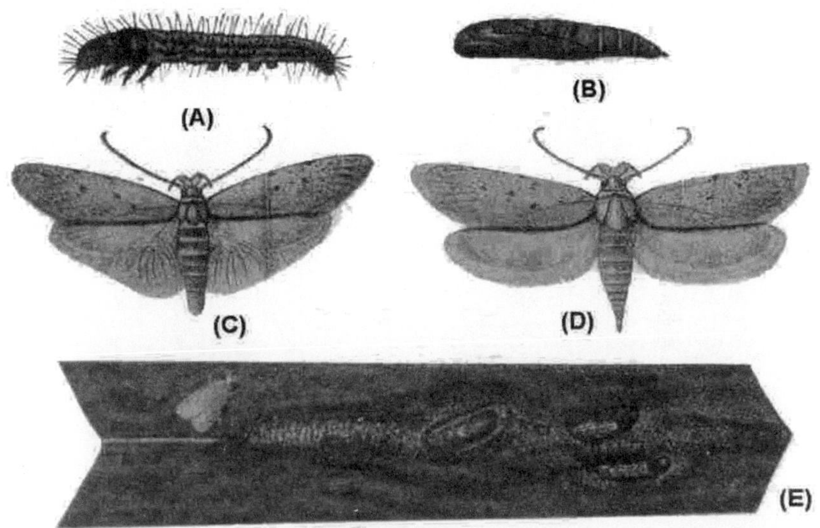

Fig. 129. *Nephantis serinopa* **A.** Caterpillar, **B.** Pupa, **C.** Female moth, **D.** Male moth, **E.** Leaf showing galleries of the caterpillar, a cocoon and a moth (Pruthi, 1969).

3. *Rhynchophorus ferrugineus* Oliv. (Fig. 130): The red weevil of coconuts belongs to family Curculionidae of the order Coleoptera. This weevil occurs throughout the palm growing districts of the country and breads in several forest trees. The beetle is about 4 cm long, including snout and varies in colour from very dark red to reddish brown with traces of up to six black spots on the pronotum. The red weevil is a serious pest of coconut palms especially those of 4 to 12 years old which may be seriously injured or killed.

Life history

The female lays eggs in wounds or cracks or soft areas in the stem and bases of leaf stalks of living palms. Egg laying may begin 3 to 5 days after the beetle emerges from the cocoon and the oviposition period of one female may last up to 8 weeks and as many as 200-500 eggs may be laid. Eggs hatch in about 4 days. The larva bores into interior of the palm stem and feeds for about a month. When full grown, it constructs a cocoon of long strips of fibres woven as a compact oval cell. They pupates inside the cocoon. The pupal stage lasts 2 to 3 weeks and immature beetle remains another week or 10 days in the cocoon before emerging. Thus under favourable conditions there may be 4 to 6 generations in a year.

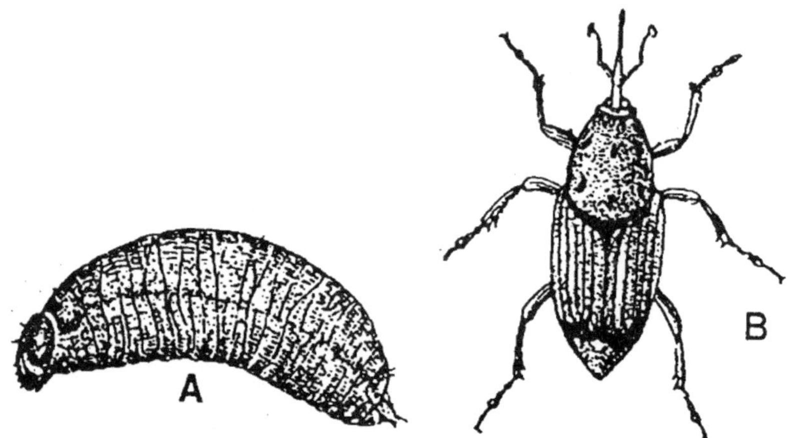

Fig. 130. Red palm weevil, *Rhyncophorus ferrugineus* **A.** Larva (Grub), **B.** Adult

Control

The insects can be killed by pouring into the live holes the emulsion of dimethoate 0.5% or lindane 1.0% or spraying on crowns every three months suppresses the pest infestation.

Some other control methods for ***Rhinoceros* beetles and red weevils of coconut** are:

Beetle : A metal rod about 75 cm long and 10 mm diameter bent into a sharp pointed hook at one end and a ring for gripping at the other end is pushed in the beetles tunnel and the beetle is hooked out. The cleaned out holes are filled with sand.

Fermenting oil cakes kept in water in an open vessel can be used to attract beetles to ground level.

Larva : Manure and compost heaps should be raked and spread out at least once every 3 months to destroy the larvae and to allow crows to pick out those overlooked.

Dead palms and fresh palm logs used for buildings and bridges should be destroyed if they have become breeding material and are producing beetles. Larvae boring in living palms should be killed by injecting a fumigant viz. carbon disulphide, calcium cyanide and ethylene dichloride, carbon tetrachloride. etc. Wounds of any kind should be tarred as *Oryctes rhinoceros* makes holes which facilitates attack of *Rhyncophorus,* it should be exterminated.

Part - III

Ecology and Behaviour

43

LIGHT

The radiant energy, termed light is one of the most important ecological factors affecting many aspects of the insect life. Several vital phenomena of the insect biology e.g. feeding, growth, development, diapause, survival and ethology in general are so profoundly influenced by light that its impact on insect's life has determined very often the distribution and speciation of these organisms in the different parts of the world.

Earlier experimental work dealing with the effects of light on insects was in relation to their sensory behaviour. However in due course of time attention has been paid to the study of light as an ecological factor on insects in relation to their growth and general metabolism. But study of the photoperiodic aspect of light on insect behaviour and development i.e. the responses of insects to daily and seasonal environmental rhythm of day light and darkness has attracted the attention of several environmental biologists (Beck, 1962, 1963 and Wilde, 1962).

Besides the indirect influence of light on insect behaviour associated with vision, this ecological factor produces various effects directly. The intensity of light is an important factor in determining the behaviour of insects. Strong intensity of light will kill any insect species. This means each insect species is adapted to an optimal light intensity in the same way as it is attuned to either an optimum temperature or humidity.

Insects which find low light intensities as their optimum are said to be ``nocturnal'' (dark active); those of medium intensities are ``crepuscular'' (dusk active) and those of higher intensities are ``diurnal'' (day active). Depending on light intensity, insect behaviour varies. For instance an insect which might show a negative response in strong light might manifest a positive reaction towards a low intensity of light. Further, there might be behavioural difference to a particular intensity of light between the two sexes of the insect species. This is very well exemplified in case of certain beetles in which only the males are attracted towards the light.

Much of the behavioural response of an insect to light intensities involves movement of insects towards or away from light, a characteristic example to indicate such a behaviour is the newly hatched codling moth larva which on emergence from the egg laid on apple tree leaf, by virtue of its positive reaction to light moves towards the apple fruits, located on the brightly illuminated parts of the tree.

Similarly, both male and female ants as well as termites, on maturation of their gonads leave the darkness of the nest and positively respond to light in the open where they embark on nuptial flight. However this positive response of light changes into a negative reaction at the end of mating while locomotive activity is an important and usual manifestation of biochemical response to light stimuli. There are certain other features of insect behaviour affected by light. One such feature is the feeding activity of insects. It has been experimentally shown that the feeding activity of *Trogoderma* larva is greater in darkness than in light when other conditions being identical. This suggests an inhibitory role played by light with respect to the feeding behaviour of this insect (Sohi, 1947; Krishna, 1960). In the case of certain insects like the larvae and adults of *Tribolium castaneum* light does not affect the feeding activity of the insect (Krishna, 1960).

Influence of light on the growth and survival of insects has attracted attention of a number of workers. Northrop (1926) was one of the few who investigated this aspect. The effect of different intensities of light in the duration of larval, pupal and imaginal life of *Drosophila* individuals which had been previously cultured in dark for 200 generations was very significant. Lower intensity reduced the larval period but moderately higher intensity increased the length of larval life. Nevertheless very high light intensity completely killed all the larvae. In the case of pupae, a lower intensity of light than needed to kill larvae was sufficient to destroy this stage and at a further decreasing intensity the length of adult life was very much reduced. Thus the influence of light has a pronounced effect on growth and general metabolism relating to the development of *Drosophila.* Experiments on the effect of different coloured rays of light on the growth of insects have also been conducted. It has been reported that in certain insects like silkworms violet rays exercise an apparent stimulating effect upon growth.

One of the important aspect of light governing insect life that has received the attention of several Entomologists and which holds a promising future is photoperiodism. It is well known that daily and seasonal duration of light has a profound effect on the production of sexual forms in Aphids, continuous short day length influencing the appearance of sexual form while long day length influencing asexual form (parthenogenetic) viviparous reproduction. Besides affecting seasonal dimorphism, photoperiodism has been shown to play an important part in growth, metamorphosis and activity rhythms of insects. Beck (1963) stated that many of the insect activities such as feeding, flying, mating and oviposition which display daily periodism are in fact an exhibition of more fundamental endogenous periodism adjusted and regulated by the environmental rhythm of photoperiod. This report of Beck is based on researches of Harker (1960, 1961), Wilde (1962) and his own (Beck, 1962, 1963) accounts for a neuroendocrine mechanism which is regulated by a centre of the brain sensitive to photoperiod, cause the profound effect on activity rhythm, growth and metamorphosis of insects.

As stated earlier, light is chiefly important in relation to behaviour and as a stimulus for those mechanisms which regulate the life cycles and keep them in step with the seasons. The essential difference between light and temperature or moisture may be illustrated by describing the behaviour of caterpillars of the spruce budworm, *Choristoneura fumiferana* in respect to light. The caterpillars tend to move towards a diffuse light; therefore they will move from the centre of a tree outward, thereby coming to the tips of the branches, where there may be fresh young leaves to eat (Wellington, 1948). Compare this with the behaviour of *Choristoneura* in a gradient of moisture. The caterpillars will move along the gradient away from a place where they are likely to be harmed by too little or too much moisture into a zone of favourable moistness. This is not so, with light, for the larvae will not be harmed by exposure either to complete darkness or to the brightest light that they are likely to meet in nature. The response to light is nevertheless a valuable adaptation, because it brings the caterpillars to a place where there is likely to be plenty of good food. This is quite characteristic of the adaptation associated with light; and for this reason Fraenkel and Gunn (1940) referred to light as a ``token" stimulus - a ``token" being a sign which represents something else. Light may indicate circumstances which are for other reasons, favourable or unfavourable, and the insect possessing the appropriate adaptation responds accordingly.

In some cases the stimulus may be provided by a gradient in illuminance (light intensity). The diurnal rhythms in illuminance and in the quality of light may be associated with rhythms in temperature, moisture, food, and so on. Of these, light may be the most consistent and reliable; the insect becomes adapted to respond to a gradient in illuminance or in the quality of the light, and light thus acts as a ``token'' stimulus, leading the animal to a place where temperature or moisture is favourable or to a place where there is an abundance of food etc. In other cases the stimulus may be provided by the changing length of day (photoperiod), this serves as a ``clock'' indicating the seasonal changes in temperature, moisture, foods and so on.

Influence of light on the inception and completion of diapause

It was mentioned dealing with diapause that some species respond to changes in their food, and others to the seasonal rhythm of temperature. But with many species the seasonal rhythm of the life cycle seems to be determined by more than merely temperature and food. Food, temperature, moisture, and air pressure all are very much less regular in their seasonal procession and are therefore likely to be less effective as ``clocks'' than photoperiod.

Kogure (1933) was able to have eggs of *Bombyx mori*, developing at any time, independently of the season and he had a ready method of telling whether a particular treatment had produced eggs with or without a diapause. He reared eggs at different photoperiods ranging from 0 to 24 hours of light each day and at different temperature ranging from 15° to 28°C. The outcome was that eggs which were exposed to less than 14 hours of light each day at 15°C gave rise to moths all of which laid pale (nondiapause) eggs, while those experiencing 16 or more hours of light each day gave rise to moths of which from 30 to 70% laid dark diapause eggs.

At 20°C the results were comparable, but the tendency towards diapause was stronger at every photoperiod and at higher temperatures (24° and 28°C) the influence of the photoperiod was entirely masked and virtually 100% of the eggs gave rise to moths laying dark eggs.

The influence of light was most marked on the developing eggs of *Bombyx mori* but the pattern of development (i.e. diapause or nondiapause in the next generation) determined during this stage was also modified by the influence of light and temperature on the larval instars. Eggs of the bi-voltine race ``showa" of the silkworm, *Bombyx mori*, were incubated at 19°C in darkness. The larvae were reared at 24°C, some in the dark for the first three instars, some in the light for the first three instars and the others were illuminated for the first, second or third instar only. The results showed that the influence of light on the early stage larva was similar to that on the egg and that the first and second instars responded more strongly than did the third.

The responses of the fourth and the fifth larval instars and the pupa to light were relatively weaker than those of the first and second larval instars. It was not evident at all in those cases where the ``diapause pattern" had been firmly imprinted during the life of the embryo. But Kogure's experiments seemed to indicate that when the response could be recognized (e.g. in larvae arising from eggs that were incubated at 15°C in the light or at 20°C in the darkness) light influenced diapause in the opposite way from eggs or young larvae, a few less of the moths which had spent the fourth and fifth larval instars and the pupal stage in the light laid dark (diapause) eggs than did the ones which had spent these stages in the dark.

The illuminance which was effective varied with the temperature, but at 15°C (the threshold for effectiveness) was about 0.014 lumens per square foot, although a significant difference from darkness was also obtained with the lowest intensity tried namely, 0.004 lumens per square foot. At 20°C the threshold was higher, lying somewhere between 0.025 and 0.5 lumens per square foot. Red and orange-yellow light with wave length exceeding 5,500 Å was equivalent to darkness and violet and blue-green light with wave lengths between 5,100 and 3,500 Å was equivalent to white light. These results are confirmed by those given by Dickson (1949) for *Grapholitha* with respect to both the abruptness of the threshold and to the particular part of the spectrum that was effective.

Very few larvae entered diapause, when they were reared in continuous light or continuous darkness at any temperature between 12° and 30°C, at intermediate temperatures between 21° and 26°C, the numbers entering diapause remained low, with a short photoperiod of less than 6 hours or a long one exceeding 14 hours but increased rapidly with photoperiods exceeding 9 hours and reached a maximum of about 100% for photoperiods between 11 and 13 hours. The influence of the photoperiod was most marked at intermediate temperatures between 21° and 26°C, at 12° and 30°C it was entirely masked even with the most influential photoperiod (i.e. 12 hours light and 12 hours darkness) virtually no larvae entered diapause at these temperatures.

The response to the photoperiod varied in a subtle way with the age of larva. Diapause was irrevocably inhibited by a long photoperiod during the first 4 days of larval life but the converse did not hold : exposure to a 12 hours photoperiod for the first 4 days of larval life failed completely to induce diapause in larvae which spent the remainder of their larval life under the influence of a longer photoperiod. But diapause was irrevocably determined for 70% of the larvae when the exposure to the 12 hours photoperiod was extended to include the first 8 days of larval life.

The pupal diapause in certain Lepidoptera appears also to respond to light. In nature the noctuid, *Diataraxia oleracea* hibernates as a pupa; but when the species was reared throughout the year in a glasshouse, a higher proportion of pupae entered diapause in the spring and autumn than in the summer. When batches of 40-50 larvae were kept at a constant temperature of 24°C and exposed to various photoperiods, more pupae entered diapause at the short photoperiods than at the long.

An undirected (unoriented) movement in response to light is called a ``photokinesis", a directed (oriented) movement in response to light is called a ``phototaxis". Apparently, kineses are less common than taxes. The larvae of the blowflies, *Calliphora erythrocephala* and *Lucilia sericata* move away from light towards darkness by alternating lateral deviations of the anterior part of the body. This sort of movement is called `Klinotaxis'. Animals with a single light receptor generally behave like this and the usual explanation offered is that the single light receptor makes comparisons of the illuminance at successive moments of time.

But most animals possess two light receptors and seem able to move to or from a source of light without these ``trial and errors" deviations of the body as they progress, e.g. the larvae of *Choristoneura fumiferana* when placed in a gradient of illuminance will move directly towards the light the larvae of *Ephestia kuhniella* in the same circumstances will move directly away from the light. When there are two lights of equal brightness, these species typically move towards or away from a position midway between the two lights. This behaviour is called ``tropotaxis", it is usually considered that the animal makes simultaneous comparisons of the brightness of the light on either side of its body (Fraenkel and Gunn, 1940).

The male firefly keeps moving towards the flash of the female, as if it retains an impression of the flash (Mast, 1912). This sort of behaviour is called ``telotaxis". Further, the insects with this capacity when exposed to two lights will move towards one of them but not along a line between the two, as in the case of tropotaxis.

The potato beetle, *Leptinotarsa decemlineata* is normally photopositive but it may become photonegative as it becomes desiccated. During dry weather the beetles move down the stem of plants into the soil, where they remain until they have been remoistened by rain. They then become photopositive again and ascend the stems of the plants where they lay their eggs.

Daily rhythms of activity

It has been observed that many species of insects are not equally active throughout the 24 hours of the day. Some are primarily nocturnal, others are diurnal and still others, which are referred to as crepuscular are active mainly at dawn and dusk.

Cockroaches move about much more at night than in the day time, when they tend to remain hidden in crevices and other dark places. In case of *Blatta orientalis* (Gunn, 1940) when the light was on from 7 AM to 7 PM and off for the rest of the 24 hours, the greatest activity was recorded in the dark half the day. In continuous light the rhythm of activity persisted for a couple of days but then disappeared. Continuous darkness had the same effect. These results indicate that the rhythm of activity is associated with the rhythm of illuminance. Though it can persist for a short time in the absence of the external rhythm.

The influence of coloured and polarized lights

Weiss (1943) exposed some 50 different species of insects to the light from coloured lamps. He kept the illuminance constant, while varying the colour from ultraviolet to infrared. When the insects were placed about 30 cm away from the colour lamps and allowed to move freely, most of them came to rest where the light was blue-green (4700-5280 Å). When the insects were put down some 2 m. away from the lights, most of them moved towards the ultraviolet light

(3650-3663Å). In some species a number of individuals also moved towards the blue-green light (4700-5280Å). The rest of the spectrum from yellow to infrared was not attractive. There are some exceptions, e.g. some butterflies, and especially firefly, which is attracted by flashes of red light with wavelength of 6,900 Å. This is the longest wave length yet recorded as visible to an insect (Wigglesworth, 1950). It was observed that the shorter wavelengths are more attractive.

Experimental evidences have shown that certain insects can detect and respond to polarized light. The proportion of the light from the sky which is polarized depends on the relative position of the sun. Frisch (1950) found that bees which have located a source of food are able to find their way to the food again and also they are able to communicate the direction and distance of the food from the hive to the other bees in the hive.

Influence of light on mating and oviposition

It has been observed that *Dacus tryoni* mates only at dusk. Before they mate, the flies behave in a characteristic way, both sexes fly actively and the males make a high pitched ``call". A similar observation was made in the flies kept in cages under constant temperature i.e. mating behaviour began at dusk. Likewise the beetle, *Aphodius howitti* mates only at dusk. The *Oncopera fasciculata* is also crepuscular. Many species of mosquitoes are known to form mating swarms during late afternoon or at dusk. In some insects oviposition may be stimulated by exposure to light of a particular illuminance or to darkness. Experimental work on *Dacus* revealed that 97% eggs were laid during periods when the lights were on. The codling moth, *Cydia pomonella*, unlike *Dacus*, laid most of its eggs during darkness i.e. about 93% of their eggs laid between sunset and dawn. The cotton boll worms, *Earias spp.* and moths of the genus *Amsacta* also lay most of their eggs during darkness or twilight (Pruthi, 1940). It has also been found that the larvae of *Tenebrio molitor*, which usually live in the dark developed more rapidly when exposed to the light from a 10 watt lamp (Cotton, 1930). Waitzinger (1933) kept some larvae of *Bombyx mori* in the dark and others in the light and found that they developed from the third to fourth instar in 5 days in the dark compared to 3.5 days in light.

Thus, the influence of light on insect's life appears to have great significance since almost all the major activities associated with the development and behaviour are governed by this factor. While the effect of intensity of light on insect behaviour has been widely studied from early times, the influence of the photoperiodic aspects of light on insect biology has recently gained much importance. Latest studies indicate a photoperiodically regulated hormonal balance that govern several of the rhythmic activities associated with insect behaviour and development.

44

HUMIDITY

The term humidity refers to water existing in the form of gas or vapour in the atmosphere. This has been one of the primary factors in determining the climate of a particular zone or region. As an ecological factor it seriously affects the distribution of animals and plants on the earth. Its influence on insects has attracted a good deal of attention from many workers and it is very well known that it is almost inseparably associated with the effect of temperature on these animals. Broadly speaking, humidity vitally affects developmental cycle, water balance and general metabolism and certain aspects of behaviour in insects.

Role of Humidity on the Developmental Cycle in Insects

The influence of humidity on the life histories of insects has been found to be very varied. According to Chapman (1931) the rice weevil failed to develop in wheat in a humidity that maintained the moisture content of the cereals below 10%. However, when the moisture content of wheat was raised to 10% or slightly above it, the development of this insect proceeded more smoothly. Another similar example to indicate the effect of humidity through the moisture content of the diet on the development of an insect is that of the grain mite which developed rapidly in flour containing 13% moisture, slowly at 12.4% moisture and died in about 2 weeks when the moisture content was reduced to 12.2%. Likewise, the bean weevil, *Callosobruchus chinensis* has been found to develop most rapidly at 90% relative humidity with marked reduction in its larval period. In certain insects however, e.g. the aphid, *Toxaptera granerium*, difference in humidity has no effect on the development of this insect. At a constant temperature of 26.6°C, the same developmental rate was maintained in this insect at relative humidity varying from 37 to 100%. This is perhaps the aphid being a sap sucking insect, the range of tolerance to the wide variations in humidity is aided by the moisture obtainable through food.

Often a certain range of humidity affects the duration of individual stages in the life cycle of an insect or it affects the survival of the adult insect. An example to indicate the former aspect is that of the pupa of the, Tachnid parasite, *Winthemia quadripunctata* which affects the pupa of the Hawk moth, *Protoparci quinquemaculatus*. The pupal instar of the parasite is so vitally affected at 0% relative humidity that the parasites fail to achieve complete development. An example to indicate the latter aspect is *Tineola* sp., where the adults lived longer at 75% relative humidity and their longevity became reduced in humidities above or below this value.

Role of Humidity on Water Balance or Metabolism

There exists strong influence of humidity on the water balance in insects. This influence is chiefly with respect to the water loss from the bodies of insects. The rate of water loss when measured in different humidities becomes a general relation to the saturation deficiency of the atmosphere. Insects living in 100% relative humidity maintain the water balance by loosing excess water from their bodies through Malpighian tubules and faeces. On the other hand, insects living in very low humidities maintain their water balance by retaining water in their bodies achieve by reabsorption of water from the hind gut either by means of rectal papillae or through the medium of reassociation of the Malpighian tubules. In both these cases the insect pass out solid insoluble uric acid as their excreta, and thus by retention of water in their bodies they escape desiccation.

In certian insects like *Tenebrio molitor*, the metabolic water appears to be one of the sources which aid the insect in maintaining water balance under varying humidities. Further, these insects gain weight in their body on account of absorption of water from the atmosphere when kept in higher humidities. A similar weight increase by the absorption of water from the atmosphere has been reported in some insect eggs such as those of the African migratory locust, *Locusta migratoria migratorioides*. Water gain or water loss in the insect's body affects several metabolic processes among which development is one.

It has been reported by several workers that low humidity causes drying out of the body fluids in the insects and higher humidity dilutes these fluids. Both are dangerous and detrimental to the insect's life. Imms (1931) stated that the effect of the humidity affecting the metabolic process in insects might be with respect to the relation of bound to free water available in the bodies of insects.

Role of Humidity on Certain Aspects of Behaviour

Variation in atmospheric moisture have been found to influence certain aspects of behaviour such as (1) feeding, (2) orientation and (3) oviposition in insects. A characteristic example to indicate the influence of humidity through the moisture content of the diet and the feeding of insects is that of the larvae and adults of *Tribolium castaneum* which feed most on wheat flour containing 14% moisture content and less on the same diet above or below this value (Krishna, 1962). Orientation of insects in relation to different humidity zones have been studied with respect to the *Tribolium confusum* by Willis and Roth (1950) and Loschiava (1952). Regarding fecundity, it has been seen that *Glossina* sp. (Tse Tse fly) have a greater egg laying capacity at 44% relative humidity with reduction in fecundity above or below this value.

The condition of the insect in relation to its behaviour in a moisture gradient

Roth and Willis (1951) observed the distribution of samples of the flour beetles, *Tribolium castaneum* and *T. confusum* in an olfactometer which gave the beetles the choice of moving into a stream of dry air (near 0% RH) or a stream of moist air, which was nearly saturated with water vapour. The beetles,which were taken from where they had been living in whole meal flour containing 10-12% of water and placed immediately in the apparatus, almost invariably came to rest in the dry air. On the other hand, beetles from the same colony, which had been kept for several days without food in air at about 0% RH before being tested in olfactometer, almost invariably came to rest in the moist air. With shorter periods of starvation is about 1.5 days with *T. castaneum* and 3.5 days with *T. confusum*, the response was intermediate, about half the beetles chose the dry air and half the moist.

Sensory organs for perceiving differences in atmospheric humidity

Moisture in air at constant temperature may be measured in units of absolute humidity, relative humidity, saturation deficit, or rate of evaporation. In certain circumstances, e.g. in a high but constant wind, the rate of evaporation, from a free water surface might be very nearly proportional to the saturation deficit of the air; but the rate of evaporation from an insect may not be so simply related to the saturation deficit. It is probable that an insect responds either to the relative humidity of the air or to its evaporating power, but not to both unless it possesses two distinct spots of receptors.

Orthokinesis, Klinokinesis and Klinotaxis

Observations on a variety of insects indicate that their behaviour in response to a gradient in humidity or to other stimuli may be classified into three types, to which the names (1) ``Orthokinesis", (2) ``Klinokinesis" and (3) ``Klinotaxis" have been given. For example *Porcellio scaber* tends to move at random with respect to the direction of the gradient in humidity, but it moves very much more rapidly in dry air than in moist.

An individual which happens, in the course of its random movements, to find its way into the region of moist air will slow down and become motionless, with the result that, in due course, most of the insect are to be found at rest at the moist end of the gradient. This sort of behaviour is called `Orthokinesis' (Fraenkel and Gunn, 1940).

The adults of the mealworm, *Tenebrio molitor* behave similarly, except that they are most active in moist air, so that they come in due course to congregate in the drier parts of the gradient (Gunn and Pielou, 1940). With the body louse, *Pediculus humanus corporis* the frequency of random turning movements increases as it approaches an unfavourable zone. The result of this is to prevent the animal from proceeding far into the unfavourable zone if the reaction is strong and immediate. If the reaction is weak, it result in a convoluted course in the unfavourable zone (Wigglesworth, 1941). This is an example of ``Klinokinesis". When the larvae of *Agriotes* meet with a change of moisture, they sometimes show a directed recoil from the dry air. This is an example of ``Klinotaxis", because it is a directed and not a random movement. It accounts for only a small part of the movements of *Agriotes* in a moisture gradient (Lees, 1943).

Survival value of insects in response to different humidity

The experimental evidences show how some insects which usually live in moist places have become adapted to recognize quite small differences in humidity and to move away from places where dryness might become a danger.

Certain species which are adapted by virtue of water conserving devices to live in dry places are equally sensitive to differences in humidity but tend to move in the opposite direction along the gradient. Two good examples are *Tenebrio* and *Tribolium*. This behaviour can doubtless be classed as an adaptation that has good survival value for species.

Water balance in terrestrial insects

It is necessary for all but a few sorts of terrestrial insects to keep the proportion of water in their bodies constant within quite narrow limits. The few special exceptions mostly belong to the resting stage in the life cycle of some species, often associated with diapause in insects.

Water may be lost by excretion, egestion, evaporation through the integument and as a waste product of respiration. Water may be gained by eating, drinking or absorption through the integument. If the source of supply is limited, as it often is, with insect, then losses from all sources must be correspondingly reduced.

The intake of water from the surroundings

Terrestrial insects obtain water in their food, and some also drink water. Bees and butterflies are commonly seen at the edges of pools, sipping water. The longevity and fecundity of some Lepidoptera are reduced when free water is not available. Muscid flies, for example, houseflies and blowflies must be supplied with water to drink if they are to be kept in cultures. Further examples are listed by Leclerq (1946) who said he had collected 500 species of all orders of insects drowned in pools in summer, victims of their thirst.

Most insects which live on dry food, such as grain meal, dried fruit etc. live without drinking. There is the odd exception of the adult spider beetle, *Ptinus pectus*, which requires a certain amount of water in addition to that which it gets with its food. Ewer and Ewer (1941) found that both longevity and fecundity were reduced if the beetles were not supplied with water to drink. The majority of insects, however, appear to obtain their water in ways other than drinking.

The food of insects varies from liquid food, such as blood and plant sap, moist food with a high water content, such as leaves, which may contain over 90% water, to dry materials, such as cereals or wood, which may contain from about 5 to 20% water. Insects which ingest large quantities of liquid food such as blood and plant sap, obtain far more water than they need in order to take in the necessary food constituents. Besides an excess of water, there is an excess of chlorides in the food of blood suckers. The excess water and chlorides have to be eliminated.

The intake of water through the integument

Some insects can absorb water through the cuticle. Larvae of the sawfly, *Cephus cinctus* (Salt, 1946) and doubtless others absorb water in this way. Eggs of grasshoppers and locusts require to absorb water from the surrounding soil during their development; so far as is known, they require it in the form of free liquid water. The passage of water into and out of the egg is not simply the result of osmotic forces but is controlled in a most remarkable way by the activity of ``hydropyle cells". These are a small group of exceptionally large cells situated at the posterior pole of the egg, immediately under a specialized part of the cuticle known as the ``hydropyle".

Water can enter and leave the egg only through the hydropyle, since the rest of the cuticle is practically impermeable to water. The hydropyle was first demonstrated in the American grasshopper, *Melanoplus differentialis* (Slifer, 1938). It has since been found to occur in other grasshoppers and locusts, e.g. the Australian plague grass hopper, *Austroicetes cruciata* (Birch and Andrewartha, 1942), the Canadian *M. bivittatus* (Salt, 1949) and the African locust, *Locustana pardalina* (Matthee, 1951).

The eggs of *Locustana* absorb water from the surroundings during their development and quite turgid. Little water is absorbed during the first few days because of the absence of hydropyle. Matthee (1951) has shown that water is absorbed through the hydropyle against an osmotic gradient.

Wigglesworth (1931) showed that the bug, *Cimex lectularius* absorbed water from unsaturated air. Breitenbrecher (1918) showed the same for the beetle *Leptinotarsa decemlineata*. The larvae of *Tenebrio molitor* absorbed water from air with a relative humidity of 88% (Mellanby, 1932); the grasshoper, *Chortophaga virdifasciata* from air with 82% relative humidity.

Conservation of metabolic water

Berger (1907) found that larvae of the meal worm, *Tenebrio molitor* maintained a constant proportion of water to dry matter in their bodies, when starved in unsaturated air and Buxton (1930) confirmed these experiments, even when starved for a month in an atmosphere of zero humidity, the larvae retained a constant water content in their bodies. Buxton argued that more water was lost in dry atmosphere as compared with a moist one and that this loss was made good by ``water of metabolism" formed during oxidation of fats and carbohydrates and retained in the body. He considered that the rate of production of metabolic water was determined by the humidity of atmosphere surrounding the insect.

Besides the meal worm, a number of other insects are known to maintain a constant proportion of water in their bodies in a dry atmosphere e.g. adults of *Lachnosterna implicata* (Sweetman, 1931), larvae of the flour moth, *Ephestia kuhniella* and the beetles, *Tribolium confusum* (Roth and Wills, 1951) and *Dermestis vulpinus* (Fraenkel and Blewett, 1944). These insects that live on dry food or in dry places must be under considerable stress to maintain an adequate amount of water in their bodies and the experiments of Buxton and others show that they do indeed possess remarkably effective mechanisms for conserving water.

Loss of water to the surroundings

Besides possessing special adaptations for obtaining water from dry surroundings, terrestrial insects have adaptations which enable them to reduce the amount of water lost to the surroundings. Water is lost in excretion, egestion as well as by respiration and by transpiration through the body surface of the several insects belonging to different orders as pointed by Wigglesworth (1948).

By Excretion and egestion

The amount of water lost with the excrement depends on the amount of water in the food and moistness of the place where the insects are living e.g. in the blood sucking bug, *Rhodnius prolixus* (Wigglesworth, 1931) excess water is eliminated through the Malphighian tubules.

In contrast to liquid feeding insects, there are many insects which live in dry surrounding on dry food. For them water is a scarce commodity and the problem is not to get rid of it but to save it. Such insects have efficient devices for reducing the amount of water lost in excretion and in faeces. Such insects conserve water by (a) reabsorption of water in the rectum or the Malpighian tubules or (b) by excretion of nitrogen as uric acid. In many insects water is reabsorbed by the rectal pads from the faeces before they leave the rectum e.g. *Tenebrio molitor* and *Pediculus humanus* etc.

By respiration and transpiration

Ramsay (1935) was able to show that transpiration from the crockroach, *Periplaneta americana* took place through both spiracles and cuticle. The Collembola, which live in moist situations, the cuticle is relatively permeable to water and the rate at which water is lost through the cuticle into dry air may be quite considerable (Davies, 1928). In others the cuticle is remarkably impermeable and most water is lost through the tracheae but this can be regulated by opening and closing mechanism of spiracles as in the larva of *Tenebrio molitor*.

Experiments of Wigglesworth and others have demonstrated the existence of a very thin wax layer forming the outer most layer of the cuticle. The water proofing properties of the cuticle, largely resides in this layer (Wigglesworth, 1948).

The eggs of the *Locustana pardalina* at the time they are laid also contain a layer of wax just under the chorion but in addition a secondary layer of wax is laid down quite early in the course of egg's development. This is secreted by the serosal cells of the developing embryo. The eggs of other desert inhabiting grasshoppers and locusts have a cuticle that is equally impermeable e.g. eggs of Australian grasshoppers, *Austroicetes cruciata* remained alive after 80 days in air dry soil. Although the eggs of the desert inhabiting grasshoppers may survive unharmed in the soil during prolong period of drought they nevertheless require moisture in the soil in order to develop. During development after the secretion of the embryonic cuticle, eggs of *Austroicetes* increase in size to a remarkable extent due to absorption of water which is regulated by the activities of the hydropyle.

The relationship between saturation deficit and survival rates

Birch (1944) determined the death rate in eggs of *Calandra oryzae* exposed to a series of constant temperatures between 15° and 32°C and a range of humidities between 20 and 90% R.H. within this range, the lethal influence of temperature could be ignored except for the lot exposed at 15°C. Experimental work has shown that eggs of *Rhizopertha dominica* are relatively more resistant to dryness than are the eggs of *Calandra oryzae*.

Drought Hardiness and acclimatization

Some insects are drought hardy throughout their lives while others become drought hardy only during certain stages of their life cycle. This is related to the level of dryness in the sorts of places where they usually live. The Australian grasshopper, *Austroicetes cruciata* is found in a region where the summer is hot and dry and it is adapted for life in this climate by virtue of a diapause in the egg stage which ensures that only the drought hardy diapausing egg shall be present during summer and that the less hardy stages (post diapause eggs, nymphs and adults) shall be present only during late winter and spring when the weather is not so dry. On the other hand there are species which without entering diapause become drought hardy as they become quiescent in response to dryness.

Lethal influence of excessive moisture

Many grasshoppers and locusts are limited by excessive moisture. The nymphs of *Melanoplus mexicanus* and *Camnula pellucida* die in large numbers during period of wet, cloudy weather when the relative humidity is close to saturation (Parker, 1930). Excessive moisture may simply drown the insects. In wet years, large number of overwintering pupae of the *Heliothis armigera* die from drowning.

Influence of moisture on fecundity

Hamilton (1950) remarked that at lower humidities the *Locusta migratoria* lived longer but failed to produce eggs below 40% RH. It was further observed that at 70% RH the locust matured sooner than in the dry air or low humidities. The length of life was also shortest at 70% RH but becoming greater as RH was increased or decreased from 70%. In a similar manner, the desert locust, *S. gregaria* also responded but egg maturation failed at low RH values, though they survived for several months. If the same desert locusts are exposed to adequate moisture, they matured eggs in the normal course and adult development completed. Further the greatest fecundity was observed at the optimum humidity which differ from insect to insect and it was

70% RH with *Locusta migratoria.* All the insects do not show a definite optimum RH for egg laying and it varies considerably from species to species (Hamilton, 1950).

The fecundity of *Calandra oryzae* increased with the moisture content of the wheat it was feeding on, from zero in wheat at 9% moisture content (34% RH) to a maximum in wheat of 14% moisture (70% RH) (Birch, 1945). Within an optimal range of temperature *Calandra* and *Rhizopertha* were able to lay eggs in wheat which was drier than the driest wheat in which eggs were laid at temperatures on either side of the optimal range. But no temperature could *Calandra* (=*Sitophilus*) lay eggs in wheat drier than 10% moisture content, the corresponding moisture content of wheat for *Rhizopertha* was 9%.

Influence of moisture on the speed of development

With some insects, development ceases almost completely in dry air, although they do not die except after prolonged exposure. Nymphs of *Locusta* provide an example of one sort of relationship between moisture and speed of development in which the speed of development is retarded at high humidities. On the other hand eggs of *Lucilia* are examples of a second type of relationships in which high humidities accelerate the speed of development. There is yet a third category, which includes those insects in which the speed of development is relatively independent of humidity e.g. Saturniid moths, *Telea polyphemus.* Other insects which have eggs like this include the bed bugs, *Cimex lectularius* and *C. rotundatus.* (Geisthardt, 1937), the firebrat, *Therombia domestica* (Sweetman, 1938), the bean weevil, *Bruchus obtectus* (Menusan, 1934) and the clothes moth, *Tineola biselliella* (Griswold, 1944). All these insects occur in dry places and their eggs have cuticles which are impermeable to water vapour. The same is true of the pupae of the moths, *Bombyx mori* (Leclerq, 1946) which develops at the same rate at RH from 20 to 100%.

Concludingly, humidity as an ecological factor, plays a vital role in the development and bionomics of the insects. The main idea derived from experiments on the humidity reaction of insects is that every insects species has an optimum humidity value in which it lives well. Above or below this value, the result will be detrimental to the insect's life. In this way humidity serves as an important ecological factor.

45

TEMPERATURE

The three environmental factors viz. temperature, humidity and light play a key role in controlling almost all life processes either directly or indirectly and all these are inter related with each other. The various vital activities like development, growth, metamorphosis, activity of several internal organs and polymorphism in locusts and aphids all are governed by the suitable temperature range for maximum activity. Any change in the tolerance zone adversely affects the proper functioning of the insect's vital organs likewise distribution and abundance of insects, formation of locust swarms, control of a pest population, maturation of reproductive organs and seasonal variations in species, aestivation and hibernation, fecundity and fertility, all are under the influence of temperature.

Since the insects are cold blooded animals their body temperature changes with the change in surrounding temperature which in turn not only control several life processes but also modifies the insect behaviour.

The range of temperature within which an insect may thrive

The extremes of temperature from the hottest to the coldest places where insects live are great. It has been claimed that the larvae of certain Diptera thrive at temperatures as high as 55°C or even higher. Brues (1939) is of the view that 52°C is about the highest temperature at which any animal is known to thrive indefinitely. At the other extreme, there is a beetle, *Astagobius angustatus*, which is known to carry on its life cycle in ice grottos, where the temperature range is between –1.7°C and 1.0°C (Allee *et al.*, 1949). But no individual species is known which can thrive over such a wide range as from 0° to 50°C. The range of temperature favourable to any particular species is related to the prevailing temperatures in the places where the insect usually lives. Not only the average temperature but also its variability will be reflected in the physiology of the particular insect. Those that live in cold places (or are active during the cold season in warm temperate zones) have a favourable range lower than that for insects from warmer zones e.g. the favourable range for the development of the eggs of *Smynthurus viridis* (Collembola) extends from 7° to 27°C (Davidson, 1931) but with *Austroicetes cruciata.* (Orthoptera) the favourable range for the development of eggs extends from 15° to 34°C (Birch, 1942).

The influence of temperature within the favourable range

When insects are allowed to move freely along a temperature gradient, they usually congregate between quite narrow limits of temperature. This narrow band of temperature has been called the ``preferred temperature" or the ``*temperature preferendum*". The behaviour of locust nymphs and adults in the field has been the subject of much study because it is generally accepted that high numbers in the outbreak area are not by themselves sufficient to produce the gregarious and migratory behaviour which is such an essential feature of the destructiveness of these insects. Given high numbers, it is still necessary to have a stimulus which will bring about ``aggregation" or crowding while the insects are in an active or excitable condition. Provided that the terrain is suitable, temperature may act in just that way for the insects tend to crowed into local situations where the temperature approaches their ``preferendum". With *Chortoicetes terminifera* this is about 42°C (Clark, 1947, 1949).

Clark studied the behaviour of early stage nymphs of this locust in an outbreak center in New South Wales (Australia). He remarked that in spite of the fact that their parents had been gregarious, the first instar nymphs were neither markedly gregarious nor excitable. They soon distributed themselves over the area in relation to the occurrence of green plants suitable for food. Most of them spent the night sheltering in the vegetation and in other places. At day break the temperature in these situations was about 2°C and the hoppers remained there motionless. They become more active as the temperature increased until at 20°C and disability due to cold could no longer be detected and their movements were normal. Between 20°C and 45°C the hoppers sought out and crowded into the warmest possible situations at ground level, taking up what Andrewartha and Birch (1954) called the basking formation. They crowded closely together often with their bodies touching and broadside to the sun.

On cool days basking groups persisted until late in the afternoon. On hot days, the basking groups would disperse as the temperature approached 45°C and the hoppers would seek more shady places, ordinarily the groups would form again later in the afternoon. This behaviour was governed by the existence of a preferred temperature at about 42°C. External influences, especially temperature plays an important causal role in crowding. A period of crowding is apparently essential for the progeny of swarms of this species to develop gregarious behaviour i.e. the capacity for mass migration. This is probably true for other locusts. The length of the necessary period of crowding varies in relation to population density and probably temperature conditions. Kennedy (1939) concluded that aggregation of *Schistocerca* was largely the result of a ``diurnal'' regime resembling in many ways to that of phase *gregaria* determined largely by temperature differences.

Influence of temperature on the rate of dispersal

It is well known that after the weevil, *Calandra oryzae* (=*Sitophilus oryzae*) has been breeding for a number of generations in a large bin of wheat, the insects will all be concentrated within a foot or two of the surface, irrespective of whether the colony started near the surface or in the depths of the bin. Birch (1946) showed that this was due to influence of temperature on the dispersal of the adult weevils. He placed small colonies of weevils on the surface and at various depths in several experimental bins of wheat.

Accordingly, as the colonies developed in the depths, they caused the temperature of the wheat immediately around them to rise. As the temperature reached and passed 32°C due to the heat of metabolism of the immobile young stages in the grains, the adults moved away to cooler places. Eventually, the layer of wheat near the surface was the only place which was cool enough to support the colony. These experiments of Birch showed that *C. oryzae* dispersed scarcely at all until the place where they were living become uncomfortably hot. Gunn *et al.* (1948) studied the behaviour of migrating swarms of the locust, *Schistocerca gregaria* in Kenya. Swarms of this locust often spend the night roosting in trees or shrubs. In the morning they descend to the ground, as the temperature rises they begin to make short low flights, but at first they are very few in the air at any one time, later the numbers making flights increase and each one stays aloft longer, so that as time passes more and more are seen in the air at once, flying over their comrades on the ground. According to Gunn *et al.* (1948) the moment at which this mass flight begins, is related to temperature but not to humidity or any other component of the environment. Swarms were occasionally observed to begin migrating when the temperature was as low as 14°-16°C. Once a swarm waited until the temperature had risen to 23°. But as a rule, migration began when the temperature was between 17° and 22°C. Migration began at a lower temperature when the maximal temperature for the previous day had been low and vice versa. In other words, the influence of temperature on the beginning of migration was modified by some stimulus (due to secretion of hormones) which was correlated with the maximal temperature of the previous day. It is much easier for locusts to become air borne and therefore to begin migrating when there are upward thermal currents near the surface of the grounds.

Influence of temperature on the speed of development

It has long been known that poikilothermic animals complete their development more rapidly in warm weather than in cool. The speed of development at different constant temperatures has been measured for many species. In nature, insects and other animals live in places where the temperature fluctuates. The life cycle may include a diapause stage which needs to be completed at a lower temperature than that which favours the more obvious processes of morphogenesis. The fluctuations may include extreme temperatures which are harmful and even short exposures to these may impair the insect's competence to develop healthily at a favourable temperature. Different stages in life cycle may have different limits to the favourable range and may respond differently to temperature within the favourable range.

In the case of *Calliphora erythrocephala* when pupae were exposed to 5°C for 8 days, the time required to complete development was reduced about 10% at each temperature except 30°C. The duration of pupal period at 18° and 23°C was also reduced by about the equivalent amount when pupae were exposed on alternate days to 5°C and 18° or 23°C (Ahmad, 1936). A continuous exposure of 16 days at 5°C was sufficient to kill about half the eggs of *Locusta migratoria;* all the eggs were dead after an exposure of 32 days, at 5°C. It is clear that 5°C is outside the limits of the tolerable range for eggs of *L. migratoria.* The degree of retardation depends on the duration of exposure at 5°C, the stage of development of embryo and the particular favourable high temperature at which development subsequently proceeds. With the rice weevil, *Calandra*, all stages may be found all the year round living together in the same handful of grains. The egg stage of *Calandra* required on the average 10.4 days to complete its development at 18.2°C and 3.6 days at 29.1°C. This indicated a proportional increase of 2.84 in the duration of the egg stage as the temperature decreased from 29.1°C to 18.2°C. For the same pair of temperatures the proportional increases in the duration of the larval, prepupal and pupal stages were respectively 4.78, 2.06 and 3.77. These are large and significant differences (Birch, 1945). The pupae of *Drosophila* developed and the adults emerged at a constant temperatures between 15°C and 33°C. At 10° and 34°C, no flies emerged although at 10°C many puparia contained fully formed flies which had failed to emerge, a few were also found in the puparia at 34°C. Influence of temperature is thus also related with the speed of development.

Influence of temperature on fecundity and the rate of egg production

The limits of the favourable range of temperature for egg production and oviposition are often of the same order as those for the development of the immature stages, but need not necessarily be similar. Thus, *Calandra oryzae* living in wheat of 14% moisture content, is able to complete its life cycle at temperatures between 15° and 34°C. Oviposition also occurs over this range but so few eggs are laid at 15°C and 34°C that it would be more realistic to consider this process to be limited to a rather narrower range about 17°-33°C (Birch, 1945). Birch (1945) studied the oviposition of two beetles viz. *Calandra oryzae* and *Rhizopertha dominica* in a carefully controlled environment and was able to measure not only the direct influence of temperature but also its interaction with the moisture content of the food, number of other insects in the environment and the age of ovipositing females.

Norris (1933) found that most females of *Ephestia kuhniella*., which had been mated with one male only, produced fertile eggs at 27° and 30°C, provided that the prepupae and pupae of both males and females had been reared at a favourable temperature lower than 27°C. As the temperature at which the prepupae and pupae were reared increased from 27° to 30°C, the proportion of females which laid no fertile egg, increased from 10-59% to 96-100%.

Cold hardiness

Kozhanchikov (1938) grouped insects in 3 classes according to their response to low temperatures. The ones in the first group can not survive for any considerable period if the temperature falls below the lower limit of the range, favourable for normal development e.g. *Calandra, Cimex* and *Locusta.* There are many of these species in which no stage in the life cycle is capable of becoming dormant at temperatures below the favourable range. In all those species that are normally active during summer resemble the subtropical and tropical forms in lacking the capacity to become dormant at low temperatures. Most member of this group reach an under cooling point and rebound at a temperature only a few degrees below 0°C. There are exceptions e.g. *Calandra and Ephestia* which are adapted to dry places or on dry food. All the members of this first group irrespective of their under cooling points, die when exposed to nonfreezing temperature and are not ``cold hardy".

The second group includes the forms that are able to become quiescent in the sense of Shelford (1927) and they may survive, inactive but healthy during prolonged exposer to cold but develop at any time if they are placed in the warmth, e.g. prepupal stage of *Agrotis segetum* and the larval stage of *Lasiocampa quercus* and pupa of *Heliothis armigera.*

The 3rd group contains the diapause stage of species from temperate climates `they hibernate in a particular stage of the life cycles. The chief difference between members of the second and third groups is that the former become ``dormant" directly in response to low temperature but with the latter dormancy has usually arisen in response to some quite different stimulus. The members of the two classes have in common a capacity to survive unharmed through quite prolonged exposure to nonfreezing low temperatures. For this reason, the members of the second and third groups might be said to be ``cold-hardy". The most cold hardy of all may even survive the freezing that accompanies the first rebound and die only after a second wave of freezing, probably associated with a second under cooling point and rebound.

Limits of the tolerable range

If an insect is exposed to an extreme high or an extreme low temperature, it will quickly die and no practical difficulties are raised when we say that death was caused by the exposure to unfavourable temperature. As the temperature becomes somewhat less extreme, death may approach more slowly, but it may still be clear that exposure to unfavourable temperature was the chief cause of death. Moving farther and farther away from the extremes, we come to a zone of moderate temperature where it is not reasonable to attribute any lethal influence to temperature. Fry (1947) called the two limits to this median tolerable zone the ``incipient lethal low temperature" and ``incipient lethal high temperature" and above and below the median tolerable zone are the ``upper lethal zone" and the ``lower lethal zone" respectively.

Influence of high temperature

It has been found that all the maggots of the fly *Anastrepha* were dead after 15 hours at 40.5°C but only half were killed at that temperature in about 11 hours. It has also been reported that deaths from heat were negligible upto about 40°C, and an exposure of 4 hours to 42°C killed about half of the larvae and at 46°C all were dead at the end of 4 hours. Buxton (1924) found certain species of Orthoptera living in the desert on bare earth with a temperature of 60°C but it is unlikely that even these desert forms could survive for very long at such a high temperature.

According to Buxton (1928) and Chapman *et al.* (1926) in the normal course of the day all the insects leave the surface of the sand when its temperature nears 50°C. This has also been observed by the author (Awasthi unpublished) in case of several insects of "Thar desert". Some climbs grasses and some enter their burrows, while others fly about some distance above the sand making hurried landings to enter their burrows.

Fecundity and fertility

Of the several internal and external factors, the temperature is one of the most important factors that controls the fecundity and fertility of insects. The term fecundity means ability to produce offspring and its value or birth rate is presented by total number of eggs laid by a female, regardless of their subsequent fate and the fertility is estimated as the total production of viable eggs giving rise young ones e.g. *Camnula pellucida* survived at 37°C for only 15.8 days but 4.2 egg pods per female were laid during that time. Whereas at 27°C the length of life was doubled (32.6 days) but only 1.0 eggs pod per female was laid (Parker, 1930). In *Locusta*, the maximum number of eggs per pod and of viable eggs was obtained at a constant temperature of 26.7°C and in *Schistocerca gregaria* at 32.2°C. In case of fluctuating temperatures however, these optima are at a considerably lower mean temperature viz. 25.1°C for *Locusta* and 23.5°C for *Schistocerca* (Hamilton, 1950).

Temperature and polymorphism

Although polymorphism in locusts, depends on several intrinsic factors (hormones) but involvements of temperature as an ecological factor is also significant. It has been reported that *Schistocerca* hoppers reared at 24°C were predominantly black but at 44°C melanin pattern was practically absent (Husain and Ahmad, 1936). The darkening of the integument at low temperature is known also in non-swarming species e.g. the hoppers of *Schistocerca abscura* are normally green, but when reared at 21°C develop black spots (Duck, 1944). The extreme temperatures have no effect on the background colour, but they certainly affect the dark pattern in a way similar to density (Dudley, 1944). The rate of growth being dependent on temperature, and on the basis of their experiments Husain and Mathur (1944) concluded that morphometrices are determined by temperature and not by density.

Influence of temperature on the incubation

The incubation time of non-diapausing eggs is affected by the temperature in the same way as all biological processes. Effect of incubation temperature depends on the hatching of locust. At the extremes of the temperatures there is a sudden reduction of percentage of hatching in *S. gregaria.* Since the best percentage obtained was eighty, some 40% hatch may be regarded as satisfactory, in which case the temperature range 23-37°C should be regarded as giving a reasonable hatch in *Schistocerca.* In *Locusta* 40% hatching was obtained by Ahmed (1936) within the range of 27-37°C and possibly below, because they hatch at the next lower temperature of 23°C was 37%. In the bug *Oncopeltus*, it was found that some high hatching percentages of eggs incubated at favourable temperatures produced nymphs and with very poor survival (Hodson and Alrawy 1958). This suggested the need for determining the viability *threshold*, depend as the lowest temperature producing a reasonably high percentage of hoppers which are able to reach the adult stage if kept under favourable conditions.

Seasonal change in form as related to the temperature

It is well known that the Lepidopterans are covered with various types of colourful hairs and scales which give them beautiful markings that serve additional purpose. They form an efficient insulating layer which helps to keep the body of insect warmer than the surrounding air. In cool climates this is necessary for efficient flight e.g. the privet hawk moth, *Sphinx ligustri* (Fig. 131). This moth can rase its body temperature to 17°C above the surrounding air but with its thorax shaved of scales and hairs, it can only achieve an excess temperature of about 8°C.

Fig. 131. The privet hawk moth, *Sphinx ligustri* resting on a tree trunk.

Several insect species particularly Lepidoptera in warm climate have different colour pattern from that of cold climatic condition. Some time this change is so drastic that leads to confusion in field identification. In several Lepidoptera, the summer forms are quite different from winter as well as spring forms which is due to temperature (see Fig. 132).

Influence of temperature on the aestivation and hibernation

Although these phenomena are controlled by several internal factors but here also temperature plays its vital role. Many insect species in hot summer or at high temperature undergo summer sleep (aestivation) likewise they hibernate in winter (winter sleep) which helps them to pass over unfavourable environmental conditions so in a way these two different and independent phenomena are under the influence of temperature low or high as the case may be.

Fig. 132. Seasonal variation in Butterflies (From Goodden, 1971)

Influence of temperature on the general metabolism

It is well known that insects have no internal regulation of body temperature. The rate of chemical processes in their metabolism are influenced directly by the temperature and are either speeded up by warmth or delayed by cold. Each species has certain limits and are killed by prolonged freezing or excessive heat. If freezing weather occurs after their eggs or larvae have begun to develop, many are killed and their populations are reduced. It has been found that in some insect eggs, larvae or pupae that over winter as resting stages with lowered metabolism either beneath the ground surface, within the plants or in the bottom of ponds and streams to escape being frozen. Certain insects trapped in ice may survive because the water in their bodies is so related to colloid that it does not freezes.

46

ECOLOGICAL ASPECTS OF DIAPAUSE

Diapause occurs as an important adaptation in many species of insects, enabling them to persist in regions from which they might otherwise be killed out by extremes of climate or to maintain high numbers in an area which might otherwise support only a few. Diapause usually occurs in that stage of the life cycle which is highly adapted to resist the rigors of the climate.

In the grasshopper, *Austroicetes cruciata*, diapause occurs in the egg stage. The diapausing egg, by virtue of a remarkably effective mechanism for preventing evaporation and yet providing for the rapid absorption of water as opportunity offers, is able to withstand prolonged exposure to dryness and thus survive the arid summer (Birch and Andrewartha, (1942).

In several insect species which live in the cool, temperate zones, diapause may occur in a stage in the life cycle which is able to withstand extreme cold or diapause may be of value to the species because it synchronizes the life cycle with the weather and thereby ensures that the active stages of the life cycle shall be present when there is an abundance of food and the weather favours rapid development, and a high survival rate.

The first experimental evidence of the phenomenon which now is termed as diapause was advanced by Duclaux (1809). He observed that eggs of the silkworm, *Bombyx mori* failed to hatch when they were kept indoors in a warm room during winter. So he kept some in an ice-box for 40 days, and then brought them into the warmth, most of them hatched. Subsequently, Duclaux took some eggs which had been in the warmth for 50 days and placed half of them at –8°C and half at 0°C. for 2 months, when they were returned to the warmth, 94% of the latter and 13% of the former hatched promptly.

Among those which had been at –8°C. many which failed to develop remained alive, because 55% of them eventually hatched in the spring after having spent the winter out of doors. Duclaux cancluded that the eggs of *Bombyx* required to be exposed to cold before they could develop and that the most suitable temperature for this was about 0°C. Wheeler (1893), coined the term ``diapause" to describe a particular stage in the embryonic development of the grasshoppers, *Xiphidium*

Henneguy (1904) used it to refer to the physiological state of dormancy or arrested development which he had observed in a number of insects and in this way diapause gained its suitable place. For a time it was used to cover almost any state of arrested development. Shelford (1929) suggested that the term diapause be restricted to cases in which development is arrested ``spontaneously" and does not respond immediately to any improvement of the environment. He suggested the term ``quiescence" for the simpler case in which development is temporarily inhibited by an unfavourable environment and may be resumed as soon as the hindrance is removed.

The bug *Rhodnius prolixus* passes through five nymphal instars, and in each instar it normally takes only one large meal of blood. If the meal is large enough (in nature it usually is) to distend the abdomen, this will cause a nervous stimulus to be carried to certain neurosecretory cells of the brain. These cells then secrete a hormone, and in due course the bug moults, passing into the next instar. Development in *Rhodnius* may be arrested in any one of three ways:–

1. Bugs may survive unfed for many weeks, and a number of bugs which received occasional small meal remained alive without developing for many months 2. Bugs which were decapitated shortly after they had engorged remained alive without developing for more than a year. 3. and, when the bugs were reared in sterile surrounding so that they contained no symbionts (*Actinomyces rhodnii*) development was arrested in the fourth or fifth instar (Wigglesworth, 1934, 1948, Brecher and Wigglesworth, 1944).

In all three cases, the bugs resumed development and moulted when a developing bug was grafted to the dormant one in such a way that the body fluids of the two could intermingle through living tissue. It is now quite clear that arrested development is due to the absence of a hormone (see: homonal control of diapause) in the first case because the requisite nervous stimulus was lacking, in the second case because the endocrine gland had been removed when the bug was decapitated, and in the third case (Wigglesworth suggests) because the hormone may not be produced in the absence of symbionts.

Wigglesworth (1948) pointed to the analogy between the arrested development of *Rhodnius* and the general case of diapause in other insects, suggesting that the immediate cause of diapause may be a failure in the secretion of hormones necessary for growth.

Variations in intensity of diapause and voltinism

With some species, every individual, in every generation, enters diapause; typically there is one generation each year. The life cycle is called ``uni-voltine" and the diapause ``obligate". In these species diapause occurs so consistently that it seems to be quite independent of any stimulus from outside the animal. On the other hand, there are many species in which one or several generations in which few or no individuals enter diapause, may be followed by a generation in which most or all individuals enter diapause; typically, there will be several generations each year. This sort of life cycle is called. ``multi-voltine", and the diapause ``facultative".

Diapause is common in the embryonic, larval and pupal stages of insects; it also occurs in nymphs and adults. Diapause may occur in diverse stages in the development of the embryo of different species, but the stage of development is usually characteristic of each species.

The influence of environment on the duration of diapause

In considering diapause it is helpful to think of development in terms of its morphological aspect, *morphogenesis*, and its physiological aspect, which by analogy, may be called *physiogenesis*.

The diapause stage, meaning that stage in the life cycle during which morphogenesis is more or less at a standstill, and diapause development to mean the physiological development, or physiogenesis, which goes on during the diapause stage in preparation for the active resumption of morphogenesis. The resumption of active morphological growth and the differentiation may be taken to indicate the completion of the diapause stage. Diapause sets in at an early stage in the morphogenesis of the embryo; it is intense and prolonged, and eggs are unusually uniform in the intensity of the diapause that is present. The influence of temperature is intricately related not only to the stage reached by the embryo but also to the history of the egg.

Diapause in the post embryonic stages usually (though not invariably) responds to low temperature also. The eggs of the lymantriid, *Euproctis* hatch during July, and the larvae reach the second or third instar by autumn. At this stage they spin a characteristic ``nest", in which they overwinter. Diapause development proceeded during winter, and by the end of February the larvae were competent to resume development if they were placed at 25°C in an atmosphere saturated with water vapour (Sanderson, 1908, Grison 1947)

Prolonged exposure to a constant low temperature promoted diapause development in the characteristic prepupal diapause in the saw flies *Cephus* (Salt, 1947) and *Gilpinia* (Prebble, 1941) and in the pupae of other insects characterized by a diapause in this stage, for e.g., the fly, *Sarcophaga* (Roubaud, 1922) and the moth *Platysamia* (Williams, 1946). The adults of *Drosophila nitens* are rather unusual. When reared in captivity, they entered a seasonal diapause during the winter, even when the temperature and illumination were kept constant. After an exposure of 10 days to 0°C, the flies were competent to resume normal reproductive activities (Bertani, 1947). It is now known that the particular range of low temperatures which favours diapause development varies with different species.

Stimuli which determine the inception of diapause

Diapause may function as an adaptation which preserves the species in a resistant stage during the unfavourable season and at the same time keeps the life cycle in step with the rhythm of the season. But a number of cases in which diapause may be induced by exposure, during an early stage of the life cycle, to a certain photoperiod have been thoroughly studied, in these cases diapause would seem to be the cause of slow growth rather than vice-versa (Andrewartha, 1952).

The influence of temperature on the inception of diapause

With the bug *Reduvius personatus* diapause may occur in any of the last three nymphal instars and usually appears in the autumn, when the temperature is falling. The larva of the beetle, *Popillia* is influenced by temperature in a similar way. When larvae were reared at 20°C, most of them experienced a prolonged diapause in the second instar; the few which escaped it in the second instar had a prolonged resting period in the third instar.

The influence of food on the inception of diapause

The food of most phytophagous species varies in quality and moisture content with the season and similar rhythmical seasonal changes may occur in food of carnivorous species, particularly if their prey exhibits a seasonal diapause. For this reason it is often difficult to decide whether an observed relationship between diapause and food is independent of temperature, light, and other components of the environment which also display a concomitant seasonal rhythm. There are, however, a number of experiments available in which these other influences have been adequately excluded and from which it seems clear that the quality of the food may be the chief stimulus determining diapause.

The changing composition of the food particularly changes in the relative proportions of fat and water, seems to be the chief cause of diapause in the larvae of *Platyedra* and *Diatraea*, which feed on annual plants. Both species are multi-voltine, with diapause likely to occur at any season of the year and apparently independently of the weather (Squire, 1937, Kevan, 1944).

The adaptive value of diapause

The firm obligate dipause which occurs in the egg stage of the grasshopper, *Austroicetes* confers upon the species three ``advantages" which are nicely related to the climate of the area where it lives; (a) it ensures that the active stages shall be present only in that season of the year which is most favourable for their development and survival (b) it ensures that the grasshopper shall spend the hot, dry summer as a diapausing egg, (c) the egg in diapause because of its remarkable resistance to water loss and its low level of respiration is the only stage in the life cycle which could resist desiccation during exposure to extremes of heat and dryness experienced during summer in this area. With species from the Northern Hemisphere the diapause stage may be especially resistant to cold. The principle is the same, and these three ``advantages" are characteristic of diapause as an adaptation in other species.

The egg of *Austroicetes*, once the cuticle and its associated wax has been laid down, becomes highly resistant to water loss and remains so while the egg continues in diapause. While the egg remain in diapause, the yolk is granular and opaque throughout. As diapause development proceeds, the posterior third of the egg, becomes filled with a non granular translucent fluid, and at the same time the egg's capacity to retain its moisture decreases (Birch and Andrewartha, 1942).

During the summer and autumn no fluid is apparent, for the small amount of liquefaction which may result from chilling during a cool night is rapidly taken up by a correspondingly small amount of growth on the part of the embryo, and the egg remains firmly in diapause. The chief hazard at this season is drought; the responses of the early stage egg are such as to ensure that it will remain firmly in diapause during summer and so remain highly resistant to desiccation.

47

KINDS OF INSECT BEHAVIOUR

Broadly speaking the manner in which an organism responds to various exogenous and endogenous stimuli is termed as behaviour. Insects mainly exhibit two kinds of behaviour viz. (1) Innate behaviour which include reflexes, kineses and taxes and (2) Learned behaviour i.e. experienced behaviour.

The innate behaviour largely consists of more or less fixed responses or series of responses to a given stimulus or the pattern of stimuli. It may be added here that the innate behaviour may be modified by experience or learning. The innate behaviour is hereditary whereas the learned behaviour is not inherited but it is acquired through interaction with the environment during the life of the individual. In several instances, there is no line of demarcation whether an observed behaviour is innate or learned. In the following pages a brief account of both the types of behaviour innate and learned is given.

INNATE BEHAVIOUR

The simplest innate behaviours found in insects are reflexes. This may involve only a part of the body as in the case of extension of proboscis or the whole body as in the case of direct reflex which occurs when an insect is placed on its dorsal side. The reflexes can be classified into two groups : 1. phasic and 2. tonic. The *phasic* reflexes are much rapid and short lived and are involved in rapid movements e.g. the extension of proboscis in flies. On the other hand the *tonic* reflexes are slow and long lived and are involved with the maintenance of posture, body turgor, muscle tone and equilibrium. The reflexes also vary in complexity, the simplest being mediated by a single affrent impulse from a receptor to an interneuron and then along an efferent neuron to an effector i.e. the reflex arc. However, most of the reflexes are very complex involving many neural connections. Reflex autonomy has been shown by ganglia of the ventral nerve cord e.g. in the silkmoth, *Bombyx mori* the oviposition, reflex which results from the ovipositor making contact with a surface, resides totally on the last abdominal ganglion. More complex innate behaviour includes various types of orientation patterns.

Orientation may be defined as ``the capacity and activity of controlling locomotion and attitude in space and time with the help of exogenous (external) and endogenous (internal) stimuli (Jander, 1963). Orientation in insects has been variously classified by different workers. For the sake of convenience, here it has been broadly classified into three types (1) kineses, (2) taxes and (3) transverse orientations which are based on the mechanisms of orientation. Kineses and taxes can be defined with respect to the type of stimulus.

Kineses : The kinesis is an unorientated or undirected action of an organism which varies according to the intensity of stimulus e.g. in *Glossina* species, there is an increase in activity in dry atmosphere relative to activity in a moist one. This behaviour causes the insect to find and remain in a humid atmosphere. The body louse, *Pediculus* makes fewer and fewer directional changes as the temperature, humidity and odour increase as the louse moves closer to a potential host.

In addition to providing specific environmental input, impulses from sensory neurons also increase overall excitability of the central nervous system turning on the insect as it were e.g. increasing temperature above the preferred range for an insect, generates increased locomotor activity resulting ultimately in the insect's moving into a cooler area. This kinesis may appear to be directed movement, the insect seems to avoid the heat by remaining in motion and turning until in a cooler place (Wigglesworth, 1972).

Orthokinesis : It is defined as a response to a stimulus in which the rate of response such as speed of movement is positively proportional to the intensity of the stimulus.

Klinokinesis : Movement resembling a random walk in which changes in direction are made when unfavourable stimuli are encountered with frequency of turning dependent upon the stimulus intensity.

Akinesis : It is a state of immobility due to lack of any stimulus. An overall lack of movement is controlled by sensory input normally by tarsal contact with a surface. Tarsal contact combined with other stimuli e.g. for dragon fly, lowered temperatures and reduced intensity of light in the evening inhibits central nervous system function in most insects (Wigglesworth, 1972).

Taxes : An oriented or directed movement of an organism towards or away from the stimulus is known as taxis. The taxes are different from the kineses in the sense that they are directed responses relative to a stimulus source. Taxes may be responses to physical or biological factors or a combination of both. The insect may move towards (positive) or away (negative) from the source of stimulus. Thus, moth's attraction to light is positive phototaxis while maggots are negatively phototactic.

The taxes may be classified according to the environmental stimuli involved e.g. phototaxis (light), geotaxis (gravity), anemotaxis (air current), rheotaxis (water current), chemotaxis (chemicals), hygrotaxis (humidity), scototaxis (darkness), thigmotaxis (contact) for details see : Wigglesworth (1965). The other taxes are klinotaxis, telotaxis, and tropotaxis.

Klinotaxis : In klinotaxic behaviour an insect moves relative to a gradient of stimulus intensity such as light source or a sound emission. Thus, a movement in a definite direction relative to a stimulus either directly towards or away from the source is known as klinotaxis.

Menotaxis : Movement that maintains a constant orientation relative to a light source e.g., the orientation with respect to a constant angle of light source is termed menotaxis.

Telotaxis : The visual fixation of an object such as prey is termed telotaxis i.e. orientation and movement directly towards a visual stimulus.

Tropotaxis : Movement whose direction is dependent on the composition of stimulation on bilateral sense organs. Attainment of orientation is directed and no deviations are required. The tropotaxis is more efficient than klinotaxis and shows the selective advantage of having paired sensory structures viz. eyes, cerci, antennae. If an antenna is removed or an eye is painted over, the insect will not orient properly. If antennae are crossed and then glued in their new position, an insect will turn towards the wrong side when orienting to an odour source.

It must be remembered, when attempting to analyze the behaviour of an insect that the mechanism involved may change from moment to moment and even at the same instant more than one mechanism may be involved.

Transverse Orientation

The transverse orientations are those in which the body is oriented at a fixed angle relative to the direction of the stimulus. The transverse orientations are of two types : (1) the light-compass reaction and (2) the dorsal light reaction (Fraenkel and Gunn, 1961). The light compass reaction consists of an orientation such that locomotion occurs at a fixed angle relative to light rays. This orientation was demonstrated in ants, caterpillars, bees, certain beetles and bugs. In the dorsal or ventral light reaction, both the long and transverse axes of the body are kept perpendicular to a directed source of light (Fraenkel and Gunn, 1961). This type of reaction is different from phototaxis because in this case, orientation is transverse to the light rays and not parallel to them e.g. both dorsal and ventral light responses are found in certain aquatic bugs. When the air bubbles trapped between the head and the antennae are removed, the back swimmer *Notonecta,* depends entirely upon a ventral light reaction for the maintenance of its normal swimming position, while the converse is true for the water boatman, *Corixa* and the creeping water bug, *Naucoris* (Rabe, 1953).

The most complex innate behaviour patterns are the fixed action patterns. These are characterized by being unlearned, species specific and adaptive. They differ from reflexes and orientations in that they require an internal readiness before they can occur and the evoking stimulus i.e. releaser is commonly not required to act during the entire course of the pattern. The releasing stimulus may be internal in the form of hormone secretion or an excitatory signal from a higher neural centre. The stimulus may be in the form of light, temperature, mechanical contact, odour etc. A variety of factors may affect the development of a state of internal readiness or motivation necessary for a given behaviour pattern to occur. These may be either internal or external in origin (Romoser, 1981).

The internal motivational factors are hormonal or neural in nature e.g. the relative concentration of juvenile hormone and the moulting hormone determine the spinning behaviour of the caterpillar of several moths and other insects. Internal receptors are commonly involved when a behaviour pattern appears spontaneously without any discernible external stimulus. An excellent example of this has been found in the blow fly, *Phormia,* in which receptors associated with the crop carry stimuli that inhibit food ingestion as long as the crop is full. No inhibitory stimuli arise from these receptors when the foregut is empty and feeding can occur. In other cases of spontaneous behviour, regular cycles of behavioural activity of internal origin determine when a particular pattern may be released e.g. crickets are active only during a certain period each day and this daily rhythm continues for several days when the insects are kept in constant darkness in the laboratory.

The external factors viz. light, temperature, substances produced by or activities of other insects of the same species and others have definite effects on motivation, in addition to serving as releasers (stimulus). Light and temperature commonly determine whether an insect is active or inactive. Every insect species has a temperature below which or above which it becomes inactive and within the range of these temperatures; specific temperature may be essential for particular activities e.g. the readiness of certain butterflies, *Pieris* etc. to copulate is determined by temperature. Likewise, the entire act of mating behaviour in the milkweed bug *Lygaeus militaris* depends on temperature. Mating takes place only at a temperature of 38°-46°C or and above as observed in this bug in the Indian desert (Awasthi and Tiwari, 1975).

It is well known that a substance secreted by one insect influences the behaviour of another insect of the same species. The term *Pheromone* is commonly used to denote such a substance. Pheromone may serve as releasers or as motivational factors or both. One of the most interesting example of a pheromone affecting behaviour is found in the honey bee. The queens produce a substance known as queen substance (secretion of the mandibular glands) which is passed from worker to worker orally and which inhibits ovary development as well as the behaviour involved with queen cell construction. Queen cells, in which new queens develop, are constructed only when a hive has lost its queen and hence the effets of the substance of her mandibular glands. Jacobson (1965, 1972, 1974) has described in considerable detail the various aspects of insect pheromones.

LEARNED BEHAVIOUR

According to Thorpe (1963) learning is that process which manifests itself by adaptive changes in individual behaviour as a result of experience. Thus learning involves the accumulation and storage (memory) of environmental information and the subsequent effects of this stored information on the behaviour of an animal, i.e. insect. Alloway (1972) defined learning as a modifiable relatively permanent change in behaviour as a result of practice. The simplest form of learning i.e. habituation is common among insects. Markl (1974) has classified the learning behaviour into (i) habituation, and (ii) latent learning.

Habituation : Different authors have defined this term in their own ways e.g. according to Dwesbury (1978) habituation may be defined as the decrease in probability or amplitude of a response occurring when a stimulus which elicits that response is presented repeatedly.

According to Gullan and Cranston (1994) habituation means reduction in the response to a stimulus with repeated exposure through modification of the central nervous systems.

Habituation (getting used to) has occurred when a stimulus that initially evoked an escape or avoidance response no longer evokes such a response. This type of learning is common in insects and has been demonstrated relative to substrate vibrations' noise, visual stimulation, chemical repellents etc. The ability of insects to learn in this fashion avoids unnecessary, energy consuming behavioural responses (Romoser, 1981).

Latent learning : It occurs without apparent reward or punishment. Social insects learn characteristics of the immediate vicinity of their nest so that they are able to find their way back from foraging flights. Olfactory conditioning may be a form of habituation or latent learning. An example of this phenomenon is a situation in which exposure to a repellent odour during a certain period of development results in the loss of repellency of this odour to adults e.g. the adults of *Drosophila melanogaster* which are normally repelled by the odour of peppermint oil. However, if the larvae have been reared in the presence of peppermint they are strongly attracted to it as adults.

Instincts : The inborn patterns of movement or automatisms, brought into operation by some appropriate conditions are termed as instincts. Instincts are normally manifested in response to a specific releaser stimulus but in the absence of such releaser, the threshold of stimulation falls and the whole elaborate response may be set off by some quite abnormal circumstances. These inborn patterns of behaviour are genetically determined or inherited, they are built into the central nervous system in the same way as morphological patterns are built into the growing body (see : Wigglesworth, 1965). Darwin (1859) has described slave making instincts and cell making instincts in detail.

Periodicity of Behaviour

Many of the cyclic activities of insects like locomotion, feeding, mating, oviposition, eclosion and others occur at regular intervals. The timing of these recurring activities is adaptive i.e. insects under natural conditions behave in ways that maximize the chances of their survival. Thus phytophagous insects hatch from eggs in synchrony with the availability of host plants, insects become dormant and cold hardy or migrate at a time that favours survival of an upcoming cold or dry season. The term periodicity is applied to such activity cycles of behaviour. The patterns of cyclic activities may be every 24 hours, monthly, or annually in coordination with lunar cycle. Periodicity in insect behaviour involves internal as well as external components. Internal or endogenous components are related to the heredity while external or exogenous components arise from the external environment.

Circadian Rhythms : These are processes that are controlled by biological clock. Most rhythms that have been identified have a period of approximately 24 hrs and are referred to as circadian (circa - about or approximately, diem- a day or daily) rhythms. The term circadian rhythm was coined by Franz Halberg (Halberg *et al.*, 1959). The diurnal species as most bees, wasps, dragon flies and butterflies are those whose activities are limited mainly to day light hours while nocturnal insects i.e. most moths and the cockroaches, crickets are active during darkness.

The crepuscular insects are most active around dusk and dawn e.g. *Drosophila* emerge from the puparium at dawn and likewise several orthopterans sing at a particular time of day or night. Circadian activity rhythms though internally controlled are subject to external modification particularly by temperature (see : Truman, 1973).

Numerous non-circadian rhythms are also known e.g. the lunar emergence of some chironomid flies (Diptera), mayflies (Ephemeroptera) and the annual emergence of some dermestid beetles (Corbet, 1966). A well known example of non circadian rhythmicity is found in the species of periodical circadas viz. *Magicicada* spp. (Homoptera), the various broods of which emerge *en-masse* at 13 or 17 year intervals.

Although rhythms are not regulated by external factors (exogenous), but some external factors are known to influence rhythms. Among the external factors that may act time cues are length of the light or dark phase of a photoperiod (i.e. a cycle consisting of a period of illumination followed by a period of relative darkness), light intensity, temperature, time at which food is available etc. (Beck, 1968).

Many exogenous environmental factors display 24 hrs periodicity which result in corresponding periodicity in insects without the involvement of biological clocks. Among these are - intensity of light, humidity, temperature, wind velocity and numerous biotic factors.

An example of how one of these factors may cause periodic behaviour is in the swarming of many species of mosquitoes, which is determind by the intensity of light. These insects continually possess the drive to swarm but do so only under appropriate conditions of light intensity and can be induced to swarm for hours if the light intensity is held artificially at the appropriate level. Under natural conditions the appropriate light intensity may be periodic e.g. that which occurs during crepuscular periods. In this situation, the mosquitoes would swarm periodically.

CONTROL OF INSECT BEHAVIOUR

Romoser (1981) remarked that an insect is continually attacked with several stimuli in space and time. Many stimuli provide information significant to the life of the insect but many do not. In order to survive, an insect must respond to the right stimuli and do the right thing at the right time escape, locate a food source, feed, locate a mate, copulate, oviposit, migrate, become dormant and so on, i.e. behaviour of an insect must be controlled. The principal factors that control the behaviour of insects are nervous and endocrine systems.

Nervous Control of Insect Behaviour

It is known from behavioural and neurophysiological studies that patterns of behaviour may occur spontaneously without any external sensory triggering. Thus, the nervous system must be capable of spontaneously generating organized patterns of behaviour from within e.g. a decapitated mantid will continuously turn and perform copulatory movements. The central nervous system in a locust with all sensory input from the head and wings removed still produces motor output nearly identical to the motor output recorded during normal flight (Wilson, 1972). These activities are considered to arise from the spontaneous, preprogrammed activity of neural pattern generators in the central nervus system.

The brain and suboesophageal ganglion both have been found to exert excitatory and inhibitory influences on internally (endogenously) controlled motor patterns. These patterns seem to reside largely in the thoracic ganglia in the flight and in this way the decerebrated mantid sends out impulses that inhibit the internal activity of the ganglia of the ventral nerve cord relative to copulation. In the case of locust, there are wind sensitive sensilla on the head which help sustain flight and also tarsal receptors which stimulate flight when contact with the substrate is lost. In addition, there are stretch receptors which aid in the regulation of flight on wing hinges and other parts of the wings and mesothorax as well as metathorax. Thus, although locust flight pattern is centrally generated, the basic pattern is modified by peripheral sensory input. Although a little is known about the neurophysiological basis of learning but removal of parts of the brain has shown that some long term memory must reside there.

Endocrine Control of Insect Behaviour

Hormones play significant roles in insect behaviour. Through acting directly on the nervous system they can cause major overall changes in patterns of behaviour. Endocrine mechanisms typically exert their influence over the longer term, coordinating behaviour with less immediate environmental demands e.g. changing seasons, changes in the availability of food etc. The determination of hormonal influences on behaviour have usually been based on one or more of the following criteria :

1. Correlation of hormone secretion and a particular behaviour pattern. 2. Application of hormone or implantation of active endocrine glands inducing a behavioural pattern sooner than it would normally occur. 3. Correlation between removal of any active endocrine gland and disappearance of a specific behaviour. 4. Restoration of the specific behaviour by hormone application or implantation of active endocrine glands. It is the last criterion that provides the strongest

evidence for hormonal control of a given behaviour. According to Riddiford and Truman (1974) two categories of hormonal effects on behaviour i.e. as modifiers or as releasers of behaviour can be recognized.

Hormones as Modifiers of Behaviour

When a hormone acts as a modifier, it changes the responsiveness of an organism. The example of a hormone acting as modifier is the influence of moulting hormone on the biting behaviour of female *Anopheles freeborni* (Beach, 1979). This mosquito does not take blood meals once overian development has been initiated. The moulting hormone, which in this case, is secreted by the ovaries inhibits biting. This was concluded on the basis that biting inhibition does not occur after removal of the ovaries and is restored by either reimplantation of ovaries or injecting moulting hormone (Romoser, 1981).

Hormones as Releasers of Behaviour

When a hormone acts as a releaser of behaviour, a specific behaviour pattern occurs within a few minutes of secretion e.g. phallic nerve stimulating hormone in the *Periplaneta americana* (Milburn *et al.*, 1960; Milburn and Roeder, 1962). This hormone can be extracted from the corpora cardiaca and when injected into male cockroaches causes the abdominal movements characteristic of copulation. The eclosion hormone secreted by the brain (i.e. neurosecretion) and corpora cardiaca in the moth *Antheraea pernyi* acts both as a releaser in triggering pre-eclosion behaviour as well as modifier by turning on adult behaviour (Truman, 1973). For more information on various aspects of insect behaviour see Awasthi (2001).

48

LANGUAGE OF BEES AND THEIR BEHAVIOUR

"The language of bees is truly perfect and their method of indicating the direction of food sources is one of the most remarkable mysteries of their complex social organization." **K. Von Frisch.**

Every one is very well aware of the dancing and humming of bees on flowers. Their different methods of dancing have different meanings and are not a useless part of their daily life. It is by means of these methods that the bees talk and by which they tell other bees about the distance and direction of the food. Besides the methods of dancing, there are other ways also by which they tell their friends about flowers full of scented nectar and ask them to go there for collecting food. Inspite of being small creatures they show a remarkable intelligence and this is one of the wonders of nature.

How the bees talk and how they tell their mates about the presence of food at a new location is quite interesting. When a bee, during its search for food, comes near an interesting flower, it dances around it for a while, collects some scented nectar from it and then returns back to its honey comb, where it dances again, telling others about the flower. During the dance, the bees nearest to the dancer become very much excited and soon join the dance, keeping their antennae close to their bodies. After a while, the dancer turns away and leaves the hive; others follow and soon a small procession reaches the feeding ground; the process being repeated many times. Thus it is clear that it is by this dancing, that one bee conveys the information about explorable feeding grounds to other bees.

The question now arises as to how the dance tells about the distance and direction of food?

Round dance

When a bee dances in simple circles. (Fig. 133A) it means, according to Frisch, "Fly out and seek in the neighbourhood of the hive". This "round dance" indicates food in near vicinity of the hive. The surface of a bee's body has the ability to hold scents for long periods and the dancer also regurgitates droplets of nectar it collected. Other bees perceive the scent on the dancer's body surface and that present in the regurgitated nectar. They start in the direction indicated by the dancer and soon search out the flower, with whose scent the two scents perceived by them tally.

It is noticeable that these dances are made only when there is a rich source of food. According to Frisch these dances also carry the basic message - "There is plenty of food and sweetness". If there is no food, then the dancer bee, after returning to the hive, does not dance.

When there are a large number of flowers rich with nectar, the dancer collects nectar from the richest flower, returns to its hive and dances vigorously; sending a large number of bees to that particular flower. When, after repeated attacks, the nectar supply of flower gets exhausted, the returning bees no longer dance on reaching their hives; with the result that no more worker bees go out to the already exhausted food source. In case nectar of the flower concerned is not much scented, bees make use of the scent gland found on their abdomens. It secretes a scented fluid, which the bees pour on the nectar, and other bees locate the flower by this particular scent, they are so familiar with.

Wagging dance

When food is at a very distant place, the bee, which comes from that feeding place, dances in a quite different manner. This particular type of dance, Frisch termed as "wagging dance" (Fig. 133B). The dancer runs a short distance in a straight line, wagging its abdomen very rapidly from side to side; turns to the left; and makes a complete angle of 180°. This process is repeated many times.

Thus, as observed by Frisch, there are two types of dances; the "round dance" indicating presence of food in near vicinity of the hive and the "wagging dance" which means that the food is at some distance sometimes many kilometres away from the beehive. As a matter of fact, the wagging dance not only announces a rich source of food at a far off distance but also tells how far away and in what direction. The actual distance and direction is indicated by the number of turns made during the dance.

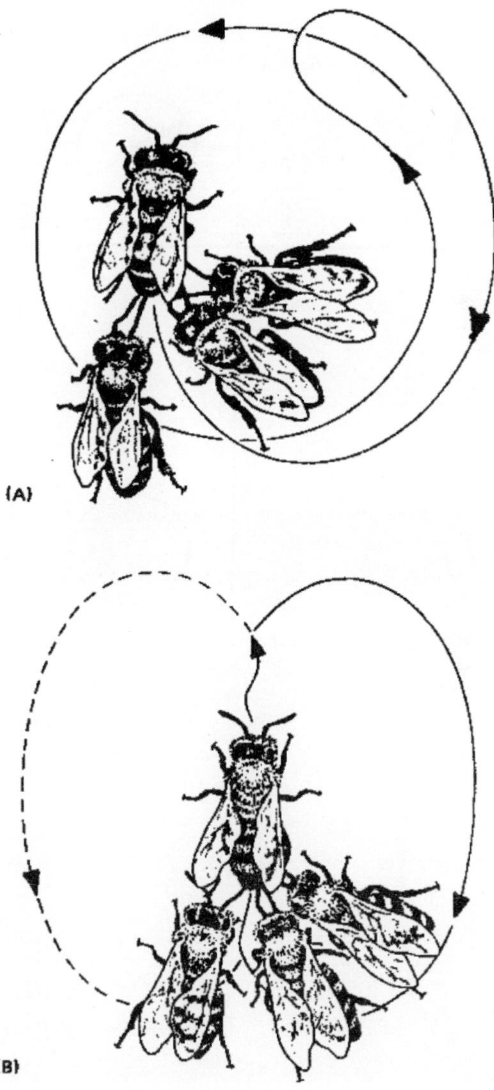

Fig. 133. Types of dances by honeybees used in communicating nectar sources. (A) round dance indicating a source near the hive, (B) wagging dance indicating direction (angle of the straight run to gravity is same as angle of nectar to sun) and distance is farther away than in the round dance (Elzinga, 1978).

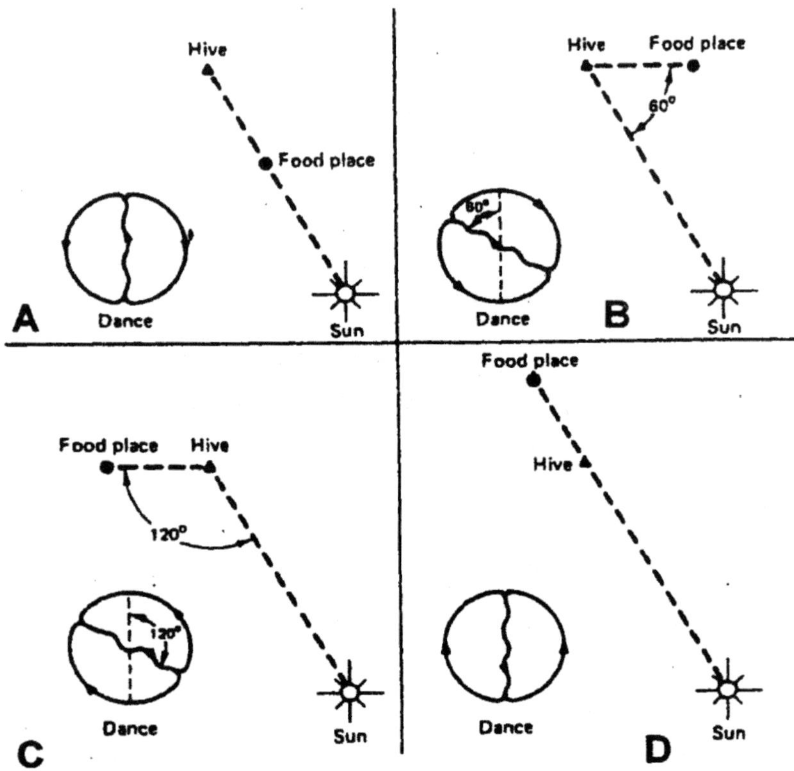

Fig. 134. The dance language of the honey bee. Four examples of indication of direction. The sun is to the south-east of the hive. The diagram in each lower left corner shows schematically the direction of the waggle dance for a given position of the feeding place.

Briefly speaking, when a rich source of food is found, its location is signalled to other forager bees and the returning dancer bee performs a ritualized dance on the vertical comb of the hive. The dance attended by other forager bees who are able to locate the food source on the basis of information provided by dancer. If the food source is less than 100 metres from hive the round dance is performed and other bees leave the hive and locate the food. If the food source is more than 100 metres from the hive, the dancer bee performs a wagging or waggle dance (Fig. 134 A-D) that contains information about the distance and direction of food, both. The distance is indicated by the duration of the straight run of the dance. Direction is indicated by transforming the direction to the food source relative to the location of the sun into an angle of the straight run of the dance relative to gravity (Fig. 134 C). The other forager bees leave the hive and locate the food source and collect the nectar.

Thus, these small creatures talk to each other in their own language. The capability of talking with each other, though in a much different language of their own is a great asset to those insects which have chosen such a high grade of social organization. That tiny animals like bees should have such remarkable capabilities, indeed makes one marvel at the wonders of nature (For details see : Frisch, 1950, 1966).

APPENDIX - I

PRINCIPAL ORDERS OF INSECTS

Class Insecta. Insects, body divided into three parts : head, thorax and abdomen, one pair of antennae, mouth parts adapted for chewing, sucking, rasping, lapping, sponging, piercing and sucking types. Thorax with 3 pairs of jointed legs, 1 pair in each pro-meso- and metathorax. Two pairs of wings variously modified, reduced or absent. Respiration through highly branched tracheal system, Excretion by Malpighian tubules, Nervous system composed of brain, suboesophageal ganglion and ganglionated ventral nerve cord. Eyes both simple and compound. Sexes separate. Usually oviparous. Development direct through nymphal instars (Hemimetabola) or indirect through larva, pupa and adult (Holometabola). Found in land and fresh water.

Sub class 1 Apterygota – Wingless insects metamorphosis absent, abdomen provided with styli and cerci.

Order 1 - Protura : Extremely small very primitive, mouth parts entognathous adapted for chewing, antennae absent, eyes and wings absent. Three pairs of 5-jointed legs. Papillate Malpighian tubules. Abdomen of 11 segments. Trachea present or absent. Metamorphosis absent, found in damp places.

Order 1. Protura

Order 2 - Collembola : Commonly known as Spring tails. Small, grey or white, antennae 4 segmented, mouth parts adapted for biting and chewing, and entognathous. Wings absent. Compound eyes, Malpighian tubules and trachea absent. Abdomen 6 segmented, springing organ (furcula) present in the 4th abdominal segment, metamorphosis absent, found in damp places.

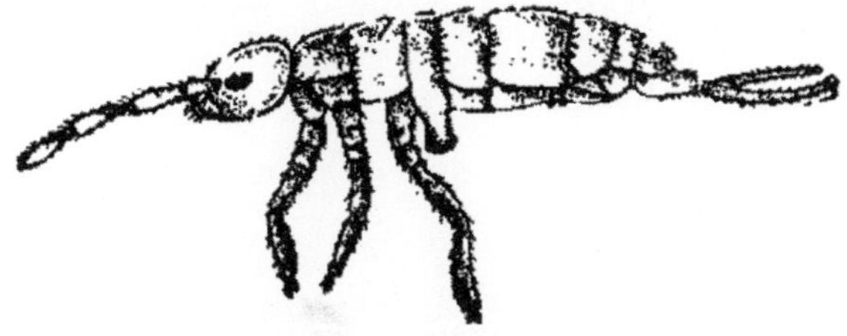

Order 2. Collembola

Order 3 : Diplura : Small in size, antennae long, eyes absent mouth parts within the head hence entognathous. Wings absent, abdomen 11 segmented cerci 2, Malpighian tubules absent, metamorphosis absent found in damp places.

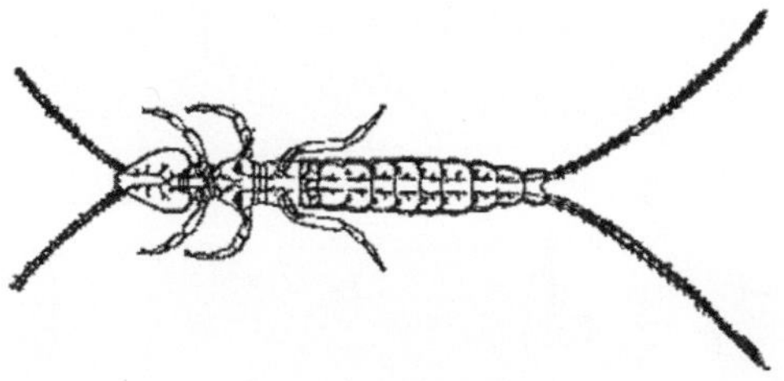

Order 3. Diplura

Order 4 : Thysanura : Bristle tails, Silver fishes. Small in size with long antennae, mouth parts biting and chewing type and ectognathous, wings absent, body covered with silvery scales, tarsi 2-4 segmented. Abdomen 11 segmented with 3 cerci at the end. Run swiftly eat starch in books, papers, clothes etc. Metamorphosis absent.

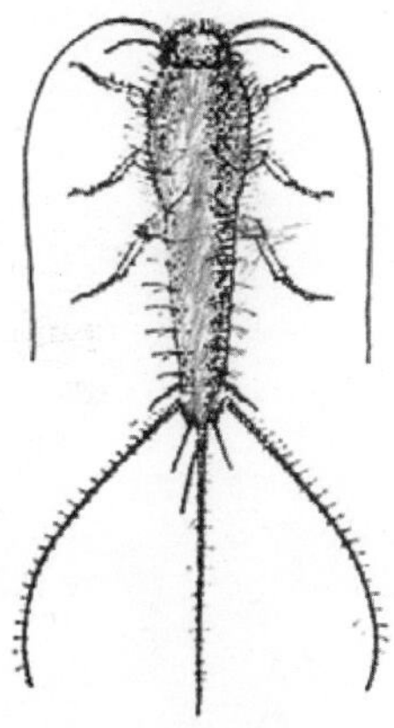

Order 4. Thysanura

Sub class 2 : Pterygota : Winged insects : Wings sometimes reduced or absent, abdominal appendages absent but cerci and genitalia present.

Division 1 Hemimetabola (Exopterygota) Metamorphosis gradual, young stages are nymphs.

Order 5 : Orthoptera : Grasshoppers, locusts and crickets. Size medium to large, mouth parts chewing type i.e. mandibulate, forewings (tegmina) narrow, hind wings membranous, broad, many veined. Hind legs modified for jumping. Abdomen with cerci and ovipositor. Mostly phytophagous (plant feeders).

Order 5. Orthoptera

Order 6 : Dictyoptera : Cockroaches and Mantids. In case of cockroaches, body, depressed, head partly under prothorax, antennae long, filiform, dark-brown or black in colour, foul smelling, nocturnal, omnivorous, spoil food and several domestic materials. In case of praying mantis, head free, prothorax long, forelegs enlarged to grasp insects as food. This order contains some household pests.

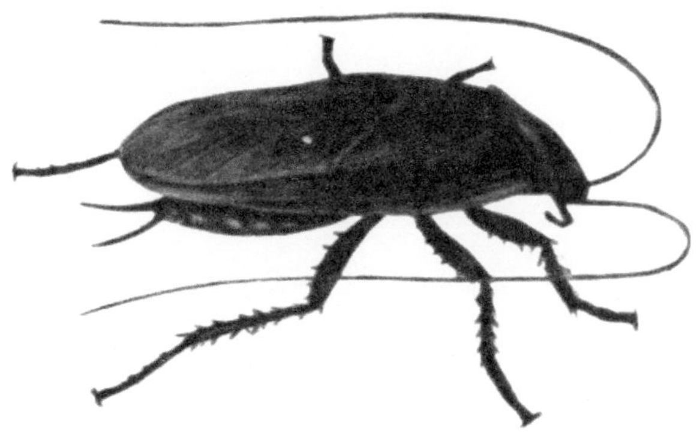

Order 6. Dictyoptera

Order 7 : Phasmida : Leaf insects and stick insects. Stick or leaf like, brown or grey or green in colour, mimicking twigs or leaves on which insects rest. Wings small or absent, mesothorax and metathorax and legs long, ovipositor small in male, cerci short and unsegmented, leaf eaters, metamorphosis slight.

Order 7. Phasmida

Order 8 : Dermaptera: Earwigs. Body elongate, mouth parts biting and chewing type, forewings short and leathery, tegmina without veins, hind wings large, semicircular fan like membranous, fold beneath the forewings at rest. Some wingless tarsi-3 segmented, last abdominal segment with 2 stout forceps, metamorphosis gradual or absent, feed on plant and small insects.

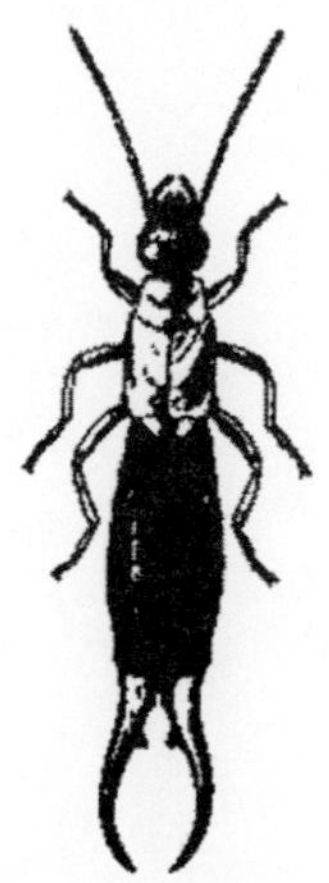

Order 8. Dermaptera

Order 9 : Plecoptera: Stone flies. Size medium, body soft, mouth parts adapted for chewing but often absent in adults, antennae long with setae, wings 4 membranous held flat on the back at rest, hind pair of wing larger, tarsi-3 segmented end of the abdomen with 2 long many jointed cerci, nymph aquatic, often with tracheal gills behind each leg, live under stone, eat small insects.

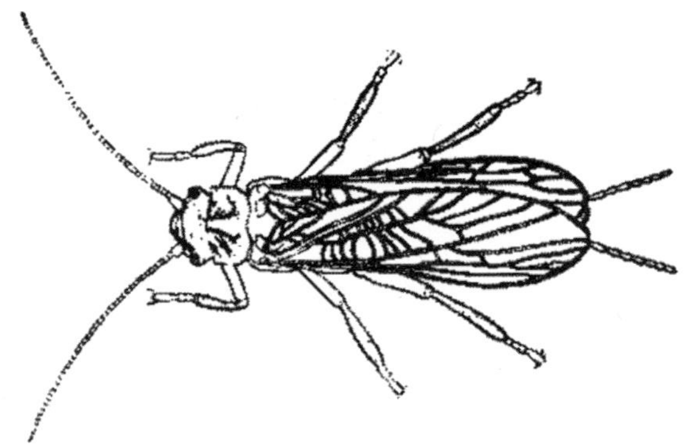

Order 9. Plecoptera

Order 10 : Isoptera: Termites. Social and polymorphic. Body soft thorax joins broadly to abdomen, mouth parts chewing type, workers and soldiers wingless, sterile blind, colourless, hence commonly known as white ants. Sexual males and females pigmented. With two pairs of similar membranous wings, which detach after nuptial flight, metamorphosis gradual or absent: Composed of three castes sexuals (king and queen), soldiers having head and mandibles large, workers excavate tunnels and build nests, collect food. Feed on cellulose. Damage all sorts of wooden materials.

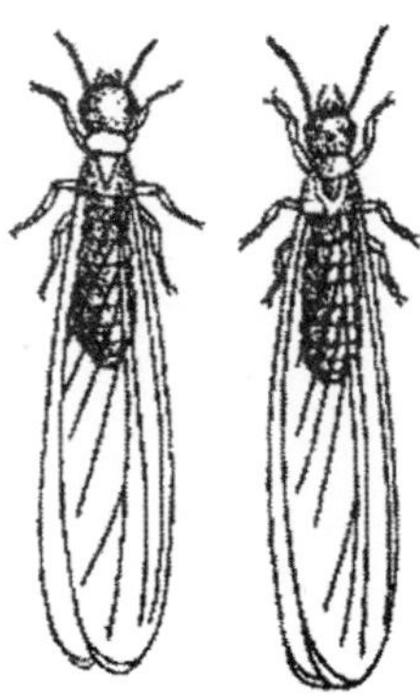

Order 10. Isoptera

Order 11: Embioptera: Web spinners. Small, body long and soft, grey in colour, having chewing types of mouth parts, forelegs with first segment of tarsi inflated containing spinning organs. Cerci- 2 segmented, males winged, females wingless, construct silken tunnels.

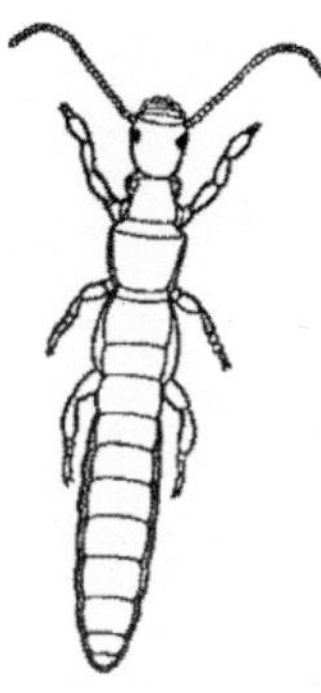

Order 11. Embioptera

Order 12 : Odonata: Dragon and damsel flies. Body brightly coloured red, yellow green etc. Mouth parts chewing type, big eyes, very prominent with 30,000 ommatidia, 2 pairs of transparent membranous wings of equal size with complex cross veins, a strong cross vein with a distinct stigma and notch, abdomen slender, elongate; strong flying, predaceous nymphs (naids) aquatic with reversible proboscis with hook like jaws, respiration by means of rectal or anal gills, metamorphosis gradual.

Order 12. Odonata

Order 13. Ephemeroptera

Order 13 : Ephemeroptera: May flies. Body soft, delicate, mouth parts adapted for chewing but vestigial, antennae short, setaceous wings 4, membranous, fore wings much larger held vertically above the body at rest, terminal end of the abdomen with 2 to 3 long cerci, nymphs aquatic with lateral tracheal gills, phytophagous, the adults live for a few hours or days, hemimetabolous metamorphosis.

Order 14: Mallophaga: Biting lice. Minute, body depressed flat, wingless, head broad, mouth parts adapted for biting, antennae short 3 to 5 segmented, eyes reduced or absent, thorax short and fused, legs short, tarsi 1-2 segmented. One or two claws for clinging to host, cerci absent, metamorphosis slight, all ectoparasite on birds and mammals, feed on feathers, hairs etc.

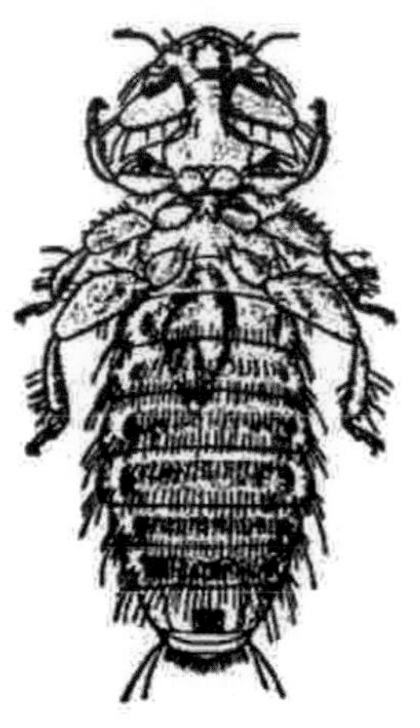

Order 14. Mallophaga

Order 15 : Anoplura : Sucking lice. Small in size, body long, depressed flat, wingless, head narrow, mouth parts modified for puncturing skin and sucking blood, eyes reduced or absent, antennae short 3-5 segmented, thorax fused, tarsi 1 segmented, each with single claw, cerci absent, all ectoparasite on mammals, feed on blood. Metamorphosis slight. Transmit several diseases.

Order 15. Anoplura

Order 16: Psocoptera: Book lice. Small in size, antennae slender, long and filiform 12-50 segmented, mouth parts chewing type, prothorax small, wings 4 or absent, tarsi 2 or 3 segmented, cerci absent, metamorphosis gradual. Feed on starch, paste, cellulose etc.

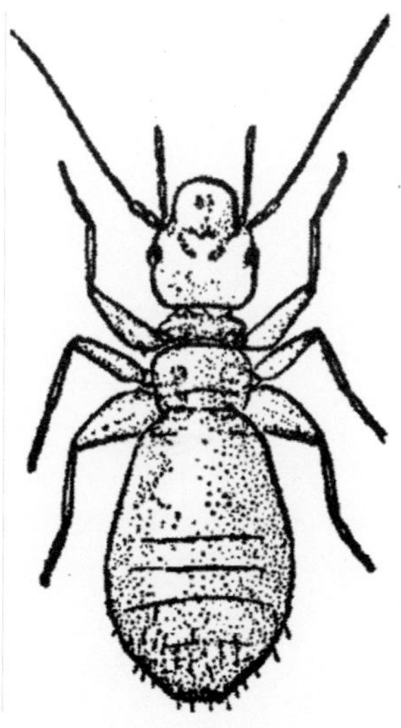

Order 16. Psocoptera

Order 17 : Heteroptera: True bugs. Medium to fairly large, mouth parts piercing and sucking type, pronotum large, triangular scutellum between wing bases, wings 4 or absent, forewings thick at bases, membranous behind cross at rest, hind wings membranous, aquatic or terrestrial, phytophagous or carnivorous.

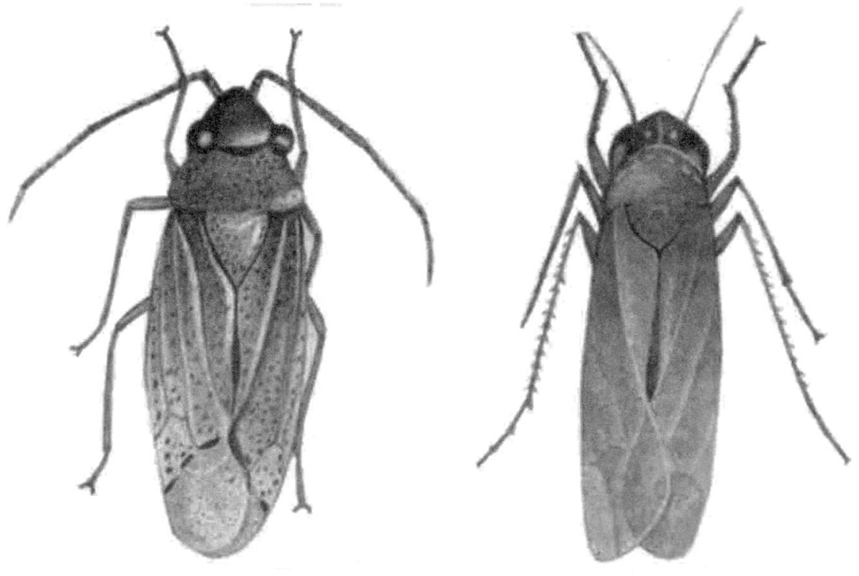

Order 17. Heteroptera Order 18. Homoptera

Order 18 : Homoptera : Aphids, cicads, scale insects. Mostly small, mouth parts piercing and sucking type bases of beak close to thorax, wings 4 membranous, feed on plant sap.

Order 19. Thysanoptera

Order 19 : Thysanoptera: Thrips. Very small, mouth parts conical, rasping sucking type, antennae 6-9 segmented wings 4, fringed by many long hairs, some without wings, tarsi 1-2 segmented mostly feed on flowers, leaves and fruits by scraping.

Division 2 : Holometabola (Endopterygota) metamorphosis complete.

Order 20 : Mecoptera: Scorpion flies. Moderate to medium sized, elongate filiform antennae, legs long and slender, mouth parts chewing type, hypognathus, wings 4 similar, membranous cerci small, some males carry terminal part of the abdomen upward like scorpion hence the name `Scorpion flies'. Larvae and adults carnivorous, larvae caterpillar-like with 3 pairs of thoracic legs and 8 abdominal prolegs, live in burrows.

Order 20. Mecoptera

Order 21 : Neuroptera : Dobsonflies, ant lions etc. Small to large antennae long, mouth parts chewing type,wings 4 similar, large membranous with many cross veins, tarsi 5 segmented, cerci absent, larvae spindle shaped with grooved sucktorial mandibles, all carnivorous, metamorphosis complete.

Order 22 : Trichoptera: Caddisflies. Small to medium sized, soft bodied, mouth parts rudimentary, setaceous antennae, and legs long, wings 4 membranous, many longitudinal and few cross veins, body and wings clothed with hairs scale like, larvae carnivorous and aquatic.

Order 21. Neuroptera

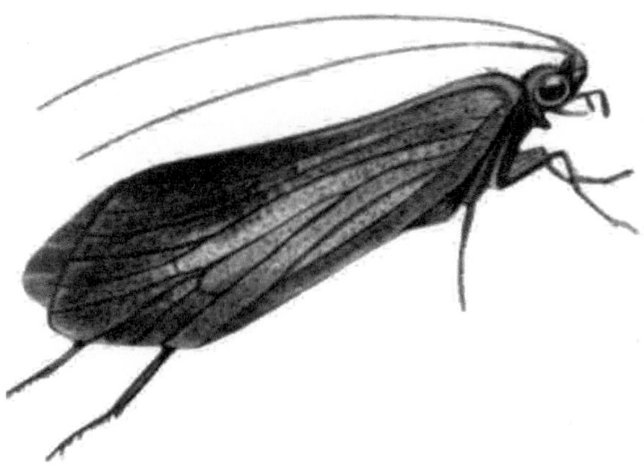

Order 22. Trichoptera

Order 23 : Lepidoptera: Butterflies and moths. Small to large in size, body, wings, legs covered with scales, mouth parts chewing type in larvae and sucktorial in adults, mandibles absent, maxillae form coiled tube or proboscis for sucking the fluids, antennae long, clubbed or pectinate, eyes large wings 4 membranous, broad with few cross veins, covered with microscopic scales, body scaly or hairy having brilliant colours, larvae worm like with 3 pairs of legs with prolegs on the abdomen, 2 silk glands on labium used to spin cocoon to contain pupa, contain many pests, some produce silk.

Order 23. Lepidoptera

Order 24 : Diptera : Two winged true flies. Forewings transparent, few veins, hind wings modified as halteres or balancers, some without wings, mouth parts piercing and sucking or sponging type, mandible absent, labium lobulated, body divisions distinct, tarsi 5 segmented, abdomen 4 to 9 segments, larva apodous (maggot), metamorphosis complete.

Order 24. Diptera

Order 25 : Siphonaptera : Fleas. Small in size, body laterally compressed, apterous, tough without wings, mouth parts piercing and sucking type, antennae short, and stout, eyes simple, or absent, legs long adapted for leaping, coxae enlarged tarsi 5 segmented, eggs laid on the host, larvae, minute, legless, feed on organic debris, pupate in cocoons, adult ectoparasite and blood sucking on birds and mammals, metamorphosis complete.

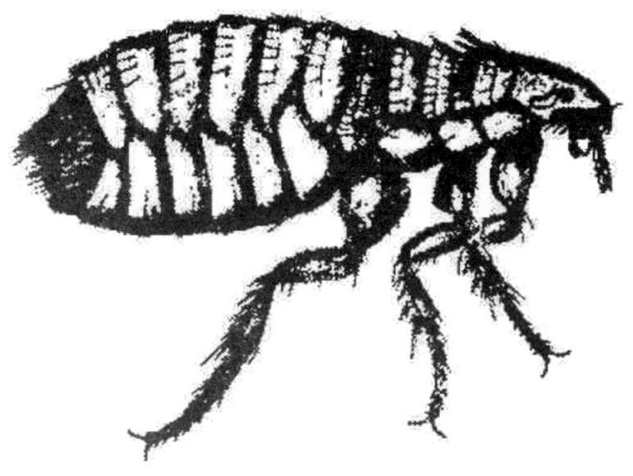

Order 25. Siphonaptera

Order 26 : Coleoptera : Beetles and weevils. Size varies from very small to large, cuticle heavy, mouth parts chewing or biting type some snout like, elytra thick, leathery, veinless; hind wings membranous, with few veins, some wingless, antennae 11 segmened, prothorax enlarged, movable, meso and metathorax united to abdomen, larvae worm like, with 3 pairs of legs, pupae rarely in cocoons, metamorphosis complete. Contain many pests of stored grains, forest-trees, vegitables etc. and some are predacious.

Order 26. Coleoptera

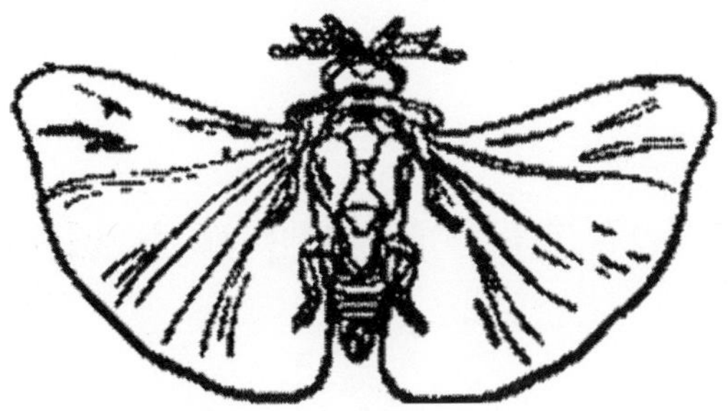

Order 27. Strepsiptera

Order 27 : Strepsiptera : Stylops. Small metathorax enlarged, mouth parts adapted for chewing or absent, males with fan shaped hind wings, forewings reduced, two clubbed halteres, live 1 or 2 days, female larva like, antennae reduced, eyes, wings, legs absent. Head and thorax fused, female and larvae endoparasitic in body of bees, wasps, Homoptera.

Order 28. Hymenoptera

Order 28 : Hymenoptera: Wasps, ants and bees. Mouth parts biting or chewing, lapping or sucking type, wings 4 membranous or absent, female with ovipositor modified for sawing, piercing or stinging, larvae caterpillar like or legless, pupae in cocoon, metamorphosis complete, mostly solitary but some social, form colonies, some parasite as larvae.

SELECTED BIBLIOGRAPHY

Agarwal, O.P. 1994. *Perspectives in Entomological Research.* Scientific Publishers (India), Jodhpur

Alford, D.V. 2007. *Pests of Fruit Crops: A Colour handbook.* Manson Publishing Ltd. London.

Alloway, T.M. 1972. Learning and Memory in Insect. *Ann. Rev. Entomol.* 17: 43-56.

Ananthakrishnan, T.N. and Raman, A. 1988. *Dynamics of Insect Plant Interaction.* Oxford and IBH Publishing Co. New Delhi.

Andrewartha, H.G. and Birch, L.C. 1954. *The Distribution and Abundance of Animals.* The University of Chicago Press, Chicago.

Atkins, M.D. 1978. *Insects in Perspective.* Macmillan Publishing Inc. New York.

Atwal, A.S. 1976. *Agricultural Pests of India and South East Asia.* Kalyani Publisher, Delhi,

Awasthi, V.B. 1968. The functional significance of nervi corporis allati-I and nervi corporis allati - II in *Gryllodes sigillatus. J. Insect Physiol.,* **14**: 301-314.

Awasthi, V.B. 1969. Effect of temperature on the neurosecretory activity in *Nezara viridula* Linn. (Heteroptera : Pentatomidae). *Experientia,* **25**: 1164-1167.

Awasthi, V.B. 1969. Studies on the neurosecretory system of the active and hibernating adults of the house cricket, *Gryllodes sigillatus* (Walk.) *Anat. Anz.,* **125** : 256-272.

Awasthi, V.B. 1972. Neurosecretory system of the adult flesh fly, *Sarcophaga ruficornis* F. (Diptera : Sarcophagidae). *Int. J. Insect Morphol. and Embryol.* **1** : 133-140.

Awasthi, V.B. 1972. Studies on the neurosecretory cells of the brain of normal and starved red cotton bug, *Dysdercus koenigii* (Fabr.) (Heteroptera : Pyrrhocoridae). *Aust. Zool.,* **17** : 34-35.

Awasthi, V.B. 1972. Studies on the neurosecretory system and retrocerebral endocrine glands of *Nezara viridula* Linn. (Heteroptera : Pentatomidae). *J. Morph.* **136** : 337-352.

Awasthi, V.B. 1973. An analytical study of the neuroendocrine system of *Anacanthotermes macrocephalus* (Isoptera : Hodotermitidae). *Zool. Beitr.,* **19** : 75-81.

Awasthi, V.B. 1973. Neurosecretory system and retrocerebral endocrine glands of *Metochus uniguttatus* Thunb. (Heteroptera : Lygaeidae). *Int. J. Insect Morphol. and Embryol.* **2** : 1-12.

Awasthi, V.B. 1973. Studies *in situ* on the neurosecretory system of *Sphaerodema rusticum* Fabr. (Heteroptera : Belostomatidae). *Zool. Beitr.,* **19 :** 151-156.

Awasthi, V.B. 1974. Neurohaemal organs in insects. *Zool. Anz.* **136**: 244-267.

Awasthi, V.B. 1975. Neurosecretory system of *Labidura riparia* (Pallas) and role of aorta as a neurohaemal organ. *J. Insect Physiol.* **21** : 1713-1719.

Awasthi, V.B. 1975. Response of neurosecretory systems of *Gryllodes sigillatus* (Walk.) and *Sphaerodema rusticum* Fabr. to salt. solution. *Zool. Jb. Anat.,* **94 :** 67-73.

Awasthi, V.B. 1975. The retrocerebral endocrine glands and associated neurohaemal organs of *Epilampra* sp. with reference to cysteine or cystine rich NSM. *Zool. Jb. Anat.,* **94** : 149-157.

Awasthi, V.B. 1976. Contribution to the neurohaemal organs in Pterygote insects. *Zeit. Mikrosk. Anat. Forsch.* **90**: 47-78.

Awasthi, V.B. 1976. Effect of nutritional factors on the appearance of neurosecretory cells and ovary development in the fleshfly, *Sacrophaga ruficornis* F. (Diptera : Sarcophagidae). *Zool. Jb. Anat.,* **95** : 1-8.

Awasthi, V.B. 1976. Neurosecretory cells in the earwig, *Euborellia annulipes* (Dermaptera : Labiduridae) and aorta as a neurohaemal organ. *Int. J. Insect Morphol. and Embryol.* **6** : 253-260.

Awasthi, V.B. 1976. The functional significance of cephalic aorta as a neurohaemal organ in *Labidura riparia* (Pallas) and other Dermaptera. *Zool. Jb. Anat.,* **95** : 109-113.

Awasthi, V.B. 1976. Ultrastructure of rectrocerebral complex of the earwig, *Euborellia annulipes* and role of aorta as a neurohaemal organ. *J. Insect Physiol.* **22** : 1181-1186.

Awasthi, V.B. 1977. Ultrastructure of corpus cardiacum of the house cricket, *Gryllodes sigillatus* (Walk.) (Orthoptera : Gryllidae). *Zool. Jb. Anat.* **98** : 476-490.

Awasthi, V.B. 1978. Studies on the neurosecretory system and retrocerebral endocrine glands of *Epilampra* sp. (Dictyoptera : Blattidae). *Zool. Beitr.,* **24** : 447-464.

Awasthi, V.B. 1979. Hormonal control of metabolism in insects. *Ind. J. Zoot.* **20**: 1-38.

Awasthi, V.B. 1979. The ultrastructure of corpus allatum of the earwig, *Euborellia annulipes* Lucas (Dermaptera : Labiduridae). *Zeit. Mikrosk. Anat. Forsch.* **93** : 981-991.

Awasthi, V.B. 1980. Neurosecretion in the frontal ganglion of *Achaea janata* Linn. and *Philosamia ricini* Hutt. (Lepidoptera). *Zeit. Mikrosk. Anat. Forsch.* **24** : 825-832.

Awasthi, V.B. 1980. Neurosecretion in the milkweed bug, *Lygaeus pandurus* Scop. (Heteroptera : Lygaeidae). *J. Hirnforsch.* **21** : 169-181.

Awasthi, V.B. 1982. The neurosecretory cells of the brain in Lepidoptera. *J. Adv. Zool.* **3** : 67-76.

Awasthi, V.B. 1983. Cardiac regulators in insects. *J. Adv. Zool.* **4**: 64-70.

Awasthi, V.B. 1985. Influence of juvenile hormone analogue on metamorphosis of *Philosamia ricini* Hutt. (Lepidoptera : Saturniidae). *Entomon,* **10**:: 165-169.

Awasthi, V.B. 1985. Present status of Insect Neuroendocrinology in India. *Ind. Rev. Life. Sci.* **5** : 59-91.

Awasthi, V.B. 2001. *Principles of Insect Behaviour*. Scientific Publishers (India), Jodhpur.

Awasthi, V.B. 2005. *Insect Neuroendocrines*. Scientific Publishers (India), Jodhpur.

Awasthi, V.B. 2007. *Agricultural Insect Pests and Their Control.* Scientific Publishers (India), Jodhpur.

Awasthi, V.B. and Singh, U.V. 1981. Dynamics of the neurosecretory cells of the brain of *Amsacta collaris* Hamp. (Lepidoptera : Arctiidae) during postembryonic development. *J. Hirnforsch.* **22** : 161-177.

Awasthi, V.B. and Singh, U.V. 1981. An analytical study of the neurosecretory cells of the brain in *Philosamia ricini* Hutt. (Lepidoptera: Saturniidae) during postembryonic development. *Experientia,* **37**: 372-375.

Awasthi, V.B. and Singh, U.V. 1982. An analytical study of the neurosecretory cells of the brain of mature larvae and diapause pupae of *Amsacta collaris* Hamp. (Lepidoptera : Arctiidae) under different temperature regimes. *Gegenbaur's Morph.* **128** : 100-106.

Awasthi, V.B. and Singh, U.V. 1982. Studies on the neurosecretory cells of the brain of *Achoea janata* Linn. (Lepidoptera Noctuidae) during postembryonic development. *Entomon.* **7** : 285-295.

Awasthi, V.B. and Singh, U.V. 1983. The anatomy and histology of the ecdysial glands of *Amsacta collaris* Hamp. (Lepidoptera: Arctiidae) during postembryonic development. *Zool. Jb. Anat.* **109** : 467-480.

Awasthi, V.B. and Singh, U.V. 1985. The neurosecretory cells of the brain in the moth, *Utetheisa pulchella* Linn. (Lepidoptera : Arctiidae) during postembryonic development. In : *Recent Advances in Developmental Biology of Animals* (Goel, S.C. and Srivastava, C.B.L., eds.). Rastogi and Co. Meerut, pp. 77.87.

Awasthi, V.B. and Singh, U.V. 1986. Histochemical analysis of the neurosecretory cells of the brain in Erisilkworm, *Philosamia ricini* Hutt. (Lepidoptera : Saturniidae) In : *Recent advances in insect Physiology, Morphology and Ecology* (Pathak, S.C. and Sahai, Y.N. eds.). Today and Tomorrow, Printers and Publishers, New Delhi. pp. 25-38.

Awasthi, V.B. and Singh, U.V. 1986. The retrocerebral endocrine glands of *Achoea janata* Linn. (Noctuidae : Lepidoptera) during post embryonic development. *Zool. Jb. Anat.,* **114** : 269-275.

Awasthi, V.B. and Singh, U.V. 1994. Studies on the ecdysial glands of *Utetheisa pulchella* Linn. (Lepidoptera : Arctiidae) during post embryonic development. In : *Perspectives in Entomological Research* (Agrawal, O.P. ed.) Scientific Publishers (India) Jodhpur.

Awasthi, V.B. and Tiwari, Y.C. 1975. Studies on the mating behaviour of *Lygaeus militaris* Fabr. (Heteroptera : Lygaeidae). Zool. Beitr. **21:** 493-499.

Ayyar, T.V.R. 1940. *Handbook of Economic Entomology for South India*, Govt. Press, Madras.

Beament, J.W.L. Treherne, J.E. and Wigglesworth, V.B. (Eds.). *Advances in Insect Physiology,* Vol. **1** to **19**. Academic Press, New York.

Beeson, C.F.C. 1941. *The Ecology and Control of the Forest Insects of India and the Neighbouring Countries.* Govt. India Press, Vasant Press, Dehra Dun.

Bellows, T.S. and Fischer, T.W. 2005. *Handbook of Biological Control.* Academic Press Elsevier, U.S.A.

Beroza, M. 1970. *Chemicals Controlling Insect Behaviour*. Academic Press, New York.

Berryman, A.A. 1988. *Dynamics of Forest Insect Populations : Pattern, causes, implications.* Plenum Press, New York, London.

Best, G.A. and Ruthven, A.D. 2005. *Pesticides: Development, Impact and Controls.* Woodland Publ. Limited, England.

Biddle, A.J. and Cattlin, N.D. 2007. *Pests, Diseases and Disorders of Peas and Beans.* A Colour Handbook, Manson Publishing Ltd., London.

Blaney, W.M. 1977. *How Insects Live.* Elsevier Publishing Project Ltd. Oxford.

Blomquist, G.J. and Vogt, R.G. 2003. *Insect Pheromones: Biochemistry and Molecular Biology.* Elsevier Academic Press, London, New York, etc.

Brian, M.V. 1983. *Social Insects : Ecology and Behavioural Biology.* Chapman and Hall, London and New York.

Briggs, G.G. 1994. *Advances in the Chemistry of Insect Control III.* Royal Society of Chemistry, Cambridge, U.K.

Bursell, E. 1970. *An Introduction to Insect Physiology.* Academic Press, New York.

Busvine, J.R. 1966. *Insect and Hygiene.* Metheun and Company Ltd. London.

Chapman, H.C. 1974. Biological Control of Mosquito larvae. *Ann. Rev. Entomol.* **19**: 33-57.

Chapman, R.F. 1972. *The Insects: Structure and Function.* ELBS edition. English University Press, London.

Chapman, R.F. 2006. *The Insects: Structure and Function.* 4th edition, Cambridge University Press.

Chapman, R.F. and Boer, G. De. 1995. *Regulatory Mechanisms in Insect Feeding.* Chapman and Hall.

Chown, S.L. and Nicolson, S.W. 2004. *Insect Physiological Ecology. Mechanisms and Patterns.* Oxford University Press.

Clausen Curtis, P. 1962. *Entomophagous Insects.* Hafner Publishing Company, New York.

Clements, A.N. 1992. *The Biology of Mosquitoes* Vol. 1, Chapman and Hall, New York and London.

Conte, Y. Le and Hefetz, A. 2008. Primer Pheromones in Social Hymenoptera. *Ann. Rev. Entomol.* **53:** 523-542.

Counce, S.J. and Waddington, C.H. 1973. *Developmental Systems: Insects.* Vol. **1** and **2**, Academic Press, New York.

Darwin, C. 1859. *The Origin of species. Mentor edition 1963.* Published by the New American Library of World Literature, Inc. New York.

Davey, K.G. 1965. *Reproduction in Insects.* Freeman Press, London.

DeBach, P. 1964. *Biological Control of Insect Pests and Weeds.* Chapman and Hall Ltd. London.

DeBach, P. 1974. *Biological Control of Natural Enemies.* Cambridge University Press, Cambridge.

DeBach, P. and David, R. 1991. *Biological Control by Natural Enemies.* Cambridge University Press.

Dent, D. 1991. *Insect Pest Management.* CAB International. Wallingford, U.K.

Dethier, V.G. 1963. *The Physiology of Insect Senses.* Methuen, London.

Dewsbury, D.A. 1978. *Comparative Animal Behaviour.* McGraw Hill Book Company, New York.

Dukas, R. 2008. Evolutionary Biology of Insect Learning. *Ann. Rev. Entomol.* **53:** 145-160.

Duncan, P.M. 1995. *Fundamentals of Insect Development*. First Indian edition. A Sagar Book House, Delhi.

Duporte, E.M. 1959. *Manual of Insect Morphology*. Reinhold Publisher. New York.

Elzinga, R.J. 1978. *Fundamentals of Entomology*. Prentice Hall of India, New Delhi.

Emden, von H.F. and Service, M.W. 2004. *Pest and Vector Control.* Cambridge University Press.

Engelmann, F. 1968. Endocrine Control of Reproduction in Insects. *Ann. Rev. Entomol.* **13:** 1-26.

Engelmann, F. 1970. *The Physiology of Insect Reproduction*. Pergamon Press, Oxford.

Engels, W. 1990. *Social Insects : An Evolutionary Approach to Castes and Reproduction*. Springer Verlag, Berlin.

Essing, E.O. 1942. *College Entomology*. Macmillan Company, New York.

Fenton, F.A. 1952. *Field Crop Insects*. Macmillan Company, New York.

Fernald, H.T. and Shepard, H.H. 1942. *Applied Entomology*. McGraw Hill Book Company, New York.

Fox, R.M. and Fox, J.W. 1964. *Introduction to Comparative Entomology*. Reinhold Publishing Company, New York.

Friend, W.G. 1958. Nutritional Requirements of Phytophagous Insects. *Ann. Rev. Entomol.* **3**: 57-74.

Frisch, K. Von 1950. *Bees, Their Vision, Chemical Senses and Language*. New York. Cornell Univ. Press.

Frisch, K. Von 1966. *The Dancing Bees*. Methuen and Company, London.

Gabe, M. 1966. *Neurosecretion*, Pergamon Press. Oxford.

Gilbert, L.I. 1964. Physiology of Growth and Development: Endocrine Aspects. In: *The Physiology of Insecta.* **1**: 149-225, Academic Press, New York.

Gilmour, D. 1965. *The Metabolism of Insects*. W.H. Freeman and Company, San Francisco.

Goodden, R. 1971. *Butterflies*. Hamlyn, London.

Graham, K. 1963. *Concepts of Forest Entomology*. Reinhold Publishing Corporation, New York.

Grover, P.M. 1937. *Lac Cultivation in India*. Indian Lac Research Institute, Namkum, Ranchi.

Gullan, P.J. and Cranston, R.S. 1994. *The Insects : An out line of Entomology*. Chapman and Hall, London.

Gwynne, D.T. 2008. Sexual conflict over Nuptial gift in Insects. *Ann. Rev. Entomol.* **53:** 83-101.

Halberg, F., Halberg, E., Barnum, C.P. and Bittner, J.J. 1959. Physiologic 24 hour periodicity in human beings and mice. The Lighting regimen and daily routine. *Amer. Assoc. Adv. Sci.* **35:** 803-878.

Heinrich, B. 1981. *Insect Thermorgulation*. John Wiley and Sons, New York.

Heller, H. and Clark, R.B. 1962. *Neurosecretion*. Academic Press, London.

Herms, W.B. and James, M.T. 1960. *Medical Entomology*. Macmillan Company, New York.

Highnam, K.C. and Hill, L. 1979. *The Comparative Endocrinology of the Invertebrates*. ELBS edition. Edward Arnold Publishers, Ltd.

Hill, D.S. 1975. *Agricultural Insect Pests of Tropics and their Control.* Cambridge University Press.

Hill, D.S. 1994. *Agricultural Entomology*. Timber Press, Portland, Oregon.

House, H.L. 1961. Insect Nutrition. *Ann. Rev. Entomol.* **6**: 13-26.

House, H.L. 1962. Insect Nutrition. *Ann. Rev. Biochem.* **31**: 653- 672.

Howse, P. Stevens, I. and Jones, O. 1995. *Insect Pheromones and their use in Pest Management*. pp. 256. Intercept Ltd. Andover Hants UK.

Howse, P.E. 1970. *Termites*. Hutchinson Univ. Lib. London.

Huffaker, C.B. 1974. *Biological Control*. Plenum Publishing Corporation, New York.

Imms, A.D. 1963. *A General Textbook of Entomology*. Revised by Richards, O.W. and Davies, R.G. 9th edition, Asia Publishing House, Bombay.

Imms, A.D. 1964. *A General Textbook of Entomology*. Revised by Richards, O.W. and Davies, R.G. Methuen Co. Ltd. London.

Jacobson, M. 1965. Insect Orientation. *Ann. Rev. Entomol.* **8:** 94-114.

Jacobson, M. 1972. *Insect Sex Pheromones*. Academic Press, New York and London.

Jacobson, M. 1974. Insect Pheromones. In: *The Physiology of Insecta.* 2nd edition. **3:** 229-276. (M. Rockstein, ed.) Academic Press, New York.

Jaw, J.H. and Regnier, F.E. 1971. Pheromones. *Ann. Rev. Biochem.* **40**: 533-548.

Johansen, O.A. and Butt, F.H. 1941. *Embryology of Insects and Myriapods.* McGraw Hill. Company, New York.

Kerkut, G.A. and Gilbert, L.I. 1985. *Comprehensive Insect Physiology, Biochemistry and Pharmacology*. Vol. **1** to **13**. Pergamon Press, Oxford.

Klowden, M.J. 2007. *Physiological Systems in Insects*. 2nd edition. Academic Elsevier Press.

Kogan, M. and Jepson, P. 2007. *Perspective in Ecological Theory and Integrated Pest Management.* Cambridge University Press. Cambridge U.K.

Krishna, K. and Weesner, F.M. (Eds.) 1969. *Biology of Termites.* Vol. **1** and **2**, Academic Press, New York and London.

Lee, R.E. and Denlinger, D.L. 1991. *Insects at Low Temperature.* Chapman and Hall, New York and London.

Lehane, M.J. 1991. *Biology of Blood Sucking Insects.* Harper Collins Academic, London.

Little, V.A. 1967. *General and Applied Entomology.* Oxford and IBH Publishing Company, Bombay and New Delhi.

Mani, M.S. 1968. *General Entomology.* Oxford and IBH Bombay and New Delhi.

Maramorosch, K. and Meinstosh, A. H. 1994. *Insect Cell Biotechnology.* Intercept Ltd. Andover, Hants, U.K.

Markl, H. 1974. Insect Behavour: Functions and Mechanisms. In: *The Physiology of Insecta.* **3:** 3-148. (M. Rockstein ed.) 2nd edition. Academic Press, New York.

McElroy, W.D. 1972. *Cell Physiology and Biochemistry.* 3rd edition Prentice Hall of India Pvt. Ltd., New Delhi.

Menn, J.J. and Beroza, M. 1972. *Insect Juvenile Hormones: Chemistry and Action.* Academic Press Inc. New York and London.

Metcalf, C.L. and Flint, W.P. 1962. *Destructive and Useful Insects: Their Habits and Control.* McGraw Hill Book Company, New York.

Metcalf, R.L. and Luckmann, W.H. 1994. *Introduction to Insect Pest Management.* 3rd edition, John Wiley and Sons Inc., New York.

Nair, K.S.S. 2007. *Tropical Forest Insect Pests : Ecology, Impact and Management.* Cambridge University, Press.

Nayar, K.K., Ananthakrishnan, T.A. and David, B.V. 1976. *General and Applied Entomology.* Tata McGraw Hill Publishing Company Limited, New Delhi.

Normann, T.C. 1976. Neurosecretion by Exocytosis. *Int. Rev. Cytol.* **46**: 1-77.

Novak, V.J.A. 1975. *Insect Hormones.* Chapman and Hall, London.

Onstad, D.W. 2008. *Insect Resistance Management: Biology, Economics and Prediction.* Elsevier-Academic Press.

Patton, R.L. 1963. *Introductory Insect Physiology.* Topan Company, Ltd., Tokyo.

Pedigo, L.P. 2003. *Entomology and Pest Management.* Pearson Education Inc.

Pfadt, Robert, E. 1962. *Fundamentals of Applied Entomology.* The Macmillan and Company, New York.

Pimentel, D. 2007. *Encyclopedia of Pest Management.* Vol. II. CRC Press, Taylor and Francis Group.

Pradhan, S. 1969. *Insect Pests of Crops.* National Book Trust of India, New Delhi.

Pradhan, S. 1983. *Insect Pests of Crops.* Revised by Jotwani, M.G. National Book Trust of India, New Delhi.

Pradhan, S. 2002. *Agricultural Entomology and Pest Control.* ICAR, New Delhi.

Price, P.W. 1975. *Insect Ecology.* John Wiley and Sons, London.

Pruthi, H.S. 1969. *Textbook on Agricultural Entomology.* Indian Council of Agricultural Research, New Delhi.

Pruthi, H.S. and Singh, M. 1950. *Pests of Stored Grains and Control.* Govt. of India Press.

Raabe, M. 1982. *Insect Neurohormones.* Plenum Press, New York.

Raabe, M. 1989. *Recent Developments in Insect Neurohormones.* Plenum Press, New York.

Reay, R.C. 1969. *Insects and Insecticides.* Oliver and Boyd, Edinburgha.

Richards, A.G. and Richards, R.A. 1977. The Peritrophic Membrane of Insects. *Ann. Rev. Entomol.* **22**: 2 19-240.

Rockstein, M. 1964. *The Physiology of Insecta.* Vol. **1** and **2**. Academic Press, New York.

Roeder, K.D. 1953. *Insect Physiology.* John Wiley and Sons Inc., New York.

Romoser, W.S. and Stoffbalk, J.G. 1973. *The Science of Entomology.* 3rd edition. WCB Publishers, Oxford, England.

Roonwal, M.L. 1979. *Termite Life and Termite Control in Tropical South Asia.* Scientific Publishers (India), Jodhpur.

Ross, H.H. 1956. *A Textbook of Entomology.* John Wiley and Sons Inc., London.

Samways, M.J. 1995. *Insect Conservation Biology.* Chapman and Hall, London.

Scharrer, B. 1977. *Peptides in Neurobiology. Historical Introduction.* H. Gainer (Ed.) Plenum Press, New York.

Schneiderman, H.A. 1972. Insect hormones and insect control. In: *Insect Juvenile Hormones: Chemistry and Action.* pp. 3-26. Academic Press Inc., New York and London.

Schoonhoven, L.M., Loon Van, J.J.A. and Dicke, M. 2006. *Insect Plant Biology.* 2nd edition. Oxford University Press.

Schowalter, T.D. 2006. *Insect Ecology: An Ecosystem Approach.* Academic Press, New York.

Shanower, T.G., Romeis J. and Minja, E.M. 1999. Insect Pests of Pigeonpea and their Management. *Ann. Rev. Entomol.* **44:** 77-96.

Sidall, J.B. 1970. Chemical Aspects of Hormonal Interactions. In: *Chemical Ecology*. pp. 281-306. Academic Press Inc. New York.

Singh, N.P. and Singh, R.A. 2002. *Scientific Crop Production with special reference to North Eastern Hill Region*. Kalyani Publishers, Ludhiana etc.

Singh, U.V. and Awasthi, V.B. 1980. The anatomy and histology of ecdysial glands of *Philosamia ricini* Hutt. (Lepidoptera : Saturniidae) during post embryonic development. *Zeit. Mikrosk. Anat. Forsch.,* **94 :** 850-864.

Singh, U.V. and Awasthi, V.B. 1981. The anatomy and histophysiology of the ecdysial glands of *Achoea janata* Linn. (Lepidoptera : Noctuidae) during post embryonic development. *Entomon,* **6** : 183-195.

Singh, U.V. and Awasthi, V.B. 1981. The neurosecretory cells and retroce-rebral endocrine glands of *Amsacta collaris* Hampson (Lepidoptera: Arctiidae) during postembryonic development. *J. Hirnforsch.* **22**: 265-278.

Singh, U.V. and Awasthi, V.B. 1982. Studies on the retrocerebral endocrine glands of the Erisilkworm, *Philosamia ricini* Hutt. (Lepidoptera : Saturniidae) during post-embryonic development. *Folia Morphologica,* 30 : 346-354.

Sinha, T.B. and Srivastava, D.C. 1970. Cellulose Digestion in *Pseudocaecilius elutus* (Psocoptera). *Enzymologia,* **39:** 84-86.

Slama, K., Ramanuk, M. and Sorm, F. 1974. *Insect Hormones and Bioanalogues*. Springer Verlag, Wien New York.

Snodgrass, R.E. 1925. *Anatomy and Physiology of the Honey bee.* McGraw Hill Company, New York.

Snodgrass, R.E. 1935. *Principles of Insect Morphology.* McGraw Hill Company, New York.

Snodgrass, R.E. 1952. *A Text Book of Arthropod Anatomy.* Cornell Univ. Press, Ithaca.

Snodgrass, R.E. 1954. *Insect Metamorphosis.* Smithson. Misc. Collns. **122** No. 9, 124 pp.

Sovage, A.H. 1956. Origin and Development of Malpighian tubules in *Schistocrca. Quart. J. Micr. Sci.* **97:** 599-615.

Speight, M.R. and Wylie, F.R. 2001. *Insect Pests in Tropical Forestry.* CABI Publishing, London.

Staal, G.B. 1971. Practical aspects of Insect Control by Juvenile Hormone. *Bull. W.H. Org.* **44:** 391-394.

Studd, J.H. 1967. *An Introduction to the Behaviour of Ants.* Edward Arnold Publ. Ltd.

Tewari, H.B. and Awasthi, V.B. 1968. The identical topography of the distribution of alkaline phosphatase and neurosecretory material in the neuroendocrine organs of *Gryllodes sigillatus* (Walk.) (Orthoptera : Gryllidae). *Gen. Comp. Endocrinol.* **10** : 330-338.

Thompson, S.N. 1999. *Nutrition and Culture of Entomophagous Insects. Ann. Rev. Entomol.* **44**: 561-592.

Thorpe, W.H. 1963. *Learning and Instincts in Animals.* 2nd edition. Methuen London.

Treherne, J.E. and Beament, J.W.L. 1965. *The Physiology of the Central Nervous System.* Academic Press, New York and London.

Truman, J.W. and Riddiford, L.M. 1974. Hormonal mechanisms underlying insect behaviour. *Adv. Ins. Physiol.* **10** : 297-352.

Uvarov, B. 1966. Grasshoppers and Locusts : *A hand book of General Acridology.* Vol. **1**. Cambridge University Press.

Wainhouse, D. 2005. *Ecological Methods in Forest Pest Management.* Oxford University Press.

Waterhouse, D.F. 1957. Digestion in Insects. *Ann. Rev. Entomol.* **2**: 1-18.

Welling, W. 1977. Dynamic Aspects of Insects, Insecticide Interactions. *Ann. Rev. Entomol.* **22**: 153-178.

West, T.F. and Hard, I.E. 1961. *Chemical Control of Insects.* Chapman and Hall Ltd., London.

Whitten, M.J. and Foster, G.G. 1975. Genetical methods of Pest Control. *Ann. Rev. Entomol.* **20:** 461-476.

Wigglesworth, V.B. 1931. Excretion in *Rhodnius. J. Exptl. Biol.* **8** : 411-457.

Wigglesworth, V.B. 1954. *The Physiology of Insect Metamorphosis.* Cambridge Univ. Press, Cambridge.

Wigglesworth, V.B. 1965. *The Principles of Insect Physiology,* ELBS Edition, Methuen and Company, London.

Wigglesworth, V.B. 1969. *Insect Physiology.* Chapman and Hall, London.

Wigglesworth, V.B. 1969. *The Control of Growth and Form.* Cornell Univ. Press, Ithaca, New York.

Wigglesworth, V.B 1970. *Insect Hormones.* Freeman. London.

Wigglesworth, V.B. 1972. *The Principles of Insect Physiology.* 7th edition, Chapman and Hall, London.

Wigglesworth, V.B. 1973. Evolution of Insect Wings and Flight. *Nature.* 246: No 5429 pp. 127-129.

Wigglesworth, V.B. 1973. The History of Insect Physiology. In : *History of Entomology*, Smith, R.F., Mittler, T.E. and Smith, C.N. (Eds.). pp. 203-328 Ann. Rev. Inc. Palo Alto, USA.

Wilde, J.De. 1962. Photoperiodism in Insects and Mites. *Ann. Rev. Entomol.* **7**: 1-26.

Wood, J.G.R. 1994. *Insects of the World.* Vol. 1 and 2 Atlantic Publishers and distributors, New Delhi, India.

Yokoyama, T. 1963. Sericulture. *Ann. Rev. Entomol.* **8:** 287-306.

INDEX

Other Related Books

ISBN: 9789388449076

ISBN: 9789388172752

ISBN: 9789388172981

ISBN: 9789388172950

ISBN: 9789388043311

ISBN: 9788172333515

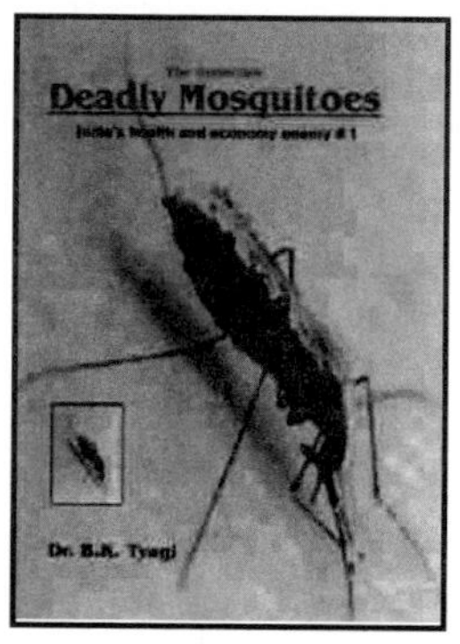

ISBN: 9788172333652

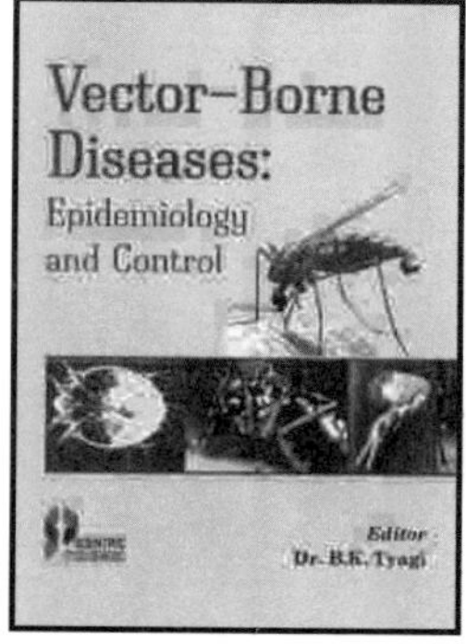

ISBN: 9788172335410

ISBN: 9789386652355

for Full Content and Sample Pages visit www.scientificpub.com

A Manual of Practical Entomology (Field and Laboratory Guide)

3rd Revised Edition

M.M. Trigunayat

A Manual of Practical Entomology (Field and Laboratory guide) is written to provide text material on different aspects of the practical syllabi of M.Sc. Entomology. All chapters are illustrative and well explained. Easy text will help students understanding the exercises. Dissections of the individual insect have been described so as to understand and follow the finest details of the anatomy of the insect concerned. Physiological and Behavioural Exercises are well presented usually in the style followed by the students. Individual Orders have been dealt with relevant necessary text and vital aspects of biology of the insect concerned. Chapters on Natural History and Rearing will inculcate interest among the students for their quest on these tiny creatures. Profusely illustrated with simple figures presentation and style gives the reader an insight to make it unique. Primarily intended for use by the post-graduate and graduate students of Entomology of the Universities in Rajasthan and elsewhere in India, it can also be used by agriculture departments, naturalists and workers in other related fields.

Contents

ISBN: 9788172339838 | 2016 | 328 pages

for Full Content and Sample Pages visit www.scientificpub.com

Principles of Insect Behaviour

2nd Revised & Enlarged Edition

V.B. Awasthi

The aim of writing this book has been to provide an account of the principles of insect behaviour for the post graduate students of Entomology and general Zoology of all the Universities and Research Institutions as well as to the research scholars engaged in the study of different aspects of Insect behaviour. It is also useful to general readers who have basic knowledge of Biology and have curiosity to know why the insects are vividly coloured, have peculiar shapes, produce sound and light; how they communicate and why bees and wasps sting when disturbed. Significance of insects in pollination of plants associated with the production of seeds, gains and fruits, the role of hormones, pheromones and other chemicals controlling various vital events of the insect behaviour has also been described. The importance of insect migration, control of insect pests by behavioural means has also been discussed. Thus, this book deals with almost all major aspects of insect behaviour. It is hoped that this book will be useful not only to the students of Entomology-Zoology and research scholars of insect behaviour, but also to naturalists who wish to know about some of the mysteries of these small creatures and their lifestyle.

Contents

ISBN: 9788172338251 | 2013 | 299 pages

Insect Neuroendocrines

V.B. Awasthi

The aim of the author has been to present an account of the neuroendocrine system of insects, the super importance of which is well established, and also to provide information as far as possible directly, through the ultrastructural illustrations. Besides, an account of some of the hormonally controlled phenomena of vital significance is given.

It is hoped that the theoretical account, and a large number of ultrastructural illustrations of some of the prominent pioneer authors included in the book, will provide ample information to the post graduate students and research scholars of insect neuroendocrinology as well as to those engaged in carrying out researches on comparative endocrinology, insect ethology, cell biology and developmental biology in various universities and acquaint them well with almost all the components of the neuroendocrine system in the active, inactive and various developmental stages, alongwith ultra-structural changes occurring in the cells and their organelles, that are associated with specific functions. I do hope that the book will stimulate the readers and serve as a ``text book cum basic research monograph" on the subject. The book also contains a list of selected bibliography for a detailed study.

ISBN: 9788172333904 | 2005 | 444 pages

Handbook of Insects:

Photographic Guide for Identification

V.B. Awasthi

Colour Book

In this book an attempt has been made to provide correct and quick identification of common insects of Asia, America, Africa and Australia for the beginners as well as graduate students of zoology and entomology by telling about the insects. Collected and to those of given in this guide and this field and class-room both for the identification. This guide will stimulate nonzoologists entomologists too who have colourful flowers and beautiful insects. This book is also useful to those who are not biologist but general peoples but watch/see insects found around them and have curiocity to know about them just like colourful flowers in a large part of the globe.

ISBN: 9789388172790 | 2019 | 202 Pages

for Full Content and Sample Pages visit www.scientificpub.com